Current Topics in Microbiology 153 and Immunology

Editors

R. W. Compans, Birmingham/Alabama · M. Cooper, Birmingham/Alabama · H. Koprowski, Philadelphia
I. McConnell, Edinburgh · F. Melchers, Basel
V. Nussenzweig, New York · M. Oldstone, La Jolla/California · S. Olsnes, Oslo · H. Saedler, Cologne · P. K. Vogt, Los Angeles · H. Wagner, Munich
I. Wilson, La Jolla/California

The Third Component of Complement

Chemistry and Biology

Edited by J. D. Lambris

With 38 Figures

Springer-Verlag
Berlin Heidelberg NewYork
London Paris Tokyo Hong Kong

JOHN D. LAMBRIS, Ph. D.

Basel Institute for Immunology
Grenzacherstr. 487,
CH-4005 Basel

ISBN 3-540-51513-5 Springer-Verlag Berlin Heidelberg New York
ISBN 0-387-51513-5 Springer-Verlag New York Berlin Heidelberg

This work is subject to copyright. All rights are reserved, whether the whole or part of the material is concerned, specifically the rights of translation, reprinting, re-use of illustrations, recitation, broadcasting, reproduction on microfilms or in other ways, and storage in data banks. Duplication of this publication or parts thereof is only permitted under the provisions of the German Copyright Law of September 9, 1965, in its version of June 24, 1985, and a copyright fee must always be paid. Violations fall under the prosecution act of the German Copyright Law.

© Springer-Verlag Berlin Heidelberg 1990
Library of Congress Catalog Card Number 15-12910
Printed in Germany

The use of registered names, trademarks, etc. in this publication does not imply, even in the absence of a specific statement, that such names are exempt from the relevant protective laws and regulations and therefore free for general use.

Product Liability: The publishers can give no guarantee for information about drug dosage and application thereof contained in this book. In every individual case the respective user must check its accuracy by consulting other pharmaceutical literature.

Offsetprinting: Saladruck, Berlin; Bookbinding: Helm, Berlin
2123/3020-543210 — Printed on acid-free paper

Preface

The third component of complement, C3, is one of the most versatile proteins and an important participant in immune surveillance and immune response pathways. Its multifunctionality is based on its ability to interact specifically with multiple serum complement proteins, cell surface receptors, and membrane-associated regulatory proteins. One of its most intriguing strategies of interaction with cell surfaces is the covalent binding of activated C3 through the internal thioester.

The field has expanded over the past 10 years and a wealth of information has accumulated. C3 from various species and many of the human C3 binding proteins have been cloned and expressed. Numerous cellular responses mediated by the different fragments of C3 have been described. The findings that C3 interacts in a ligand-receptor-like fashion with proteins of nonself origin such as the gC of herpes simplex virus, a 70-kDa protein from *Candida albicans*, proteins from Epstein-Barr virus, etc. has opened a new field of investigation. The papers assembled in this volume summarize the wealth of data on the various aspects of the C3 interactions; together they bring to the reader new information on the chemistry, molecular genetics, biology, and pathophysiology of C3 and C3-binding proteins. Emphasis is given to structural features as they relate to functions.

Spring 1989   JOHN D. LAMBRIS,
                                    HANS J. MÜLLER-EBERHARD

Table of Contents

J. E. Volanakis: Participation of C3 and Its Ligands in Complement Activation 1

S. R. Barnum, G. Fey, and B. F. Tack: Biosynthesis and Genetics of C3 23

D. J. Becherer, J. Alsenz, and J. D. Lambris: Molecular Aspects of C3 Interactions and Structural/Functional Analysis of C3 from Different Species 45

R. P. Levine and A. W. Dodds: The Thioester Bond of C3 73

D. T. Fearon and J. M. Ahearn: Complement Receptor Type 1 (C3b/C4b Receptor; CD35) and Complement Receptor Type 2 (C3d/Epstein-Barr Virus Receptor; CD21) 83

H. Rosen and S. K. A. Law: The Leukocyte Cell Surface Receptor(s) for the iC3b Product of Complement . . 99

D. M. Lublin and J. P. Atkinson: Decay-Accelerating Factor and Membrane Cofactor Protein 123

D. P. Vik, P. Muñoz-Cánoves, D. D. Chaplin, and B. F. Tack: Factor H 147

M. P. Dierich, H. P. Huemer, and W. M. Prodinger: C3 Binding Proteins of Foreign Origin 163

T. E. Hugli: Structure and Function of C3a Anaphylatoxin 181

R. B. Sim and S. J. Perkins: Molecular Modeling of C3 and Its Ligands 209

D. Bitter-Suermann and R. Burger: C3 Deficiencies 223

J. Alsenz, J. D. Becherer, B. Nilsson, and J. D. Lambris: Structural and Functional Analysis of C3 Using Monoclonal Antibodies 235

Subject Index 249

Indexed in Current Contents

List of Contributors

You will find their addresses at the beginning of the respective contributions

AHEARN J. M.
ALSENZ J.
ATKINSON J. P.
BARNUM S. R.
BECHERER J. D.
BITTER-SUERMANN D.
BURGER R.
CHAPLIN D. D.
DIERICH M. P.
DODDS A. W.
FEARON D. T.
FEY G.
HUEMER H. P.
HUGLI T. E.

LAMBRIS J. D.
LAW S. K. A.
LEVINE R. P.
LUBLIN D. M.
MUÑOZ-CÁNOVES P.
NIELSSON B.
PERKINS S. J.
PRODINGER W. M.
ROSEN H.
SIM R. B.
TACK B. F.
VIK D. P.
VOLANAKIS J. E.

Participation of C3 and its Ligands in Complement Activation

J. E. VOLANAKIS

1 Introduction 1
2 Overview of Complement Activation 2
3 Participation of C3 in Complement Activation 4
4 C3 Convertase of the Alternative Pathway 4
4.1 The Noncatalytic Subunit $C3_{H_2O}$/C3b 4
4.2 The Catalytic Subunit Factor B 6
4.3 Factor D 10
4.4 Properdin 11
4.5 Assembly of the Alternative Pathway C3 Convertase 12
5 C5 Convertase of the Alternative Pathway 13
6 C5 Convertase of the Classical Pathway 14
7 Summary 15
References 15

1 Introduction

The complement system comprises a group of proteins in the blood which upon activation generate fragments and protein-protein complexes expressing biological activities. The biochemistry, function, and genetics of the system have been recently reviewed (CAMPBELL et al. 1988; MÜLLER-EBERHARD 1988). Complement activation proceeds in a sequential cascase-like fashion that is similar to the activation of other humoral effector systems in the blood, such as the coagulation, the fibrinolytic, and the kinin-generating system. Complement-derived biologically active products mediate a variety of important functions including increased vascular permeability, chemotaxis of phagocytic cells, activation of inflammatory cells, opsonization of foreign particles and cells, and direct killing of foreign cells (reviewed in MÜLLER-EBERHARD and MIESCHER 1985). Thus, complement plays a major role in host defense against pathogens. Accumulating evidence suggests complement activation products also function as growth and/or differentiation factors for B cells and possibly have additional effects on other cells and tissues.

Division of Clinical Immunology and Rheumatology, Department of Medicine, University of Alabama at Birmingham, Birmingham, AL 35294, USA

The complement system consists of more than 30 distinct proteins (Table 1). In their native state these proteins are either serum soluble or associated with cell membranes. Functionally, complement proteins can be categorized as those participating in the activation sequences (the classical and the alternative pathway), those regulating the activation and the activities of the system, and those serving as receptors for active fragments. Certain complement proteins overlap these physicochemical and functional categories. The introduction of recombinant DNA methods in complement research in the early 1980s has contributed to the rapid acquisition of information regarding the primary structure, function, biosynthesis, and genetics of complement proteins and has paved the way for a structural definition of reactive sites and active centers.

2 Overview of Complement Activation

Despite the complexity implied by multiple interacting proteins, complement activation is characterized by relative simplicity and economy of design. The most important activities in terms of host defense are derived from two proteins, C3 and C5, which

Table 1. Proteins of the complement system

Prevalent form in native state	Functional group		
	Participating in activation sequence	Regulatory	Receptors
Serum soluble	C1q, C1r, C1s, D C4, C2, C3, B C5, C6, C7, C8, C9	C1INH C4bp, H, I, P C3a/C5a INA S protein	
Membrane associated		CR1, MCP DAF HRF	C1qR CR1, CR2, CR3, CR4, CR5 C3a/C4aR, C5aR

Established symbols have been used for most complement proteins. In addition, the following generally accepted abbreviations have been used: INH, inhibitor; C4p, C4b-binding protein INA, inactivator; R, receptor, e.g., CR1, complement receptor type 1; DAF, decay-accelerating factor; MCP, membrane cofactor protein; HRF, homologous restriction factor

are structurally homologous and probably represent gene duplication products (DE BRUIJN and FEY 1985; WETSEL et al. 1987). Additional biologically active fragments are derived from C4, another structural homolog of C3 (BELT et al. 1984) and perhaps also from C2 and factor B. Expression of activity requires cleavage of C3 and C5 by highly specific proteases, termed convertases (Fig. 1). There are two C3 and two C5 convertases. One of each is assembled during activation of complement by the classical or the alternative pathway. C3 convertases are bimolecular while C5 convertases

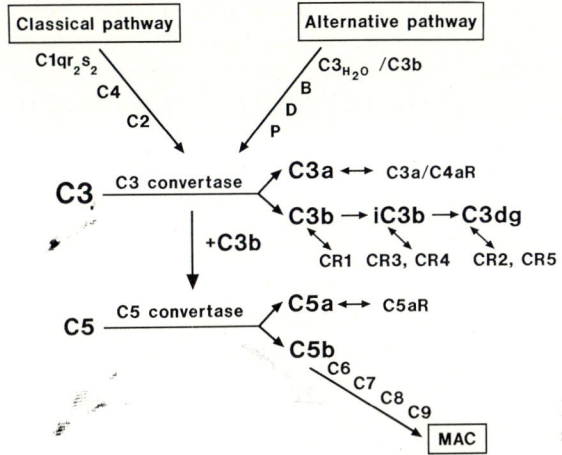

Fig. 1. Activation of the complement system. ↔, Ligand-receptor interactions

trimolecular protein complexes. The two activation pathways utilize different proteins to form these enzymes. In addition, the assembly of the convertases is initiated by different activating substances in the two pathways. However, the resulting enzymes have identical substrate and peptide bond specificity, giving rise to identical biological products.

Characteristic of the simplicity and economy of design of complement activation is the fact that C5 convertases are derivatives of C3 convertases (Fig. 1), and that C3 and C5 are activated by their respective convertases in similar fashion. In each case, a single peptide bond near the NH_2-terminus of the α-chain is cleaved to generate a small peptide, C3a or C5, and a large two-polypeptide chain fragment, C3b or C5b. Each of these four fragments expresses biological activities important to host defense. C3a, C5a, and C3b carry out their functions by interacting with specific receptors on the surface of effector cells (reviewed by FEARON and WONG 1983). Briefly, C3a, C5a, and also C4a, a homologous peptide generated during activation of C4, stimulate release of histamine from mast cells and blood basophils, contract smooth muscle, and increase vascular permeability. The structure and function of C3a is detailed in the article by HUGLI (this volume). C5a has additional functions evoking neutrophil responses, including chemotaxis, release of lysozomal enzymes, generation of oxygen radicals and increased adherence. C3b has multiple biologic activities that are mediated through interaction with a specific receptor, CR1, present on most blood cells.

The biologic consequences of C3b-CR1 interactions include clearance of immune complexes and phagocytosis and are discussed in the article by FEARON and AHEARN in this volume. C3b in complex with factor H, CR1, or membrane cofactor protein (MCP) is cleaved by factor I, a serine proteinase, to iC3b (DAVIS and HARRISON 1982), which is recognized by specific receptors, CR3 and CR4 (p150/95), present on leukocytes, resulting in enhanced phagocytosis. The structure and function of CR3 and CR4 are described in the article by ROSEN and LAW (this volume). Further cleavage of the α'-chain of iC3b by I in the presence of CR1 (Ross et al. 1982; MEDOF et al. 1982) generates C3c and C3dg. The latter fragment is recognized by specific receptors, CR2, on B lymphocytes and also by distinct receptors, CR5, on neutrophils. CR2 is discussed in the article by FEARON and AHEARN (this volume). The biologic activity of C5b does not depend on interaction with specific receptors. Instead, C5b initiates the assembly of a large protein-protein complex, termed membrane attack complex (MAC; MÜLLER-EBERHARD 1988), by interacting sequentially with C6, C7, C8, and C9. The resulting complex interacts directly with the lipid bilayer of membranes through hydrophobic domains of the participating proteins and eventually forms a transmembrane channel which leads to cell death.

The potential pathogenic potential of complement activation is controlled efficiently by several regulatory proteins (reviewed by VOLANAKIS 1988) that act at points of enzymatic amplification of the activation sequence and also at the level of effector molecules. They effectively control the extent of activation of the complement cascade and also protect the cells of the host from the pathogenic potential of complement activation products.

3 Participation of C3 in Complement Activation

From the above brief overview it is evident that C3 is the pivotal protein of the complement system (BARNUM 1989). It is the most abundant complement protein in blood (130 mg/dl) composed of two disulfide-linked polypeptide chains, α (115 kDa) and β (75 kDa). Clearly, cleavage of the Arg-77–Ser-78 bond near the NH_2-terminus of the α-chain of C3 is the key event in complement activation not only in terms of the multiple biologic activities expressed by C3a, C3b, and the further proteolytic degradation fragments of C3b, but also in terms of the multiple functions of C3b in the activation sequence per se. As shown in Fig. 1, C3b participates in the formation of the C3 and C5 convertases of the alternative pathway and of the C5 convertase in the classical pathway. Molecular interactions leading to the formation of these enzymes and the structure of participating proteins is discussed below.

4 C3 Convertase of the Alternative Pathway

4.1 The Noncatalytic Subunit $C3_{H2O}$/C3b

Formation of the alternative pathway C3 convertase is intimately related to certain unique structural and functional features of C3. The structure of C3 is discussed in the article by BECHERER et al. (this volume) and the biosynthesis and genetics of C3 in the article by BARNUM et al. (this volume). Structural features relevant to the assembly of C3 convertase are considered here briefly. The complete primary structure of human C3 has been deduced from the nucleotide sequence of overlapping cDNA clones (DE BRUIJN and FEY 1985). The molecule is encoded as a single 1663 amino acid long polypeptide chain, which includes a 22-residue signal peptide and four Arg residues linking the COOH-terminus of the β-chain to the NH_2-terminus of the α-chain. The Arg linker is removed during posttranslational modification leading to the two-polypeptide chain structure of the mature protein.

An additional important posttranslational modification takes place before secretion of C3 (IIJIMA et al. 1984). It involves the formation of a thioester bond between Cys-1010 and Gln-1013 of the pro-C3 polypeptide chain (TACK et al. 1980; THOMAS et al. 1982). The mechanism leading to the formation of the thioester bond is unknown, however it has been shown that in C4, which has a similar thioester bond, this process precedes proteolytic processing of the single polypeptide chain into the three chains of mature C4, but that it follows core N glycosylation (KARP 1983). VAN LEUVEN (1982) has suggested that the thioester bond forms during protein folding by the action of a transglutaminase-like active site in the polypeptide chain. Further folding of the polypeptide chain results in burying of the thioester bond in the hydrophobic interior of the molecule where it is protected from water and thus, relatively stable. The chemistry and function of protein thioester bonds are discussed in the article by LEVINE and DODDS (this volume). Under physiologic conditions the thioester bond in native C3 undergoes hydrolysis at slow rates with a half-life of 230 h giving rise to $C3_{H2O}$ (PANGBURN and MÜLLER-EBERHARD 1980), which displays a free sulfhydryl at Cys-1010 and a Glu at residue 1013 (numbering of pro-C3). Hydrolysis of the thioester results in conformational changes (ISENMAN et al. 1981) accompanied by loss of certain functional activities and acquisition of others. Among the latter is the ability of $C3_{H2O}$ to form Mg^{2+}-dependent complexes with complement factor B.

Formation of the $C3_{H2O}B(Mg^{2+})$ complex is considered to represent the first step towards formation of the so-called "initiation" C3 convertase of the alternative pathway (PANGBURN et al. 1981). In a second step, the serine proteinase, factor D catalyzes the cleavage of the Arg-228–Lys-229 peptide bond in B resulting in the release of the NH_2-terminal, 30 kDa, Ba fragment of B and the formation of the $\overline{C3_{H2O}Bb}$ complex which expresses C3 convertase activity. This series of reactions initiated with the hydrolysis of the thioester bond in native C3 and concluding with the formation of the initiation C3 convertase and cleavage of C3 into C3a and C3b is considered to occur in the blood continuously at slow rates (NICOL and LACHMAN 1973). Thus, a constant supply of small amounts of freshly generated C3b is available at all times. Both the $\overline{C3_{H2O}Bb}$ C3 convertase and the products of its catalytic action, C3a and C3b, are quickly inactivated by the control proteins, C3a/C5a INA and factors H and I.

Cleavage of C3 by a C3 convertase results in a pronounced conformational change in C3b (ISENMAN and COOPER 1981) associated with an extremely labile (metastable) thioester bond apparently resulting from its exposure on the surface of the molecule (LAW et al. 1980; SIM et al. 1981; TACK et al. 1980). The metastable thioester bond of C3b has a short half-life estimated at 60 µs and is highly reactive towards nucleophiles. The reactive carbonyl serves as acyl donor resulting in the formation of ester or amide bonds on reaction with hydroxyl or amino groups, respectively. Thus, C3b becomes covalently attached to the surface of neighboring cells or proteins displaying reactive nucleophiles. Alternatively, the metastable thioester bond reacts with H_2O to give fluid-phase C3b, which like $C3_{H2O}$ can form an unstable fluid-phase C3 convertase and thus contribute to the continuous physiologic low-level cleavage of C3. The fate of surface-bound C3b seems to be entirely dependent on the nature of the surface. C3b bound to the surface of a nonactivator of the alternative pathway, e.g., host's red cells, reacts preferentially with factor H which acts as a cofactor for the enzyme, factor I.

Cleavage of the α-chain of C3b by I results in iC3b which is subsequently cleaved further to C3c, released in the fluid phase, and C3dg, which remains bound to the surface. Additional control proteins CR1, decay-accelerating factor (DAF), and MCP on the membrane of host cells also interact with bound C3b preventing its interaction with B, dissociating C3b-bound B, and in the case of CR1 and MCP acting as cofactors for I. The structure and function of factor H is discussed in the article by VIK et al. (this volume); DAF and MCP are discussed in the article by LUBLIN and ATKINSON (this volume). In contrast, C3b bound to the surface of an activator of the alternative pathway has a higher avidity for factor B than for H (FEARON and AUSTEN 1977a, b) and thus can form a C3 convertase. Activators include various polysaccharides, lipopolysaccharides, bacteria, viruses, fungi, parasites, tumor cells, and certain immunoglobulins in complex with antigen (reviewed in (MÜLLER-EBERHARD and SCHREIBER 1980). The exact chemical features allowing discrimination between activators and nonactivators and their mode of action are not completely understood. Experiments by FEARON (1978) demonstrated that removal of sialic acid from the surface of sheep erythrocytes resulted in decreased binding of H to erythrocyte-bound C3b, thus transforming the cells from nonactivators to activators of the alternative pathway. Similar effects of sialic acid have been described for other cell surfaces (KAZATCHKINE et al. 1979; OKADA et al. 1982; EDWARDS et al. 1982). Subsequent experiments (reviewed in PANGBURN and MÜLLER-EBERHARD 1984) indicated sialic acid is probably not the only chemical moiety determining the properties of a surface with respect to alternative pathway activation.

The C3 convertase formed on the surface on an activator of the alternative pathway is termed "amplification" convertase because it forms part of a positive feedback loop, forming additional C3 convertase. Its assembly proceeds through steps identical to those described for the initiation C3 convertase. Factor B binds to $\overline{C3b}$ in the presence of Mg^{2+}; it is then cleaved by factor D giving rise to the $\overline{C3bBb}$ complex. An additional protein, properdin, plays an important role in the upregulation of the amplification C3 convertase. Properdin binds to the surface-bound $\overline{C3Bb}$ complex resulting in its stabilization (FEARON and AUSTEN 1975; MEDICUS et al. 1976). Binding of properdin may also protect the amplification C3 convertase from the action of the regulatory proteins, factors H and I. However, eventually the amplification C3 convertase comes under control with the release of Bb in an inactive from and the degradation of C3b to iC3b.

4.2 The Catalytic Subunit Factor B

Complement factor B is a 90-kDa single-polypeptide chain glycoprotein, structurally and functionally homologous to complement component C2. The genes for both proteins are located within the major histocompatibility complex locus on the short arm of human chromosome 6 (CARROLL et al. 1984; WHITEHEAD et al. 1985) and are less than 500 bp apart (CAMPBELL and BENTLEY 1985). The 5' end of the C2 gene lies approximately 600 kb from the 5' end of the gene encoding the HLA-B antigen and the 3' end of the factor B gene approximately 30 kb from the 5' end of the C4A gene (DUNHAM et al. 1987; CARROLL et al. 1987). The genes encoding tumor necrosis factors α and β were mapped between the C2 and the HLA-B genes and a novel gene, termed RD, between the factor B and the C4A genes (LÉVI-STRAUSS et al. 1988).

The complete primary structure of factor B has been determined from cDNA and protein sequencing (MOLE et al. 1984). The polypeptide chain consists of 739 amino acid residues with a calculated M_r of 83000. It exhibits 39% residue identity with C2 (BENTLEY 1986; HORIUCHI et al. 1989) which, along with the close proximity of the corresponding genes, indicates a gene duplication event at a distant evolutionary time. The polypeptide chain of B contains four sites for potential N-linked glycosylation and the mature protein 8.6% carbohydrate (TOMANA et al. 1985). Transmission electron micrographs have revealed a three-domain globular structure for B (SMITH et al. 1984a; UEDA et al. 1987). The gene segment coding for each of the three domains appear to have been derived from three unrelated gene superfamilies (Fig. 2). Thus, factor B represents an example of a "mosaic" protein (DOOLITTLE 1985) which is also true for several other complement proteins, including C1r, C1s C2, C7, C8α, C8β, C9, and factor I (reviewed in CAMPBELL et al. 1988).

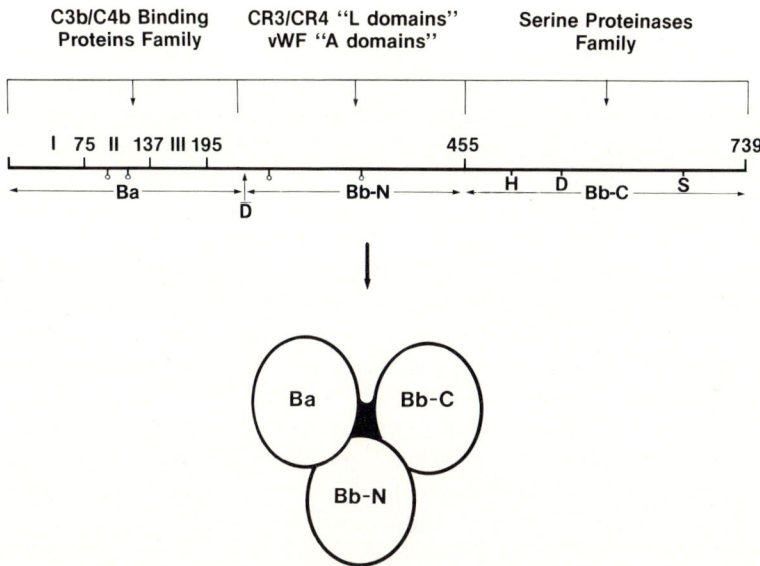

Fig. 2. Diagrammatic representation of the mature polypeptide chain of factor B. *Top*, the probable derivation of the gene segments encoding each of the three domains from three unrelated gene superfamilies; *thick line*, polypeptide chain: ↓, Possible N-glycosylation site; *H, D, S*, active site residues His, Asp, and Ser, respectively; I, II, III, consensus repeat units. *Bottom*, a schematic representation of the electron microscopic appearance of the molecule

The NH_2-terminal domain of factor B, termed Ba, contains three direct consensus repeats, each approximately 60 amino acid residues long. Each repeat is encoded by a separate exon (CAMPBELL et al. 1984). Interest in the structure and function of these repeats stemmed from their presence in variable numbers in C2, CR1, CR2, H, MCP, DAF, and C4p, all of which bind fragments of C3 and/or C4. Members of this newly described gene superfamily (reviewed in REID et al. 1986) also include proteins not known to bind C3 and/or C4 fragments such as β_2-glycoprotein I, clotting factor XIII, and the interleukin 2 receptor. Each of these proteins contains a different number of consensus repeats from two in C1r (JOURNET and TOSSI 1986) to 30 in CR1 (KLICKSTEIN et al. 1987). The consensus repeat units are contiguous, and in the complement regulatory proteins they represent the major structural element. The most characteristic shared structural feature among repeats is the presence of four invariable disulfide-linked half-cystine residues. It has been determined that in β_2-glycoprotein I (LOZIER et al. 1984) and in C4bp (JANATOVA et al. 1989) the first and third half cystines and the second and fourth are disulfide linked giving rise to a compact triple-loop structure. It seems likely that the same bonding pattern is present in the consensus repeats of all other proteins in the superfamily, including the three repeats of factor B.

UEDA et al. (1987) demonstrated that Fab fragments of a monoclonal antibody recognizing an epitope on the Ba domain inhibited in a dose-dependent fashion binding of intact B to red cell-bound C3b. In addition, PRYZDIAL and ISENMAN (1987) showed that Ba could inhibit the formation of the $\overline{\text{C3bBb}}$ convertase. They also demonstrated a specific, metal-ion independent interaction between Ba and C3b by using a cross-linking reagent. Taken together, these data indicate that Ba contains a binding site for C3b. Similar data have indicated the corresponding NH_2-terminal domain of C2, C2b, contains a similar binding site for C4b (NAGASAWA and STROUD 1977, KERR 1989; OGLESBY et al. 1988). The topology and chemical nature of these sites are unknown; however, it seems reasonable to assume they are contained within one of the consensus repeats. Several considerations indicate that the presence of this binding site on Ba cannot fully account for the initial binding of B to C3b or $C3_{H2O}$. First, Mg^{2+} is required for binding of B to C3b, and the Mg^{2+} binding site has been localized on Bb (FISHELSON et al. 1983). Second, neither Ba nor Bb in isolated form have substantial affinity for C3b. Third, Bb remains bound to C3b after cleavage of B by D. Electron micrographs of the C3 convertase, $\overline{\text{CoVFBb}}$, formed with the cobra analog of C3b, CoVF, have clearly shown Bb attached to CoVF through one of its two globular domains (SMITH et al. 1982). It thus seems likely that initial binding of B to $C3b/C3_{H2O}$ depends on two low-affinity binding sites, one on Ba and the other on one of the other two domains of B. Mg^{2+} apparently acts as an allosteric effector of the latter site.

The middle domain of factor B (residues 229–454; Fig. 2, Bb-N) exhibits amino acid sequence homology to a 187 amino acid residue long region near the NH_2 terminus of the α-chain of CR3 (Mac-1; CORBI et al. 1988) and to a homologous region on the α-chain of CR4 (p150/95; CORBI et al. 1987). Both of these leukocyte adhesion glycoproteins have binding affinity for iC3b, and interestingly they both require the presence of divalent cations for binding iC3b (Ross et al. 1983; MICKLEM and SIM 1985). CR3 and CR4 are structurally homologous to extracellular matrix receptors such as the vitronectin receptor, the fibronectin receptor, and glycoprotein IIb and are therefore considered to be members of a gene superfamily of cell-cell and cell-matrix receptors termed integrins (HYNES 1987). However, no other integrin has on its α-chain a region similar to those found in CR3 and CR4, which led to its designation as L domain to indicate its presence on leukocyte integrins only. A domain termed "A domain," homologous to the middle domain of factor B, is present in three imperfect tandem repeats in the polypeptide chain of von Willebrand factor (vWF) (SADLER et al. 1986). vWF is a glycoprotein playing an essential role in hemostasis. The region of vWF containing the A domains is believed to contain binding sites for collagen and platelet glycoprotein Ib (GIRMA et al. 1986).

The middle domain of C2 (residues 218–444) is homologous to that of factor B, but its homology to the L domains of CR3 and CR4 and to the A domains of vWF is not as strong as that of factor B. The middle domain of factor B and of C2 have been proposed (SMITH et al. 1982; PRYZDIAL and ISENMAN 1987) to contain binding sites for C3b and C4b, respectively, but direct supporting evidence is missing. However, studies on the oxidation of C2 by I_2, which results in stabilization of the $\overline{C4b2a}$, C3 convertase of the classical pathway (POLLEY and MÜLLER-EBERHARD 1967), demonstrated that the effect of I_2 is due to oxidation of the single free thiol of Cys-241 within this domain (PARKES et al. 1983). The relative stability of the C4b2aoxy convertase is apparently the result of a higher affinity binding of C2aoxy to C4b as compared to C2a as indicated from a slower rate of dissociation from the complex. Conversely, reduction of the free thiol abolishes the hemolytic activity of C2. It is of further interest that guinea pig C2, which forms a more stable convertase than human C2 (KERR and GAGNON 1982), has an Ala residue substituting for Cys-241 of human C2. Taken together these data are consistent with the suggestion (SMITH et al. 1982) that the middle domain of B contains a binding site for $C3b/C3_{H2O}$.

The carboxyl-terminal domain of factor B (Fig. 2, Bb-C) is homologous to serine proteinases, including highly conserved segments around the active site residues and the substrate binding site. Both intact B and fragment Bb express esterolytic activity, and the active site has been mapped with peptide thioester substrates homologous to the sequences at the cleavage/activation site of C3 and C5 (KAM et al. 1987). Bb was shown to be about ten fold more reactive towards these synthetic substrates than B and approximately 1000-fold less reactive than trypsin. This low catalytic efficiency is consistent with the mode of action of factor B in the activation sequence of the alternative pathway. Proteolytic activity against C3 is expressed by Bb only in the context of a complex with $C3_{H2O}$ or C3b. Comparison to other serine proteinases indicates the main distinctive structural feature of the carboxyl-terminal domain of factor B is the absence of a free NH_2-terminal residue and of the highly conserved amino terminal region that is characteristic of all members of this large family of homologous enzymes. In all other eukaryotic serine proteinases the positively charged α-NH_2-terminal residue is generated during activation from their zymogen form and plays an essential role in the molecular rearrangement that results in the enzymatically active conformation of the catalytic center (STROUD et al. 1975). In the apparent absence of a free NH_2-terminus, assumption of the catalytically active form of Bb must be achieved through a different mechanism probably involving C3b. A 33-kDa fragment obtained from a partial digest of B with porcine elastase and consisting essentially of the COOH-terminal domain of the protein was shown to exhibit serine esterase as well as hemolytic activity (LAMBRIS and MÜLLER-EBERHARD 1984). Interestingly, this fragment also displayed binding affinity for C3b.

4.3 Factor D

Complement factor D is the enzyme that catalyzes the cleavage of C3b-bound B, thus completing the assembly of the C3 convertase of the alternative pathway, $\overline{C3bBb}$. Factor D is a 24-kDa single-polypeptide chain serine proteinase. The serine proteinase nature of the enzyme was initially demonstrated by its irreversible inhibition by diisopropyl fluorophosphate (FEARON et al. 1974) and subsequently confirmed by amino acid sequence homologies with other members of the serine proteinase family of enzymes (VOLANAKIS et al. 1980; DAVIS 1980). The almost complete primary structure of factor D has been deduced from amino acid sequencing (JOHNSON et al. 1984; NIEMANN et al. 1984). The polypeptide chain consists of 222 amino acid residues with a calculated M_r of 23750. The primary structure of D exhibits about 40% identity with the B chain of human plasmin and an average of 35% identity with the pancreatic enzymes kallikrein, trypsin, chymotrypsin, and elastase. A much stronger homology (62%) is observed between factor D and a mouse protein, termed adipsin, synthesized by adipocytes and secreted in the blood (COOK et al. 1985, 1987). Recent studies (ROSEN et al. 1989) have indicated that adipsin has factor D-like functional properties i.e., it can cleave factor B into Ba and Bb only in the presence of C3b, and it can substitute for factor D in hemolytic assays. It thus seems likely that adipsin is mouse factor D.

Factor D isolated from serum of normal individuals or from urine of patients with Fanconi's syndrome (VOLANAKIS and MACON 1987) lacks the amino-terminal activation peptide that characterizes other serine proteinase zymogens. In addition, it exhibits esterolytic activity against peptide thioesters, homologous to the sequence of the factor B activation/cleavage site (KAM et al. 1987). However, the catalytic efficiency of D is 10^3–10^4 times lower than that of $\overline{C1s}$ and trypsin. The low esterolytic activity of purified factor D is compatible with the apparent absence of a zymogen for the enzyme in blood (LESAVRE and MÜLLER-EBERHARD 1978). A similar absence of a structural zymogen was noted in biosynthetic studies using U937 and HepG2 cells which secreted only active D (BARNUM and VOLANAKIS 1985a, b). A partial cDNA clone for factor D was isolated from a U937 library (MOLE and ANDERSON 1987). The nucleotide sequence of the insert has not been reported. However, it was reported to contain 17 amino acid residues at the NH_2-terminus not present in the sequence of D isolated from serum. It is not clear whether this sequence represents a leader or an activation peptide. However, it is interesting to note that an activation peptide coded for in the adipsin mRNA is apparently cleaved off before secretion of the enzyme by adipocytes or by mammalian cells transfected with adipsin cDNA (ROSEN et al. 1989; B. SPIEGELMAN, personal communication).

In the absence of a structural zymogen, other mechanisms are apparently contributing to the regulation of D activity in blood, including the extremely restricted specificity of the enzyme and its rapid catabolic rate. Studies in our laboratory (VOLANAKIS et al. 1985; SANDERS et al. 1986) have demonstrated that factor D is filtered through the glomerular membrane and catabolized in the proximal renal tubules, resulting in a fractional catabolic rate of 59.6% per hour (PASCUAL et al. 1988). This rapid catabolic rate maintains very low plasma levels of the enzyme (1.8 ± 0.4 µg/ml; BARNUM et al. 1984) and thus may contribute to the regulation of its activity. In fact, factor D is the limiting enzyme in the activation sequence of the alternative pathway (LESAVRE and MÜLLER-EBERHARD 1978). The mode of action of D may also contribute significantly to the regulation of its enzymatic activity. Active D cleaves its single known substrate, factor B, only in the context of the Mg^{2+}-dependent $C3_{H2O}$/C3bB complex. Active site mapping of D with peptide thioesters (KAM et al. 1987) and active site inhibitors (C. M. KAM et al. unpublished data) revealed some interesting functional features. As mentioned above, D was found to express esterolytic activity against Arg thioesters, but its catalytic efficiency was three to four orders of magnitude lower than that of $\overline{C1s}$ and trypsin. One of the most interesting findings was that extension of the substrate to include a P_3 or P_4 residue corresponding to the cleavage site of B resulted in loss of esterolytic activity. In contrast to its low reactivity with peptide thioester substrates, D reacted with active site inhibitors at rates comparable to those measured for trypsin. For example, APMSF inhibited D with a $K_{obs}/[I]$ of $110 M^{-1} s^{-1}$ as compared to $150 M^{-1} s^{-1}$ measured for trypsin. On the basis of these results, it has been proposed that the active center of D as it exists in serum exhibits a zymogen-like conformation, characterized by an obstructed binding site, and that the active conformation is induced by the substrate, $C3_{H2O}$/C3bB.

4.4 Properdin

Properdin is the final protein participating in the assembly of the amplification C3 convertase of the alternative pathway. It is necessary for the formation of a stable C3 convertase and therefore for efficient activation of the pathway. Properdin was first described by PILLEMER and his associates (1954) as an important component of host defenses against pathogens in a series of experiments demonstrating the existence of an alternative pathway for complement activation. Human properdin is a glycoprotein consisting of cyclic oligomers of a polypeptide chain of approximate M_r of 50000 (MINTA and LEPOW 1974; MEDICUS et al. 1980; DISCIPIO 1982). A partial amino acid sequence of human properdin has been published (REID and GAGNON 1981) and a cDNA clone coding for mouse properdin has been isolated and sequenced (GOUNDIS and REID 1988). The deduced amino acid sequence of mouse properdin indicated a 441-residue-long polypeptide chain with two sites for potential N glycosylation. The most interesting structural feature of properdin is the presence of six tandem consensus repeats, each approximately 60 amino acid residues in length, occupying the middle of the polypeptide chain. Alignment of the repeats to maximize homologies indicates 35 of the average 60 residues in each repeat are conserved, including six Cys, six Pro, and three Trp residues. Consensus repeats homologous to those found in properdin are also present in the adhesive glycoprotein thrombospondin (LAWLER and HYNES 1986) and in the complement proteins C6 (D. N. CHAKRAVARTI, personal communication), C7 (DISCIPIO et al. 1988), C8α (RAO et al. 1987), C8β (HOWARD et al. 1987), and C9 (DISCIPIO et al. 1984). Interestingly, the three thrombospondin consensus repeats are within a 70-kDa chymotryptic fragment known to contain binding sites for matrix proteins, including type V collagen, laminin, and fibronectin (MUMBY et al. 1984). It is of further interest that the properdin consensus repeat exhibits structural homology to conserved regions near the COOH-terminus of the polypeptide chains of circumsporozoite proteins from *Plasmodium falciparum*, *P. knowlesi*, and *P. vivax* (GOUNDIS and REID 1988).

An electron microscopic study of human properdin (SMITH et al. 1984b) revealed the protein to be polydisperse, consisting of cyclic dimers, trimers, tetramers, and higher oligomers. Trimers represented the most abundant form accounting for 45% of the observed oligomers followed by dimers (30%) and tetramers (10%). This distribution is in reasonably good agreement with that determined for properdin oligomers in human serum (PANGBURN 1989). Properdin protomers appeared as long, flexible rods exhibiting a reproducible sharp bend near the middle. No isolated monomers were detected, and no redistribution of individual oligomers was observed, indicating a high avidity of association between binding sites (SMITH et al. 1984b).

As mentioned above, the function of properdin relates to its ability to bind to the $\overline{\text{C3bBb}}$, C3 convertase resulting in a decreased rate of dissociation of Bb and thus stabilization of the enzyme (FEARON and AUSTEN 1975; MEDICUS et al. 1976). The primary binding specificity of properdin is for the C3b subunit of the convertase. Complexes of properdin and C3 in the fluid phase have been demonstrated (CHAPITIS and LEPOW 1976), and binding of properdin to red cell-bound C3b (EC3b) can be shown. An affinity constant of $2.9 \times 10^7 \, M^{-1}$ with a 1:1 stoichoimetry was determined for the binding of properdin to zymosan-bound C3b (DiSCIPIO 1981). The binding site for properdin has been mapped within a 34 amino acid residue segment of the α-chain of C3 (DAOUDAKI et al. 1988). The affinity of properdin for zymosan-bound C3b was found to increase by approximately four fold in the presence of factor B (DiSCIPIO 1981). Similarly, the affinity of properdin is higher for EC3bBb than for EC3b (FARRIES et al. 1988). Thus, it seems likely that Bb also has binding sites for properdin, although complexes between B or Bb and properdin have not been demonstrated.

4.5 Assembly of the Alternative Pathway C3 Convertase

A hypothetical model for the assembly of the alternative pathway C3 convertase, based on the information reviewed above, is shown in Fig. 3. Similar models have been proposed previously (PRYZDIAL and ISENMAN 1987; OGLESBY et al. 1988). Factor B is depicted as a three-domain structure based on its electron microscopic appearance (SMITH et al. 1984a; UEDA et al. 1987). Initial binding of B to activator-bound C3b depends on two low-affinity binding sites, one on the NH_2-terminal and the other on the middle domain of B. Mg^{2+} apparently acts as an allosteric effector of the latter binding site (FISHELSON et al. 1983). The precise topology of the corresponding sites on C3 has not been determined. Studies by BURGER et al. (1982) demonstrated that monoclonal antibodies reacting with the α-chain of C3 and the C3c fragment inhibited the binding of B but not that of H to surface-bound C3b. Anti-C3d monoclonals had a reverse effect, i.e., they inhibited binding of H but not of B. However, C3c has no binding affinity for B. More recently, O'KEEFE et al. (1988) obtained a C3 fragment, termed C3o, by digesting C3 with a protease derived from cobra venom. C3o is similar to C3c except for an additional ten amino acids (residues 955–964, pro-C3 numbering) at the carboxyl terminus of the NH_2-terminal fragment of the α'-chain of C3c. C3o was able to interact with B in the presence of Mg^{2+} in a manner similar to C3b, i.e., C3o-bound B could be cleaved by D into Ba and Bb. It thus seems likely that the region 955-Glu-Gly-Val-Gln-Lys-Glu-Asp-Ile-Pro-Pro-

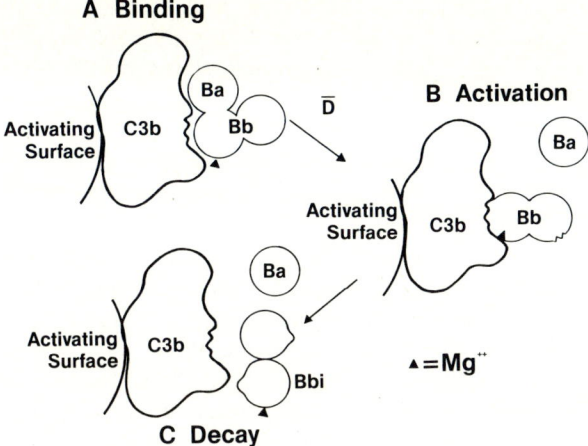

Fig. 3. Hypothetical model for the assembly (A), activation (B), and decay (C) of the C3 convertase of the alternative pathway of complement activation

964 of the α-chain of C3 represents one of the presumed two-factor B binding sites. A thermodynamic study of the binding of factor B to fluid-phase C3b or $C3_{H_2O}$ (PRYZDIAL and ISENMAN 1988) indicated that, under physiologic conditions, hydrophobic interactions are dominant in C3bB or $C3_{H_2O}B$ complex formation. Ionic forces are also likely to contribute to the interactions since low ionic strength enhances complex formation (DISCIPIO 1981). The association constant for binding of B to C3b was determined to be $2.5 \times 10^6 \, M^{-1}$, whereas a five fold lower affinity was measured for the $C3_{H_2O}B$ interaction (PRYZDIAL and ISENMAN 1988).

$C3b/C3_{H_2O}$-bound B interacts with D inducing the active conformation of the enzyme which in turn catalyzes the cleavage of the Arg-228–Lys-229 bond in B. Cleavage of B induces a transient conformational change in Bb resulting in higher binding affinity for C3b, sequestration of Mg^{2+} (FISHELSON et al. 1983), and expression of proteolytic activity for C3 (Fig. 3). The transient conformation can be stabilized by properdin. Decay of the C3 convertase is due to dissociation of Bb which is accelerated by binding of H, CR1, DAF, and MCP to C3b. The conformation of dissociated Bb in Fig. 3 is depicted as different from that in the native molecule based on the electron microscopic appearance of Bb which is characterized by projections and angular surfaces (UEDA et al. 1987). In addition, the finding of 99% loss of C3-cleaving activity of dissociated, as compared to C3b-bound, Bb (FISHELSON and MÜLLER-EBERHARD 1984) indicates a diffent conformation of the active center.

5 C5 Convertase of the Alternative Pathway

Cleavage of C3 by the amplification C3 convertase formed on the surface of an alternative pathway activator results in the deposition of many C3b molecules in the vicinity of the convertase and eventually leads to the formation of a $\overline{C3bBbC3b}$

complex which expresses C5 convertase activity (DAHA et al. 1976; MEDICUS et al. 1976). It was originally thought that the second C3b molecule in the trimolecular complex was bound covalently to the surface of the activator in the immediate vicinity of the C3 convertase. The second C3b molecule was shown to provide a binding site for C5 which allowed its cleavage by Bb into C5a and C5b (VOGT et al. 1978; ISENMAN et al. 1980). The C3b-C5 interaction was found to exhibit a stoichiometry of 1:1 and an association constant of 4.8×10^5 M^{-1} for fluid-phase C3b (ISENMAN et al. 1980) and 5.7×10^6 M^{-1} for bound C3b (DiSCIPIO 1981). Recently, KINOSHITA et al. (1988) demonstrated the second C3b molecule in the $\overline{C3bBbC3b}$ complex is covalently attached through an ester bond to the α'-chain of the first, surface-bound C3b molecule. Thus, the alternative pathway C5 convertase can be described as a trimolecular complex in which the catalytic subunit, Bb is bound noncovalently to a covalently linked C3b dimer. The association constant for binding of C5 to the C3b dimer is $1.5-2.4 \times 10^8$ M^{-1}, significantly higher than that for monomeric C3b. This finding suggests (KINOSHITA et al. 1988) that C5 binds to both C3b molecules in the complex through two relatively low affinity sites resulting in an increased avidity for the C3b-C3b complex and, thus, selective binding of C5 to the convertase.

6 C5 Convertase of the Classical Pathway

In the classical pathway the C5 convertase forms by a mechanism similar to that in the alternative pathway, i.e., C3b binds to the C3 convertase resulting in a switch of substrate specificity from C3 to C5. The classical pathway C3 convertase is a bimolecular complex, $\overline{C4bC2a}$, assembled on the surface of activators through steps similar to those described for the alternative pathway (reviewed in REID and PORTER 1981; MÜLLER-EBERHARD 1988). The main difference between the two pathways is that in the classical pathway the activating enzyme, $\overline{C1s}$, generates both the noncatalytic and the catalytic subunits of the convertase, C4b and C2a, respectively. Further, $\overline{C1s}$ circulates in the blood in enzymatically inactive zymogen form, C1s, as part of Ca^{2+}-dependent complex $C1qr_2s_2$ of three proteins. Activation of C1s by $\overline{C1r}$ follows binding of the recognition protein C1q to a classical pathway activator and the ensuing autocatalytic activation of C1r (LOOS 1982; COLOMB et al. 1984; SCHUMAKER 1987). $\overline{C1s}$ cleaves C4, a protein structurally homologous to C3, into C4a and C4b resulting in the covalent binding of C4b to the activator through a transacylation reaction involving the reactive carbonyl group of the internal thioester of C4 and hydroxyl or amino groups, depending on the C4 isotype, on the surface of the activator (LAW et al. 1984; ISENMAN and YOUNG 1984). C2, a homolog of factor B, binds to surface-bound C4b in a Mg^{2+}-dependent reaction and is cleaved by $\overline{C1s}$ into C2b, released in the fluid-phase and the two-domain C2a which remains bound to C4b, representing the catalytic domain of both the C3 and the C5 convertase. Similarly to Bb, C2a expresses proteolytic activity for C3 only in the context of a complex with C4b. Cleavage of C3 by the $\overline{C4b2a}$ C3 convertase results in deposition of C3b molecules on the surface of the activator. These C3b molecules can initiate the assembly of the amplification C3 convertase depending on the chemical nature of the activating surface. Formation of C5 convertase requires the covalent binding of

C3b to the α'-chain of activator-bound C4b through an ester bond (TAKATA et al. 1987). The C4b-C3b ester bond is relatively unstable with an estimated half-life at 37 °C of 7.9 h. An association constant of 2.1×10^8 M^{-1} was measured for the binding of C5 to C4b-C3b dimers, suggesting that both C4b and C3b may provide binding sites for C5 (TAKATA et al. 1987).

7 Summary

C3, the most abundant complement protein in blood, plays a central role in the activation sequence of the complement system as well as in host defense. Expression of the multiple functions of C3 requires its cleavage by highly specific enzymes termed C3 convertases. C3 in a conformationally altered form, $C3_{H2O}$, resulting from the slow spontaneous hydrolysis of the internal thioester bond of native C3, initiates the assembly of a C3 convertase which continuously cleaves C3 in the blood at slow rates generating a constant supply of small amounts of C3b. When an activator of the alternative complement pathway is present, C3b becomes covalently attached to its surface via an ester or amide bond. Activator surface-bound C3b initiates the assembly of an "amplification" C3 convertase, $\overline{C3bBb(P)}$, which can efficiently activate C3 and generate additional convertase complexes on the surface of the activator. C3b generated by an amplification or classical pathway C3 convertase can also bind covalently to the noncatalytic subunit, C3b or C4b, respectively, resulting in the generation of a C5 convertase, an enzyme catalyzing the cleavage/activation of C5. In terms of participation in host defense, several fragments of C3, including C3a, C3b, iC3b, and C3dg, mediate a number of important functions such as increased vascular permeability, enhancement of phagocytosis, elimination of immune complexes, and perhaps also proliferative responses and/or differentiation of B cells.

Acknowledgement. The expert secretarial assistance of Ms. Beth McDowell is gratefully acknowledged.

References

Barnum SR (1989) C3 structure and molecular genetics. In: Sim RB (ed) Biochemistry and molecular biology of complement. MTP Press Lancaster
Barnum SR, Volanakis JE (1985a) In vitro biosynthesis of complement protein D by U937 cells. J Immunol 134: 1799–1803
Barnum SR, Volanakis JE (1985b) Biosynthesis of complement protein D by HepG2 cells: a comparison of D produced by HepG2 cells, U937 cells and blood monocytes. Eur J Immunol 15: 1148–1151
Barnum SR, Niemann MA, Kearney JF, Volanakis JE (1984) Quantitation of complement factor D in human serum by a solid-phase radioimmunoassay. J Immunol Methods 67: 303–309
Belt KT, Carroll MC, Porter RR (1984) The structural basis of the multiple forms of human complement component C4. Cell 36: 907–914
Bentley DR (1986) Primary structure of human complement component C2. Homology to two unrelated protein families. Biochem J 239: 339–345

Burger R, Denbel U, Hadding U, Bitter-Suermann D (1982) Identification of functionally relevant determinants on the complement component C3 with monoclonal antibodies. J Immunol 129: 2042–2050

Campbell RD, Bentley DR (1985) The structure and genetics of the C2 and factor B genes. Immunol Rev 87: 19–37

Campbell RD, Bentley DR, Morley BJ (1984) The factor B and C2 genes. Philos Trans R Soc Lond [Biol] 306: 367–378

Campbell RD, Law SKA, Reid KBM, Sim RB (1988) Structure, organization, and regulation of the complement genes. Annu Rev Immunol 6: 161–195

Carrol MC, Campbell RD, Bentley DR, Porter RR (1984) A molecular map of the human major histocompatibility complex class III region linking complement genes C4, C2 and factor B. Nature 307: 237–241

Carrol MC, Katzman P, Alicot EM, Koller BH, Geraughty DE, Orr HT, Strominger JL, Spies T (1987) Linkage map of the human major histocompatibility complex including the tumor necrosis factor genes. Proc Natl Acad Sci USA 84: 8535–8539

Chapitis J, Lepow IH (1976) Multiple sedimenting species of properdin in human serum and interaction of purified properdin with the third component of complement. J Exp Med 143: 241–257

Colomb MG, Arlaud GJ, Villiers CL (1984) Activation of Cl. Philos Trans R Soc London [Biol] 306: 283–292

Cook KS, Groves DL, Min HY, Spiegelman BM (1985) A developmentally regulated mRNA from 3T3 adipocytes encodes a novel serine protease homologue. Proc Natl Acad Sci USA 82: 6480–6484

Cook KS, Min HY, Johnson D, Chaplinsky RJ, Flier JS, Hunt CR, Spiegelman BM (1987) Adipsin: a circulating serine protease homolog secreted by adipose tissue and sciatic nerve. Science 237: 402–405

Corbi Al, Miller LJ, O'Connor k, Larson RS, Springer TA (1987) cDNA cloning and complete primary structure of the alpha subunit of a leukocyte adhesion glycoprotein, p 150, 95. EMBO J 6: 4023–4028

Corbi AL, Kishimoto TK, Miller LJ, Springer TA (1988) The human leukocyte adhesion glycoprotein Mac-1 (complement receptor type 3, CD11b) α subunit. Cloning, primary structure, and relation to the integrins, von Willebrand factor and factor B. J Biol Chem 263: 12403–12422

Daha MR, Fearon DT, Austen KF (1976) C3 requirements of alternative pathway C5 convertase. J Immunol 117: 630–634

Daoudaki ME, Becherer JD, Lambris JD (1988) A 34-amino acid peptide of the third component of complement mediates properdin binding. J Immunol 140: 1577–1580

Davis AE (1980) Active site amino acid sequence of human factor D. Proc Natl Acad Sci USA 77: 4938–4942

Davis AE, Harrison RA (1982) Structural characterization of factor I — mediated cleavage of the third component of complement. Biochemistry 21: 5745–5749

de Bruijn MHL, Fey GH (1985) Human complement component C3: cDNA coding sequence and derived primary structure. Proc Natl Acad Sci USA 82: 708–712

DiScipio RG (1981) The binding of human complement proteins C5, factor B, $\beta_1 H$ and properdin to complement fragment C3b on zymosan. Biochem J 199: 485–495

DiScipio RG (1982) Properdin is a trimer. Mol Immunol 19: 631–635

DiScipio RG, Gehring MR, Podack ER, Kan CC, Hugli TE, Fey GH (1984) Nucleotide sequence of cDNA and derived amino acid sequence of human complement C9. Proc Natl Acad Sci USA 81: 7298–7302

DiScipio RG, Chakravarti DN, Müller-Eberhard HJ, Fey GH (1988) The structure of human complement C7 and the C5b-7 complex. J Biol Chem 263: 549–560

Doolittle RF (1985) The genealogy of some recently-evolved vertebrate proteins. Trends Biochem Sci 10: 233–237

Dunham I, Sargent CA, Trowsdale J, Campbell RD (1987) Molecular mapping of the human major histocompatibility complex by pulsed-field gel electrophoresis. Proc Natl Acad Sci USA 84: 7237–7241

Edwards MS, Kasper DL, Jennings AF, Baker CJ, Nicholson-Weller A (1982) Capsular sialic acid prevents activation of the alternative complement pathway by type III, group B streptococci. J Immunol 128: 1278–1283

Farries TC, Lachmann PJ, Harrison RA (1988) Analysis of the interaction between properdin and factor B, components of the alternative pathway C3 convertase of complement. Biochem J 253: 667–675

Fearon DT (1978) Regulation by membrane sialic acid of β1H-dependent decay-dissociation of amplification C3 convertase of the alternative complement pathway. Proc Natl Acad Sci USA 75: 1971–1975

Fearon DT, Austen KF (1975) Properdin: binding to C3b and stabilization of the C3b-dependent C3 convertase. J Exp Med 142: 856–863

Fearon DT, Austen KF (1977a) Activation of the alternative complement pathway due to resistance of zymosan-bound amplification convertase to endogenous regulatory mechanisms. Proc Natl Acad Sci USA 74: 1683–1687

Fearon DT, Austen KF (1977b) Activation of the alternative pathway with rabbit erythrocytes by circumvention of the regulatory action of endogenous control proteins. J Exp Med 146: 22–33

Fearon DT, Wong WW (1983) Complement ligand-receptor interactions that mediate biological responses. Annu Rev Immunol 1: 243–271

Fearon DT, Austen KF, Ruddy S (1974) Properdin factor D: characterization of its active site and isolation of the precursor form. J Exp Med 139: 355–366

Fishelson Z, Müller-Eberhard HJ (1984) Residual hemolytic and proteolytic activity expressed by Bb after decay-dissociation of $\overline{C3bBb}$. J Immunol 132: 1425–1429

Fishelson Z, Pangburn MK, Müller-Eberhard HJ (1983) C3-convertase of the alternative complement pathway. Demonstration of an active stable C3b,B(Ni) complex. J Biol Chem 258: 7411–7415

Girma JP, Kalafatis M, Pietu G, Lavergne J-M, Chopek MW, Edgington TS, Meyer D (1986) Mapping of distinct von Willebrand factor domains interacting with platelet GPIb and GPIIb/IIIa and with collagen using monoclonal antibodies. Blood 67: 1356–1366

Goundis D, Reid KBM (1988) Properdin, the terminal complement components, thrombospondin and the circumsporozoite protein of malarial parasites contain similar sequence motifs. Nature 335: 82–85

Horiuchi T, Macon KJ, Kidd VJ, Volanakis JE (1989) cDNA cloning and expression of human complement component C2. J Immunol (in press)

Howard OMZ, Rao AG, Sodetz JM (1987) Complementary DNA and derived amino acid sequence of the β subunit of complement protein C8: identification of a close structural and ancestral relationship to the α subunit and C9. Biochemistry 26: 3565–3570

Hynes RO (1987) Integrins: a family of cell surface receptors. Cell 48: 549–554

Iijima M, Tobe T, Sakamoto T, Tomita M (1984) Biosynthesis of the internal thioester bond of the third component of complement. J Biochem 96: 1539–1546

Isenman DE, Cooper NR (1981) The structure and function of the third component of human complement. I. The nature and extent of conformational changes accompanying C3 activation. Mol Immunol 18: 331–339

Isenman DE, Young JR (1984) The molecular basis for the difference in immune hemolysis activity of the Chido and Rodgers isotypes of human complement component C4. J Immunol 132: 3019–3027

Isenman DE, Podack ER, Cooper NR (1980) The interaction of C5 with C3b in free solution: a sufficient condition for cleavage by fluid phase C3/C5 convertase. J Immunol 124: 326–331

Isenman DE, Kells DI, Cooper NR, Müller-Eberhard HJ, Pangburn MK (1981) Nucleophilic modification of human complement protein C3: correlation of conformational changes with acquisition of C3b-like functional properties. Biochemistry 20: 4458–4467

Janatova J, Reid KBM, Willis AC (1989) Disulfide bonds are localized within the short consensus repeat units of complement regulatory proteins; C4b-binding protein. Biochemistry (in press)

Johnson DMA, Gagnon J, Reid KBM (1984) Amino acid sequence of human factor D of the complement system. Similarity in sequence between factor D and proteases of non-plasma origin. FEBS Lett 166: 347–351

Journet A, Tosi M (1986) Cloning and sequencing of full-length cDNA encoding the precursor of human complement component C1r. Biochem J 240 : 783–787

Kam C-M, McRae BJ, Harper JW, Niemann MA, Volanakis JE, Powers JC (1987) Human complement proteins D, C2, and B. Active site mapping with peptide thioester substrates. J Biol Chem 262: 3444–3451

Karp DR (1983) Post-translational modification of the fourth component of complement. Effect of tunicamycin and amino acid analogs on the formation of the internal thiol ester and disulfide bonds. J Biol Chem 258 : 14490–14495

Kazatchkine MD, Fearon DT, Austen KF (1979) Human alternative complement pathway: membrane-associated sialic acid regulates the competition between B and β1H for cell-bound C3b. J Immunol 122: 75–81

Kerr MA (1980) The human complement system. Assembly of the classical pathway C3 convertase. Biochem J 189: 173–181

Kerr MA, Gagnon J (1982) The purification and properties of the second component of guinea pig complement. Biochem J 205: 59–67

Kinoshita T, Takata Y, Kozono H, Takeda J, Hong K, Inoue K (1988) C5 convertase of the alternative complement pathway: covalent linkage between two C3b molecules within the trimolecular complex enzyme. J Immunol 141: 3895–3901

Klickstein LB, Wong WW, Smith JA, Weis JH, Wilson JG, Fearon DT (1987) Human C3b/C4b receptor (CR1). Demonstration of long homologous repeating domains that are composed of the short consensus repeats characteristic of C3/C4 binding proteins. J Exp Med 165: 1095–1112

Lambris JD, Müller-Eberhard HJ (1984) Isolation and characterization of a 33000-Dalton fragment of complement factor B with catalytic and C3b binding activity. J Biol Chem 259: 12685–12690

Law SK, Lichtenberg NA, Levine RP (1980) Covalent binding and hemolytic activity of complement proteins. Proc Natl Acad Sci USA 77: 7194–7198

Law SKA, Dodds AW, Porter RR (1984) A comparison of the properties of two classes, C4A and C4B, of the human complement component C4. EMBO J 3: 1819–1823

Lawler J, Hynes RO (1986) The structure of human thrombospondin, an adhesive glycoprotein with multiple calcium-binding sites and homologies with several different proteins. J Cell Biol 103: 1635–1648

Lesavre P, Müller-Eberhard HJ (1978) Mechanism of action of factor D of the alternative complement pathway. J Exp Med 148: 1498–1509

Lévi-Straus M, Carroll MC, Steinmetz M, Meo T (1988) A previously undetected MHC gene with an unusual periodic structure. Science 240: 201–204

Loos M (1982) The classical complement pathway: mechanism of activation of the first component by antigen-antibody complexes. Prog Allergy 30: 135–192

Lozier J, Takahasi N, Putnam FW (1984) Complete amino acid sequence of human plasma β_2-glycoprotein I. Proc Natl Acad Sci USA 81: 3640–3644

Medicus RG, Götze O, Müller-Eberhard HJ (1976) Alternative pathway of complement: recruitment of precursor properdin by the labile C3/C5 convertase and the potentiation of the pathway. J Exp Med 144 : 1076–1093

Medicus RG, Esser AF, Fernandez HN, Müller-Eberhard HJ (1980) Native and activated properdin: interconvertability and identity of amino- and carboxy-terminal sequences. J Imunol 124: 602–606

Medof ME, Prince GM, Mold C (1982) Release of soluble immune complexes from immune adherence receptors on human erythrocytes is mediated by C3b inactivator independently of β1H and is accompanied by generation of C3c. Proc Natl Acad Sci USA 79: 5047–5051

Micklem KJ, Sim RB (1985) Isolation of complement-fragment-iC3b-binding proteins by affinity chromatography. The identification of p 150, 95 as an iC3b-binding protein. Biochem J 231: 233–236

Minta JO, Lepow IH (1974) Studies on the subunit structure of human properdin. Immunochemistry 11: 361–368

Mole JE, Anderson JE (1987) Cloning the cDNA for complement factor D: evidence for the existence of a zymogen for the serum enzyme (Abstract). Complement 4: 196

Mole JE, Anderson JE, Davison EA, Woods DE (1984) Complete primary structure for the zymogen of human complement factor B. J Biol Chem 259: 3407–3412

Müller-Eberhard HJ (1988) Molecular organization and function of the complement system. Annu Rev Biochem 57: 321–347

Müller-Eberhard HJ, Miescher PA (eds) (1985) Complement. Springer, Berlin Heidelberg New York

Müller-Eberhard HJ, Schreiber RD (1980) Molecular biology and chemistry of the alternative pathway of complement. Adv Immunol 29: 1–53

Mumby SM, Rangi GJ, Bornstein P (1984) Interactions of thrombospondin with extracellular matrix proteins: selective binding to type V collagen. J Cell Biol 98: 646–652

Nagasawa S, Stroud RM (1977) Cleavage of C2 by C1 into the antigenically distinct fragments C2a and C2b: demonstration of binding of C2b to C4. Proc Natl Acad Sci USA 74: 2998–3001

Nicol PAE, Lachman PJ (1973) The alternative pathway of complement activation. The role of C3 and its inactivator (KAF). Immunology 24: 259–275

Niemann MA, Bhown AS, Bennett JC, Volanakis JE (1984) Amino acid sequence of human D of the alternative complement pathway. Biochemistry 23: 2482–2486

Oglesby TJ, Accavitti MA, Volanakis JE (1988) Evidence for a C4b-binding site on the C2b domain of C2. J Immunol 141: 926–931

Okada N, Yasuda T, Okada H (1982) Restriction of alternative complement pathway activation by sialoglycolipids. Nature 299: 261–263

O'Keefe MC, Caporale LH, Vogel C-W (1988) A novel cleavage product of human complement component C3 with structural and functional properties of cobra venom factor. J Biol Chem 263: 12690–12697

Pangburn MK (1989) Analysis of the natural polymeric forms of human properdin and their functions in complement activation. J Immunol 142: 202–207

Pangburn MK, Müller-Eberhard HJ (1980) Relationship of a putative thiolester bond in C3 to activation of the alternative pathway and the binding of C3b to biological targets of complement. J Exp Med 152: 1102–1114

Pangburn MK, Müller-Eberhard HJ (1984) The alternative pathway of complement. Springer Semin Immunpathol 7: 163–172

Pangburn MK, Schreiber RD, Müller-Eberhard HJ (1981) Formation of the initial C3 convertase of the alternative complement pathway. Acquisition of C3b-like activities by spontaneous hydrolysis of the putative thioester in native C3. J Exp Med 154: 856–867

Parkes C, Gagnon J, Kerr MA (1983) The reaction of iodine and thio-blocking reagents with human complement components C2 and factor B. Biochem J 213: 201–209

Pascual M, Steiger G, Estreicher J, Macon K, Volanakis JE, Schifferli JA (1988) Metabolism of complement factor D in renal failure. Kidney Int 34: 529–536

Pillemer L, Blum L, Lepow IH, Ross OA, Todd EW, Wardlaw AC (1954) The properdin system and immunity. I. Demonstration and isolation of a new serum protein, properdin, and its role in immune phenomena. Science 120: 279–285

Polley MJ, Müller-Eberhard HJ (1967) Enhancement of the hemolytic activity of the second component of human complement by oxidation. J Exp Med 126: 1013–1025

Pryzdial ELG, Isenman DE (1987) Alternative complement pathway activation fragment Ba binds to C3b. Evidence that formation of the factor B-C3b complex involves two discrete points of contact. J Biol Chem 262: 1519–1525

Pryzdial ELG, Isenman DE (1988) A thermodynamic study of the interaction between human complement components C3b or $C3(H_2O)$ and factor B in solution. J Biol Chem 263: 1733–1738

Rao AG, Howard OMZ, Ng SC, Whitehead AS, Colten HR, Sodetz JM (1987) Complementary DNA and derived amino acid sequence of the α subunit of human complement protein C8: evidence for the existence of a separate α subunit messenger RNA. Biochemistry 26: 3556–3564

Reid KBM, Gagnon J (1981) Amino acid sequence studies of human properdin. N-terminal sequence analysis and alignment of the fragments produced by limited proteolysis with trypsin and the peptides produced by cyanogen bromide treatment. Mol Immunol 18: 949–959

Reid KBM, Bentley DR, Cambell RD, Chung LP, Sim RB, Kristensen T, Tack BF (1986) Complement system proteins which interact with C3b or C4b. A superfamily of structurally related proteins. Immunol Today 7: 230–234

Rosen BS, Cook KS, Yaglom J, Groves DL, Volanakis JE, Damm D, White T, Spiegelman B (1989) Adipsin has complement factor D activity: an immune-related defect in genetic and acquired obesity. (to be published)

Ross GD, Lambris JD, Cain JA, Newman SL (1982) Generation of three different fragments of bound C3 with purified factor I or serum I. Requirements for factor H vs CR1 cofactor activity. J Immunol 129: 2051–2060

Ross GD, Newman SL, Lambris JD, Devery-Pocius JE, Cain JA, Lachman PJ (1983) Generation of three different fragments of bound C3 with purified factor I or serum. II. Location of binding sites in the C3 fragments for factor B and H, complement receptors, and bovine conglutinin. J Exp Med 158: 334–352

Sadler JE, Shelton-Inloes BB, Sorace JM, Titani K (1986) Cloning of cDNA and genomic DNA for human von Willebrand factor. Cold Spring Harbor Symp Quant Biol 51: 515–523

Sanders PW, Volanakis JE, Rostand SG, Galla JH (1986) Human complement protein D catabolism by the rat kidney. J Clin Invest 77: 1293–1304

Schumaker VN (1987) Activation of the first component of complement. Annu Rev Immunol 5: 21–42

Sim RB, Twose TM, Paterson DS, Sim E (1981) The covalrnt-binding reaction of complement component C3. Biochem J 193: 115–127

Smith CA, Vogel CW, Müller-Eberhard HJ (1982) Ultrastructure of cobra venom factor-dependent C3/C5 convertase and its zymogen, factor B of human complement. J Biol Chem 257: 9879–9882

Smith CA, Vogel CW, Müller-Eberhard HJ (1984a) MHC class III products. An electron microscopic study of the C3 convertases of human complement. J Exp Med 159: 324–329

Smith CA, Pangburn MK, Vogel CW, Müller-Eberhard HJ (1984b) Molecular architecture of human properdin, a positive regulator of the alternative pathway of complement. J Biol Chem 259: 4582–4588

Stroud RM, Krieger M, Koeppe RE, Kossiakoff AA, Chambers JL (1975) Structure-function relationships in the serine proteases. In: Reich E, Rifkin DB, Shaw E (eds) Proteases and biological control. Cold Spring Harbor, New York, pp 13–42

Tack BF, Harrison RA, Janatova J, Thomas ML, Prahl JW (1980) Evidence for the presence of an internal thioester bond in third component of human complement. Proc Natl Acad Sci USA 77: 5764–5768

Takata Y, Kinoshita T, Kozono H, Takeda J, Tanaka E, Hong K, Inoue K (1987) Conalent association of C3b with C4b within C5 convertase of the classical complement pathway. J Exp Med 165: 1494–1507

Thomas ML, Janatova J, Gray WR, Tack BF (1982) Third component of human complement: localization of the internal thiolester bond. Proc Natl Acad Sci USA 79: 1054–1058

Tomana M, Niemann M, Garner C, Volanakis JE (1985) Carbohydrate composition of the second, third and fifth components and factor B and D of human complement. Mol Immunol 22: 107–111

Ueda A, Kearney JF, Roux KH, Volanakis JE (1987) Probing functional sites on complement protein B with monoclonal antibodies. Evidence for C3b-binding sites on Ba. J Immunol 138: 1143–1149

van Leuven F (1982) Human alpha-2-macroglobulin. Structure and function. Trends Biochem Sci 7: 185–187

Vogt W, Schmidt G, von Buttlar B, Dieminger L (1978) A new function of the activated third component of complement: binding to C5, an essential step for C5 activation. Immunology 34: 29–40

Volanakis JE (1988) Structure, molecular genetics, and function of complement control proteins: an update. In: The year in immunology, vol 3. Karger, Basel, pp 275–290

Volanakis JE, Macon KJ (1987) Isolation of complement protein D from urine of patients with Franconi's syndrome. Anal Biochem 163: 242–246

Volanakis JE, Bhown AS, Bennett JC, Mole JE (1980) Partial amino acid sequence of human factor D: homology with serine proteases. Proc Natl Acad Sci USA 77: 116–119

Volanakis JE, Barnum SR, Giddens M, Galla JH (1985) Renal filtration and catabolism of complement protein D. N Engl J Med 312: 395–399

Wetsel RA, Ogata RT, Tack BF (1987) Primary structure of the fifth component of murine complement. Biochemistry 26: 737–743

Whitehead AS, Colten HR, Chang CC, Demars R (1985) Localization of the human MHC-linked complement genes between HLA-B and HLA-DR by using HLA mutant cell lines. J Immunol 134:641–643

Biosynthesis and Genetics of C3*

S. R. BARNUM[1], G. FEY[2], and B. F. TACK[2]

1 Introduction 23
2 C3 Biochemistry and Structure 23
3 C3 Biosynthesis 25
4 Sites of C3 Biosynthesis 26
5 C3 Transcription and Alternative Transcripts 29
6 C3 and the Acute-Phase Response 29
7 C3 Gene Structure 31
7.1 Chromosomal Location 35
7.2 C3 Polymorphism and Deficiency 36
8 Conclusions 37
References 37

1 Introduction

The biosynthesis, regulation of synthesis, and genetics of C3 have been extensively studied (for reviews see COLTEN 1976; COLE and COLTEN 1988; FEY et al. 1983). In fact, perhaps more is known about these aspects of C3 biology than about any other complement component. This extensive characterization is due primarily to the central role of C3 in complement activation and regulation as well as in the host immune response. Certain practical aspects have also contributed to the diverse background on C3, including its early characterization relative to many other complement proteins and its ready availability due to its high serum concentration.

2. C3 Biochemistry and Structure

Human C3 is a glycoprotein with a mean serum concentration of 1.2–1.3 mg/ml (KOHLER and MÜLLER-EBERHARD 1967). The molecular weight of C3 is approximately 183,000, as calculated from the derived amino acid sequence (DEBRUIJN and FEY 1985)

---

[1] Dept. of Microbiology University of Alabama at Birmingham, UAB Station THT/444 Birmingham, AL 35294 USA.
[2] Department of Immunology, Research Institute of Scripps Clinic, La Jolla CA 92037, USA.
* This research was supported by NIH grants AI 19222 (to B.F.T.) and AI 19551 (to G.H.F.).

and carbohydrate analysis (TOMANA et al. 1985; HASE et al. 1985; HIRANI et al. 1986). It is composed of two polypeptide chains, α and β, with molecular weights of 115,000 and 75,000 respectively (as determined by sodium dodecyl sulfate polycrylamide gel electrophoresis) (SDS-PAGE). Carbohydrate composes approximately 1.7% of C3, with the only two types of sugar identified being mannose and N-acetylglucosamine (TOMANA et al. 1985; HASE et al. 1985; HIRANI et al. 1986). Each chain contains N-linked high-mannose type carbohydrate (Fig. 1). The sites of glycosylation are residues 63 (β-chain) and 917 (α-chain; DE BRUIJN and FEY 1985). The α-chain sugar structures are composed of $(Man)_9(GlcNAc)_2$-Asn and $(Man)_8(GlcNAc)_2$-Asn, while the β-chain contains $(Man)_6(GlcNAc)_2$-Asn and $(Man)_5(GlcNAc)_2$-Asn (HASE et al. 1985; HIRANI et al. 1986). Other biochemical parameters of C3 are shown in Table 1.

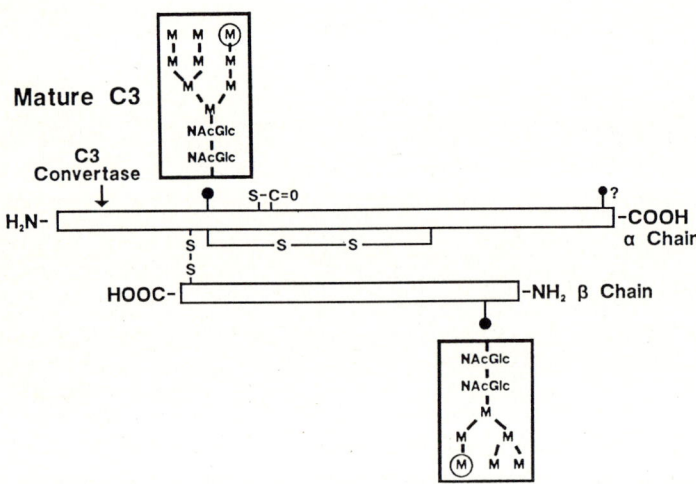

Fig. 1. A schematic representation of mature human C3. The ● symbol represents the positions of carbohydrate attachment; *question mark* near the symbol on the α-chain represents a putative attachment site. *Insets* show the structure of the high-mannose oligosaccharides found on the α- and β-chains of C3

Table 1. C3 biochemical characteristics

M_r	~183000
pI	~5.9
Carbohydrate content	1.7%
Extinction coefficient	9.7 ($\varepsilon_{1\,cm}^{1\%}$ at 280 nm)
Serum concentration	1.2–1.5 mg/ml
Fractional catabolic rate	1.7%/h

Derived from DE BRUIJN and FEY 1985; MÜLLER-EBERHARD and SCHREIBER 1980; ALPER and PROPP 1968; KOHLER and MÜLLER-EBERHARD 1967; TACK and PRAHL 1976; RUDDY et al. 1975; BEHRENDT 1985

C3 contains a reactive β-cysteinyl-γ-glutamyl thiolester bond which is formed between the side chains of neighboring cysteine and glutamine residues. Thiolester structure, reactivity, and mechanism of transacylation in C3, C4, and $α_2$-macroglobulin have been extensively studied (for review see TACK 1983; SOTTRUP-JENSEN 1987). The thiolester is an important and essential feature of C3 host defense functions. Activation of C3, generating the C3b fragment, results in the transient reactivity of the thiolester, allowing C3b to attach covalently to bacteria, cell surfaces, and immune complexes. This binding of C3 effects the complement host defense functions of opsonization and immune surveillance. (HOSTETTER et al. 1984). For additional information on the C3 thiolester see LEVINE and DODDS (this volume).

3 C3 Biosynthesis

C3 in humans is synthesized as a single-chain precursor (MORRIS et al. 1982a) similar to its structural homologs C4 and C5 (MORRIS et al. 1982b). The precursor contains a 22 amino acid leader peptide presumably cleaved during translocation into the endoplasmic reticulum (DE BRUIJN and FEY 1985). The mouse leader peptide is 24 amino acids in length and is highly homologous to its human counterpart (approximately 70%; LUNDWALL et al. 1984). In the human precursor, the α- and β-chains are connected by a tetra-arginine linker as deduced from the cDNA sequence (DE BRUIJN and FEY 1985). This molecule, termed prepro-C3, is diagrammed in Fig. 2. The arginine linker is similar to those seen in human C5 (RPRR; LUNDWALL et al. 1985; WETSEL et al. 1988) and C4 (β-α-chain, RKKR; WHITEHEAD et al. 1983; BELT et al. 1984). This linker appears to be an evolutionarily conserved feature since it is also present in mouse C3 (RRRR; WETSEL et al. 1984), C4 (β-α-chain, RNKR; OGATA et al. 1983), and C5 (RSKR; WETSEL et al. 1987). The main feature of the linker structures is the preponderance of arginine and lysine residues, however there are functional noncharged (serine, glutamine) and nonpolar (proline) substitutions. The linker is removed shortly before secretion since metabolic labeling studies demonstrate that no mature C3 (two chains) is visible intracellularly up to 4 h after labeling but is detected extracellularly in 30 min, and that no precursor C3 is detected extracellularly (MORRIS et al. 1982a; WARREN et al. 1987). Precursor C3 has been demonstrated to be processed in a similar fashion in other species, such as guinea pig (GOLDBERGER et al. 1981), mouse, and rat (STRUNK et al. 1988).

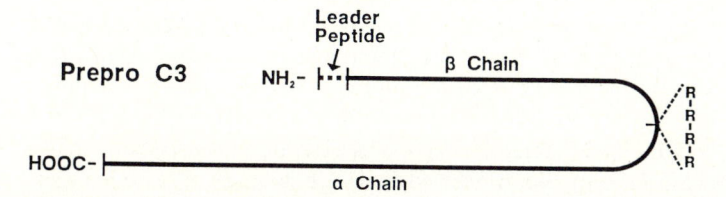

Fig. 2. Schematic view of prepro-C3. The single-chain C3 precursor is diagrammed to show the leader peptide and tetra-arginine linker region. *R*, arginine

After entering the endoplasmic reticulum, human C3 is transported to the Golgi complex where it is glycosylated at two of three Asn-X-Thr/Ser sites with high-mannose carbohydrate (Hubbard and Ivatt 1981). It is interesting to note that in mice only the α-chain is glycosylated (Fey et al. 1980; Bednarczyk and Capra 1988), a finding in keeping with the sequencing results of Lundwall et al. 1984) in which no potential glycosylation sites were identified in the β-chain. In addition, the guinea pig C3 β-chain also appears not to be glycosylated (Thomas and Tack 1983).

Between transport from the Golgi complex to the plasma membrane, C3 is cleaved to the mature two-chain form, probably by a cathepsin-like protease with a specificity for dibasic residues, as has been seen for neuropeptide hormone precursor processing (Steiner et al. 1984). Studies with the mouse macrophage line J774.2 suggest that the cleavage occurs at neutral pH, or that the enzyme(s) involved are active at a broad pH range. These same studies also demonstrated that treatment with the carboxylic ionophore monensin inhibits C3 processing and intracellular transport (Bednarczyk and Capra 1988). Since monensin acts on the transport pathway at the level of the Golgi complex (Tartakoff 1983), these data substantiate kinetic studies in both human and murine cell lines demonstrating that pre-C3 cleavage occurs late in intracellular transport (Morris et al. 1982a, b; Bednarczyk and Capra 1988).

The biosynthesis of the thiolester bond in C3 is an intracellular posttranslational event (Iijma et al. 1984). Cell-free translation studies using rabbit liver mRNA yields a product nonreactive with methylamine but reactive with iodoacetimide, indicating the presence of a free thiol group. Addition of liver homogenate to the cell free product results in a methylamine-reactive product, demonstrating that the required enzyme(s) is present intracellularly. The particular enzyme (possibly a transglutaminase) or enzymes involved and their subcellular location have not yet been identified.

4 Sites of C3 Biosynthesis

Four major sites of synthesis have been identified for human C3: hepatocytes, monocytes, and epithelial (fibroblasts) and endothelial cells (see Table 2). Raji cells have recently been suggested to be a fifth site, but further characterization of C3 biosynthesis by these cells is required (discussed below). It is interesting to note that murine astrocytes (neutral equivalent to a macrophage) also synthesize and secrete C3 and appear to increase synthesis in response to lipopolysaccharide (LPS) stimulation (Lévi-Strauss and Mallat 1987). If human astrocytes also synthesize and secrete C3, it would suggest a role for complement in neural immunology.

The liver is the primary source (greater than 90%) of serum complement (Alper and Rosen 1976), and the synthesis of most complement components by hepatocytes has been amply documented (Colten 1976; Cole and Colten 1988). Primary hepatocyte cultures have been employed (Alper et al. 1969; Colten 1972), and C3 biosynthesis could be documented in fetal liver tissue as early as 14 weeks or detected in fetal serum between 5 and 10 weeks (reviewed in Colten 1976). Recent studies, however, have employed the hepablastoma-derived HepG2 and Hep3B2 cell lines (Knowles et al. 1980), and C3 biosynthesis by both cell lines has been established

Table 2. C3 biosynthethic sites

Cell or tissue type	Species
Liver (Parenchymal cell)	Human, mouse, rat, rabbit, guinea pig, chicken, quail, cobra, trout, lamprey
Monocyte/macrophage	
Primary monocyte cultures	Human, guinea pig
U937 (monocyte cell line)	Human
J774.2 (monocyte cell line)	Mouse
Endothelial derived from capillary and umbilical cord	Human
Epithelial	
Aveolar type II	Mouse, rat
Primary and transformed cultures	Human
Liver	Rat
Astrocytes (brain)	Mouse
Raji (transformed B cell)	Human

See text for references

(MORRIS et al. 1982a, b; DARLINGTON et al. 1986). C3 produced by HepG2 cells is functionally active as determined by hemolytic assay and appears antigenically and functionally identical to C3 purified from serum. C3 is synthesized as a single-chain precursor in HepG2 cells and is subsequently processed to the mature two-chain polypeptide on or shortly after secretion (MORRIS et al. 1982a, b). Kinetics of C3 secretion by this cells line, as assessed by pulse-chase metabolic labeling methods, are comparable to those for other complement components (MORRIS et al. 1982b; GOLDBERGER et al. 1984; BARNUM and VOLANAKIS 1985) with the mature protein present extracellularly in 30 min or less.

Many species, including primitive vertebrates, have been shown to synthesize functionally active C3 or a C3 homolog which retains the gross structural features of human C3. In those cases where it has been examined, the amino acid sequence homology is quite high between human C3 and C3 of lower vertebrates. Since these C3 homologs were identified from serum, their likely biosynthetic site is the liver. Among the other species known to produce C3 are rats (STRUNK et al. 1975), rabbits (KUSANO et al. 1986), quail (KAI et al. 1983), cobra (EGGERTSEN et al. 1983), *Xenopus* (SEKIZAWA et al. 1984), rainbow trout (NONAKA et al. 1984a), and lamprey (NONAKA et al. 1984b). Interestingly, the lamprey C3 has three polypeptide chains rather two, making it structurally more similar to C4. Perhaps this suggests that C4 is the more primitive of the two components. For more information on the structure and function of C3 from different species, see the article by BECHERER et al. (this volume).

Of extrahepatic synthetic sites for complement, cells of the monocyte/macrophage series are the most studied (COLTEN 1976; COLE and COLTEN 1988), partly because of their availability. Primary cultures are predominantly used although the monocyte-like cell line U937, established from a histiocytic lymphoma has also been utilized (SUNDSTROM and NILSSON 1976). C3 biosynthesis by monocytes (EINSTEIN et al. 1977;

WHALEY 1980; ZIMMER et al. 1982; STRUNK et al. 1983; LAPPIN et al. 1986; STRUNK et al. 1985; ST JOHN SUTTON et al. 1986) and U937 cells (NICHOLS 1984; MINTA and ISENMAN 1987) demonstrates that the molecule is structurally similar to serum C3, is functionally active, and appears to be secreted with kinetics similar to that for that for hepatocyte C3. Similar findings have been documented in the mouse macrophage cell line J774.2 (BEDNARCZYK and CAPRA 1988) and mouse peritoneal macrophages (FEY et al. 1980).

Differences have been noted in the C3 synthetic rates of monocytes versus macrophages, with macrophages synthesizing three to four times more C3 than monocytes (LAPPIN et al. 1986). It is likely that this reflects a maturational difference since U-937 cells (which are monocytic and therefore less mature) synthesize considerably less C3 (MINTA and ISENMAN 1987) than adherent monocytes (which are essentially spread and activated macrophages; HAMILTON et al. 1987). This is further supported by studies in which stimulation of U-937 cells with phorbol esters such as phorbol myristate acetate (PMA) promoted differentiation into more macrophage-like cells and increased the amount of C3 synthesized (MINTA and ISENMAN 1987).

In addition to monocytes as an extrahepatic site of complement synthesis, endothelial and epithelial cells also synthesize C3. UKEI and coworkers (1987) quantitated C3 production by cultured human capillary endothelial cells by enzyme-linked immunosorbent assay and calculated a secretory rate of 250 ng $C3/10^6$ cells per 5 days. The purity of these cell cultures was assessed by presence of factor VIIIR Ag and morphology and demonstrated a nearly pure endothelial cell preparation. This is a significant point when working with primary cultures since contamination by other cell types, particularly macrophages, could easily augment C3 production. The synthetic rate seen for these cells appears lower than that seen with primary monocyte cultures, in which a rate of over 400 ng C3/ml per 5 days was quantitated (HAMILTON et al. 1987). However it is difficult to be precise on this point since the number of cells used in the latter experiment were not determined. Immunoprecipitation of endothelial cell supernatants from cells metabolically labeled with [^{35}S]-methionine followed by SDS-PAGE and autoradiography, revealed bands with molecular weights comparable to those of serum C3. The supernatants were not assessed for functional activity (UKEI et al. 1987). Similar studies by WARREN and coworkers (1987) demonstrated C3 production by cultured human umbilical vein endothelial cells (however the purity of these cell preparations was not documented).

Epithelial cells also synthesize and secrete C3. WHITEHEAD and colleagues (1981) screened human primary and transformed fibroblasts, T and B cell lines and various cervical and colon carcinoma cell lines for C3 synthesis using a C3-specific monoclonal antibody. Only primary and transformed fibroblasts and the cervical carcinoma cell line D98/AH-2 synthesized C3 as determined by immunoprecipitation of [^{35}S]-methionine-labeled cell culture supernatants. The functional activity of fibroblast C3 was not determined. Pulmonary type II aveolar epithelial cells have recently been shown to synthesize C3, but the functional activity was not assessed (STRUNK et al. 1988). Other species capable of synthesizing and secreting C3 from epithelial tissue are mice and rats (STRUNK et al. 1988; GUIGUET et al. 1987). Taken together these data demonstrate that local complement production by these cell types plays a major role in host defense at both inflammatory and mucosal sites since these cells along with macrophages are capable of producing all the components necessary for complement activation by either the classical or alternative pathways (for review on macrophages see JOHNSON and HETLAND 1988).

Recently the synthesis of C3 by Raji cells has been demonstrated by W. LERNHARDT and colleagues (personal communication). C3 immunoprecipitated from [^{35}S]-methionine-labeled cells has the same apparent molecular weight and structure as serum C3. No further work has been completed as yet with regards to activity or secretion kinetics. The data conflict with those of WHITEHEAD and coworkers (1981) who were unable to immunoprecipitate C3 from Raji cell supernatants. Further studies are required to resolve this discrepancy.

5 C3 Transcription and Alternative Transcripts

Northern blot analysis of C3 mRNA demonstrates that the message is approximately 5.2–5.3 kb in size for all tissue types known to synthesize C3. This is comparable to the 5.1-kb (DOMDEY et al. 1982) and 5.2–5.3-kb (WHEAT et al. 1987) size reported for mouse liver C3. It should be noted that these analyses utilized agarose gel electrophoresis, which may not completely denature all RNA species due to secondary structure features, and therefore the sizes reported are not absolute. The hepatocyte cell lines HepG2 and Hep3B2 produce a single C3 message at 5.2–5.3 kb in size from poly(A)$^+$ mRNA (WARREN et al. 1987; G. DARLINGTON and R. STRUNK, personal communication). Peripheral blood monocytes (WARREN et al. 1987; STRUNK et al. 1985) and the monocyte-like cell line U937 (W. LERNHARDT, personal communication) produce a similar size message at 5.2 kb. Endothelial cells from umbilical cord veins synthesize a 5.2 kb message in an amount one-fifth that of monocytes and 1% that of hepatocytes (WARREN et al. 1987). These findings suggest differential regulation of C3 expression in these tissue types and that regulation is exerted at least partly at a pretranslational level. Interestingly, Raji cells, a transformed B cell line, also produce a 5.2-kb message and a truncated message approximately 1.8 kb in size (W. LERNHARDT, personal communication). The truncated message corresponds to the C3 α-chain based on Northern blotting with C3 α-chain probes and is likely to be analogous to truncated C3 messages produced in murine T-cell hybridoma, lymphoma, and macrophage cell lines (LERNHARDT et al. 1986). The truncated message could arise from alternative splicing, alternative initiation of transcription, or a second truncated C3 gene. It is also possible that this message is an aberrancy of a transformed cell line.

6 C3 and the Acute-Phase Response

Aside from tissue-specific regulation, C3 synthesis is also modulated during the acute phase and in response to a variety of stimulatory agents. The serum concentration of C3 increases approximately 50% during the acute phase; thus it is only marginally elevated in comparison to C-reactive protein, which may increase 500- to 1000-fold (KUSHNER 1982; KOJ 1985). The specific mediator or mediators responsible for the increase in serum C3 concentration are not known. However, several hepatocyte-specific mediators have been described, e.g., interleukin (IL) 1, tumor necrosis factor (TNF), glucocorticoids, hepatocyte-stimulating factors including IL-6 and COLO HSF I-III, and γ-interferon (reviewed in KOJ 1986; FEY and GAULDIE 1988). Several studies have directly assessed various mediators in augmenting C3 biosynthesis in hepatocytes, monocytes, and epithelial cells (see Table 3). LPS or lipid A stimulation either of peripheral blood monocytes or in the monocyte-like cell line U937 increased C3 synthesis 2 to 30-fold with no apparent change in secretion kinetics (NICHOLS 1984; STRUNK et al. 1985; ST JOHN SUTTON et al. 1986). Monocytes from cord blood appear refractory to LPS stimulation, which is most likely due to maturational differences (ST JOHN SUTTON et al. 1986). These results are significantly different from the 50% increase in serum C3 concentration during the acute phase, suggesting that the regulation of C3 production in monocytes is independent of that in hepatocytes, and that local monocyte C3 production plays an important role in host defense. Other mediators such as PMA and γ-interferon also increased C3 synthesis by U937 cells (NICHOLS 1984; MINTA and ISENMAN 1987). However, HAMILTON and coworkers (1987) found C3 synthesis to be downregulated in peripheral blood monocytes treated with γ-, β- or α-interferon. The discrepancy in these results may be due to an aberrancy in the U937 cell line or may reflect maturational differences between the cells employed. Interestingly, PMA induced a greater number of cells to produce C3, increasing the population of C3-producing cells from 5% to 30% (MINTA and ISENMAN 1987). Whether this was a direct or an indirect effect remains to be clarified. The hepatocyte cell line Hep3B2 increases C3 synthesis in response to conditioned monocyte medium, IL-1, and TNF (DARLINGTON et al. 1986). The mechanism involved is unclear, but the increase in C3 mRNA suggests that regulation is pretranslational. Rat epithelial

Table 3. Regulation of C3 synthesis

Modulator	Change in C3 synthesis	Cell type
Acute phase	50% inc.	Liver, monocyte
LPS, lipid A	2–30 × inc.	Monocytes, U937
PMA (phorbol ester)	5 × inc.	U937
Interferon		
γ	2 × inc.	U937
	~50% dec.	Monocytes
α	~50% dec.	Monocytes
β	~20% dec.	Monocytes
IL-1	10 × inc.	Hep3B2 (liver cell line)
	~40% inc.	Epithelial (primary culture)
TNF	10 × inc.	Hep3B2
Testosterone	8–10 × inc.	Serum levels in male mice
Hydrocortisone	9 × inc.	Rat hepatoma cell lines
PHA-stimulated monocyte culture medium (IL-6?)	~20 × inc.	Hep3B2

See text for references. inc.: increase; dec.: decrease

cell lines synthesize increased amounts of C3 (2.5-fold) when incubated with conditioned medium from mononuclear cells stimulated with phytohemagglutinin (GUIGUET et al. 1987). Preliminary characterization of the mononuclear cell factor or factors involved demonstrates that the stimulation is effected by protein molecules of two sizes, 15–20 and 25–60 kDa. This size is consistent with the size of rat IL-6 (W. NORTHEMANN et al., unpublished findings), suggesting IL-6 mediated stimulation of C3 synthesis. IL-1 alone increases C3 synthesis by the rat epithelial cell lines approximately 40% but was not as effective as conditioned media (GUIGUET et al. 1987). It is interesting to note that the level of increased synthesis effected by IL-1 is in the same range as that seen during the acute-phase response.

IL-6 and the related hepatoctye-stimulating factors probably play a more significant role in modulating the acute-phase response than any other modulator either singly or in combination with other modulators (for review see FEY and GAULDIE 1989). Since most of the major cell types that produce and have receptors for IL-6 also synthesize C3, it is likely that IL-6 serves both an autocrine and a paracrine role in the increase in C3 synthesis during the acute-phase response.

The role of hormones in C3 synthesis in humans is poorly understood although it has been reported that C3 serum levels increase late in pregnancy (PROPP and ALPER 1968). The effect of hormones on C3 levels in mice has been more closely studied, and these studies demonstrate that serum C3 levels in male mice are eight to ten times higher than in females (CHURCHILL et al. 1967). This is in keeping with androgen regulation found for many mice complement and noncomplement proteins. Other studies with rat hepatoma cell lines showed marked increase in C3 synthesis in response to hydrocortisone (STRUNK et al. 1975).

7 C3 Gene Structure

The central role played by C3 in the complement system has led us (BARNUM et al. 1989) to begin examination of its intron/exon organization with the aim of better understanding both structure/function aspects of C3 and the evolutionary development and relationships of C3 to C4, C5, and α_2-macroglobulin. Three cosmid clones isolated from an amplified human liver cosmid library (FEY et al. 1984) have been partially characterized. These clones, termed *cos*1–3, have insert sizes of 34, 36, and 29 kb, respectively. All three of the clones overlap as determined by Southern blot analysis using M13 template probes (derived from sequencing the C3 cDNA) and contain sequence only in the α-chain. Two of the cosmid clones (*cos*1 and -2) were shown by a combination of sequencing and Southern blot analysis to cover the α'-chain region of the C3 gene. Accordingly, these two clones were used to identify the intron/exon junctions, partially map the positions of the exons within the gene, and estimate the size of this region of the C3 gene.

The α'-chain of C3 is encoded by 24 exons beginning with a split serine codon (immediately 3' to the AGG coding for the carboxy-terminal arginine residue of C3a) and ending with 3' nontranslated sequence (Table 4). The exons range in size from

52 bp up to 213 bp with an average size of 115 bp. Exons with this average size would code for 38.3 amino acids, which is fewer than the reported average of 44.6 amino acids for internal exons (TRAUT 1988). The last exon is incompletely defined but contains 157 bp of coding sequence and currently over 200 bp of nontranslated sequence (data not shown), extending the previously reported sequence for this region (DEBRUIJN and FEY 1985) by over 200 bp. The size of each exon and its location in the cDNA sequence of C3 (DEBRUIJN and FEY 1985) are shown in Table 4. Also shown in Table 4 are the splice donor and acceptor sequences demarcating the 5' and 3' ends of the 23 introns in the C3 α'-chain gene region. The consensus of the donor sequences was GTGAGT with a frequency of 100%, 96%, 52%, 83%, 91%, and 34%, respectively. This consensus sequence is in good agreement with the GTAAGT sequence described in the literature (BREATHNACH et al. 1978; MOUNT 1982; GREEN 1986; PADGETT et al. 1986). Guanine residues were found more frequently than adenine in position 3 of the consensus sequence; this is not a significant difference since both residues are predominant in this position (MOUNT 1982).

There was, however, one exception to the generally invariant GT dinucleotide consensus sequence. This was a cytosine residue substituting for the nearly invariant thymine residue in position 2 of the consensus sequence located at the beginning of

Table 4. Sizes of the C3 α'-chain exons, location in the cDNA sequence and intron/exon junction sequences

Exon number	Location in cDNA	Length (bp)	Splice acceptor	Splice donor
1	2306–2414	109	tcctgcag/GTA	TGG/gtaagg
2	2415–2501	87	ccatccag/AAT	AGG/gtgaga
3	2502–2643	142	ccgagcag/GAT	AAG/gtgggt
4	2644–2856	213	cctggcag/GTG	GTG/gtgagt
5	2857–2922	66	acgcgtag/CCG	GTG/gtgagt
6	2923–3010	88	acctgcag/AAG	AAG/gtgaga
7	3011–3214	204	tgctgcag/GGA	AGG/gtgggc
8	3215–3291	77	ctctgcag/GGT	CTG/gtgagt
9	3292–3450	159	cactccag/GCT	ATT/gtaaga
10	3451–3549	99	tatctcag/GGT	ACC/gtaagt
11	3550–3706	157	tctcatag/AGC	AAG/gtgagg
12	3707–3870	164	ctttgcag/ATA	CAG/gcaagt
13	3871–4029	159	ggttctag/GCC	GAG/gtacag
14	4030–4089	60	ctctgcag/ACC	TCG/gtaagg
15	4090–4180	91	cttaccag/GTG	CAG/gtaaaa
16	4181–4232	52	gtttctag/AAA	CAG/gtaaga
17	4233–4320	87	ctatccag/GTA	CAG/gtatga
18	4321–4410	91	ttccccag/CTG	AAG/gtaagg
19	4411–4516	106	cccctcag/GTC	TGG/gtgagc
20	4517–4606	90	gcccgcag/AGG	AGG/gtgagt
21	4607–4689	83	ccctgcag/AGA	ATG/gtgagt
22	4690–4774	85	ccccacag/TGT	CAG/gtcagg
23	4775–4910	136	cactccag/GCT	CAA/gtgagt
24	4911–5049	157	ccctccag/CCT	
	Consensus	–	ccctgcag/GNT	AAG/gtgagt

N, any nucleotide

intron 12. This substitution is rarely seen (fewer than 1% of over 400 genes analyzed; PADGETT et al. 1986) but has been observed in the adult chicken α-globin gene (DODGSON and ENGEL 1983), the mouse α-A-crystallin gene (KING and PIATIGORSKY 1983), and the rabbit β-globin gene (WIERINGA et al. 1983). The conversion of the GT dinucleotide to GC is apparently a functional substitution since the transcripts for these genes are processed normally. Conversion of the guanine to an adenine or removal of the consensus sequence does however prevent proper splicing (FELBER et al. 1982; TREISMAN et al. 1982; WIERINGA et al. 1983, 1984). The 3' intron consensus sequence was CCCTGCAG with a frequency of 58%, 50%, 54%, 58%, 46%, 83%, 100%, and 100%, respectively (Table 4). There were no unusual sequences found in this region of the introns, which are generally pyrimidine-rich and end in the invariant AG dinucleotide (BREATHNACH et al. 1978; MOUNT 1982).

The structural mapping of that region of the C3 gene encoding the α'-chain gene has been completed. A map of the region is shown in Fig. 3; also shown here are the relative positions of the cosmid clones used to derive the map. All 23 introns have been sequenced or their size estimated by restriction mapping and Southern blot analysis (Table 5). Ten introns have been completely sequenced and are surprisingly short for a gene of this size (estimated to be approximately 35–40 kb). These introns range in size from 85 to 242 bp with an average of 140 bp in length. The remaining introns are estimated to be from approximately 250 bp to over 4 kb in length with an average of 1450 bp in length. The C3 α'-chain introns are predominantly type 0 and type 1 introns (42% each), where type 0 introns interrupt between codons, and type 1 interrupt after the first nucleotide of a codon. The remaining 16% are type 2 introns (interruption after the second nucleotide of a codon). This deviates slightly from the recently reported average composition of 54%, 27%, and 18% for types 0, 1, and 2 introns, respectively (TRAUT 1988). The combination of Southern blot analysis and sequencing demonstrates that the C3 α'-chain gene region is approximately 23 kb in length. This is nearly as large as the mouse C3 gene (WIEBAUER et al. 1982), suggesting that the full-length human C3 gene is larger than its murine counterpart.

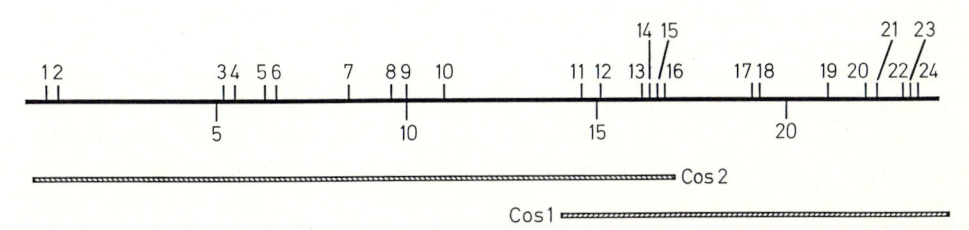

Fig. 3. Physical map of the C3 α'-chain gene. The positions of the 24 exons are shown above the line; distance in kilobases is shown below the line. The orientation of the C3 cosmid clones, *cos1* and *cos2*, relative to the α'-chain are shown in the lower portion of the figure

C3 binds to more components of the complement system than any other single complement protein. Among the components that bind or interact with C3 or its proteolytic fragments are B, P, I, C5, H, CR1, CR2, CR3, and CR4. Several of these components bind exclusively to the α-chain and in some cases the specific regions of

Table 5. Sizes of the human C3 α'-chain intron

Intron number	Length (bp)	Intron number	Length (bp)
1	242	12	~1000
2	~4200	13	155
3	94	14	124
4	~700	15	138
5	126	16	~2300
6	~1750	17	111
7	~1000	18	~1700
8	~250	19	~900
9	~900	20	196
10	~3500	21	~600
11	~470	22	85
		23	128

Intron sizes shown with the symbol ~ were estimated by Southern blotting. The remaining introns were completely sequenced

interaction have been mapped (for reviews see LAMBRIS and MÜLLER-EBERHARD 1986; LAMBRIS 1988; BECHERER et al., this volume). Correlation of protein domains with binding or cleavage sites in C3 as delineated by intron/exon junctions has revealed a number of interesting findings. Each of the binding sites mapped in the α'-chain of C3 (LAMBRIS et al. 1985; WRIGHT et al. 1987; DAOUDAKI et al. 1988; BECHERER and LAMBRIS 1988; LAMBRIS et al. 1988) appear to be within a single given exon. These binding sites and their relationship to the structure of the α'-chain gene region are shown in the upper portion of Fig. 4. The sites for the complement receptors CR1–CR3 are contained within exons, 1, 12, and 16, respectively. The binding site for the regulatory protein factor H is also in exon 12 while the properdin binding site is located in exon 18. If the sequences encoded by these exons are sufficient to create the necessary binding sites, then these data would support the hypothesis that exons are large enough to specify functional modules in proteins (TRAUT 1988).

The lower portion of Fig. 4 shows the approximate location of some of the various proteolytic cleavage sites found in the C3 α'-chain relative to the location of the intron/exon junctions. The most 5' exon (exon 1) starts immediately after the C3a coding sequence, suggesting that C3a is encoded by a separate exon or exons. This cleavage site correlates with the C5 convertase cleavage site at the genomic level as determined by partial characterization of C5 gene structure (S. R. BARNUM, R. A. WETSEL, D. NOACK and B. F. TACK, unpublished results). The 3' end of exon 6, which delineates the NH_2-terminal end of the C3d fragment, correlates with the proposed complement protein I cleavage sites at this location. None of the other enzyme cleavage sites (kallikrein, trypsin, elastase, or others) appear to correspond exactly to any intron/exon junction sites, although the elastase and the most carboxy-terminal I cleavage site are within ten nucleotides of junction sites. This is not surprising, however, since these cleavage sites are not biologically relevant for C3 function. As previously determined for C4, the thiolester region of C3 is encoded within a single exon (YU et al. 1986). Limited comparison of C3 α'-chain intron/exon organization with that of human C5 (S. R. BARNUM, R. A. WETSEL, D. NOACK, and B. F. TACK,

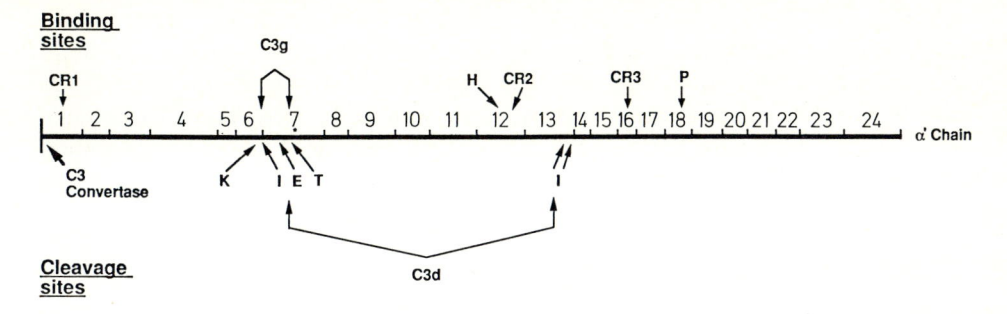

Fig. 4. The location of the binding sites of complement receptors and other complement molecules and of the proteolytic cleavage sites in the C3 α'-chain. The approximate location of the binding sites of complement receptors 1–3 (*CR1–CR3*), complement factor H (*H*), and properdin (*P*) are shown in the upper portion of the figure in relation to intron/exon junctions. The lower portion of the figure shows the approximate location of the proteolytic cleavage sites in the C3 α'-chain in relation to the intron/exon junction location. The position of the C3g and C3d fragment s are shown for orientation purposes. *Asterisk* denotes the location of the thiolester region. *I*, Complement protein I; *E*, elatase; *T*, trypsin; *K*, kallikrein

unpublished results; WETSEL et al. 1988) and rat α_2-macroglobulin (M. HATTORI et al., unpublished results) reveals a distingly different genomic structure for each molecule (data not shown). This suggests that, although these molecules share extensive sequence homology and structural similarities, their genomic structures have been diverging for some time. We are currently sequencing and mapping clones covering the β-chain and the remaining portion of the C3 α-chain to complete the genomic analysis of C3.

7.1 Chromosomal Location

The human C3 gene is located on chromosome 19. This was demonstrated by studies in which the expressed protein product segregated with this chromosome and was confirmed by blotting experiments in which a 1.39-kb C3 genomic DNA probe hybridized to the DNA prepared from the appropriate human-mouse somatic cell hybrids (WHITEHEAD et al. 1982). Thus C3 is unlinked to the major histocompatibility region (MHC) class III genes (containing C4, C2, and B) since this region is located on chromosome 6 in humans (CARROLL et al. 1984). In contrast, the mouse C3 gene is located on the same chromosome as the class III MHC genes (chromosome 17) but is unlinked to this region (PENALVA DA SILVA et al. 1978; NATSUMME-SAKAI et al. 1979). In rats the C3 gene has been localized to chromosome 9 (SZPIRER et al. 1988). The genes for other structurally related proteins such as C5 and α_2-macroglobulin are also unlinked to the C3 gene, being located on chromosomes 9 and 12, respectively (WETSEL et al. 1988; KAN et al. 1985). It was concluded from the chromosome localization studies that there is only one C3 gene in humans in the haploid genome, or that if multiple copies exist, they would have to display exactly the same characteristic 12-kb *Eco*RI restriction fragment (WHITEHEAD et al. 1982). Recent work has more accurately mapped the location of the gene by excluding the gene from the distal long arm of chromosome 19 (19q133–19qter; WIEACKER et al. 1983).

Human C3 is part of an extensive linkage group on chromosome 19 including the Lewis blood group, myotonic dystrophy, the ABH secretor system, the Lutheran blood group, the insulin receptor, the low-density lipoprotein receptor, the genes for plasma lipid transport (apolipoproteins C1, C2, and E), peptidase D, glucose 6-phosphate isomerase, and the chorionic gonadotropin B subunit. The exact orientation of these genes along the chromosome appears, however, to require additional localization studies (SHAW et al. 1986; LUSIS et al. 1986).

7.2 C3 Polymorphism and Deficiency

Two major electrophoretic variants of C3 have been identified ($C3^S$ and $C3^F$) as slow- and fast-migrating forms of the protein. These genes are inherited in an autosomal codominant fashion indicating that they are two allelic variants (WIEME and DEMULENAARE 1967; ALPER and PROPP 1968; ALPER 1973; ALPER and ROSEN 1984). The slow variant is predominant in all major races (Caucasian, Negro, and Oriental) with a frequency of 0.79 in Caucasians compared to 0.20 for the fast form. Other races have considerably lower frequency of the fast variant (American Negro) or very little detectable polymorphism (Oriental; RITTNER and SCHNEIDER 1988). A wide number of protein variants have also been characterized (26 all together), and all are recognized as allelic gene products, the sum of which make up a gene frequency of 0.001 (RITTNER and SCHNEIDER 1988). Of the variants best characterized to this point by two-dimensional electrophoresis, all the charge differences have been localized to the β-chain (CARROLL and CAPRA 1979; FEY et al. 1983). Restriction fragment length polymorphisms have also been defined using a variety of enzymes. These polymorphisms have been described in several ethnic groups (DAVIES et al. 1983; FEY et al. 1983).

Complete C3 deficiency has been recognized for some time, although such cases are rare (ALPER et al. 1972; ROSS and DENSON 1984). Of the several cases described, numerous clinical syndromes have been associated with the deficiency, ranging from fever to nephritis (ALPER and ROSEN 1981, 1984; ROSS and DENSON 1984; COLTEN 1985). These individuals often suffer from rheumatic disease and systemic lupus erythematosus (reviewed in ROSS and DENSON 1984). Interestingly, women present more frequently than men with C3 deficiency with a female to male ratio of 6.1:1. This disequilibrium probably reflects a selection bias, although it may also suggest that C3 deficiency in men is more frequently lethal (ROSS and DENSON 1984). Homozygous-deficient individuals usually have less than 1/1000 the serum C3 levels of normals and in general are markedly susceptible to infection by pyogenic micro-organisms leading to frequent episodes of pneumonia, meningitis, and other conditions. The primary etiological agents are pneumococcus, streptococcus and *Hemophilus influenzae*. In addition, these individuals also have reduced or nonexistent C3 hemolytic activity, fail to opsonize particles, and have lost chemotactic and bactericidal activity (BALLOW et al. 1975). Monocytes from homozygous-deficient individuals produce approximately 25% the normal amount of C3 produced by monocytes from normals, again suggesting that C3 regulation in monocytes is independent of that in hepatocytes. C3 from these monocytes is morphologically and functionally identical to that produced by normals (EINSTEIN et al. 1977). As might be expected, heterozygous-deficient individuals are normal in most respects even though they have approximately half the normal serum level of C3. However they may present with variable susceptibility to infection (PUSSELL et al. 1980). The molecular basis for the hetero- or homozygous deficiencies remains unclear. However, point mutations or deletions of regulatory or structural portions of the gene are possible explanations. Some individuals apparently suffer from C3 deficiency secondary to hypercatabolism due to a primary defect in factor I synthesis (thereby leading to wasteful consumption of alternative pathway components, including C3). Direct evidence linking these observations is however lacking (COLTEN 1985). For more information on C3 deficiencies, see the article by BITTER-SUERMANN and BURGER (this volume).

8 Conclusions

Despite the fact that C3 biosynthesis has been documented in several cell types, and that the level of synthesis can be modified by numerous biological mediators, the relative biosynthetic contribution of the various cell types to the in vivo inflammatory response or during the acute-phase response is unclear. Also unclear is the regulation of C3 biosynthesis between the C3-producing tissues in both normal and inflammatory states. The gene structure is still under investigation in our laboratory along with studies to understand the molecular basis for C3 deficiency. These studies are essential in analyzing C3 structure/function relationships and for expression work. Clearly, C3 will serve as an interesting model for tissue-specific regulation, and undoubtedly both *cis* and *trans* regulatory elements described for other genes (reviewed in Maniatis et al. 1987) will be found to play a role in C3 regulation. We and others are addressing these questions, but much remains to be done. Obviously C3 still poses a number of interesting questions, and the answers to these questions will significantly enhance knowledge of complement structure/function relationships and regulation of complement in the inflammatory state.

Acknowledgements. The authors thank Drs. G. DARLINGTON, M. HATTORI, W. LERNHARDT, and R. STRUNK for sharing unpublished findings. We are also grateful to Bonnie Towle for secretarial assistance.

References

Alper CA (1973) Genetics and the C3 molecule. Vox Sang 25: 1–8
Alper CA, Propp RP (1968) Genetic polymorphism of the third component of human complement (C'3). J Clin Invest 47: 2181–2191
Alper CA, Rosen FS (1976) Genetics of the human complement system. Adv Hum Genet 7: 141–188
Alper CA, Rosen FS (1981) Complement deficiencies in humans. In: Clinical immunology update. Elsevier, New York, pp 59–75
Alper CA, Rosen FS (1984) Inherited deficiencies of complement proteins in man. Springer Semin Immunopath 7: 251–261

Alper CA, Johnson AM, Birtch AM, Moore FD (1969) Human C3: evidence for the liver as the primary site of synthesis. Science 163: 286–288

Alper CA, Colten HR, Rosen FS, Rabsonn AR, Macnab GM, Gear JSS (1972) Homozygous deficiency of C3 in a patient with repeated infections. Lancet II: 1179–1181

Ballow M, Shira JE, Harden L, Yang SY, Day NK (1975) Complete absence of the third component of complement in man. J Clin Invest 56: 703–710

Barnum SR, Volanakis JE (1985) Biosynthesis of complement protein D by HepG2 cells: a comparison of D produced by HepG2 cells, U937 cells and blood monocytes. Eur J Immunol 15: 1148–1151

Barnum SR, Amiguet P, Amiguet-Barras F, Fey G, Tack BF (1989) Complete intron/exon organization of DNA encoding the α' chain of human C3. J Biol Chem 264: 8471–8474

Becherer JD, Lambris JD (1988) Identification of the C3b-receptor binding domain in third component of complement. J Biol Chem 263: 14586–14591

Bednarczyk JL, Capra JD (1988) Posttranslational processing of the murine third component of complement. Scand J Immunol 27: 83–95

Behrendt N (1985) Human complement component C3: characterization of active C3 S and C3 F, the two common genetic variants. Mol Immunol 22: 1005–1008

Belt KT, Carroll MC, Porter RR (1984) The structural basis of the multiple forms of human complement C4. Cell 36: 907–914

Breathnach R, Benoist C, O'Hare K, Gannon F, Chambon P (1978) Ovalbumin gene: evidence for a leader sequence in mRNA and DNA sequences at the exon-intron boundaries. Proc Natl Sci USA 75: 4853–4857

Carroll MC, Capra JD (1979) Molecular weight variations in the polypeptide subunits of the third and fourth components of complement derived from various mammalian species. Fed Proc 38: 1291

Carroll MC, Campbell RD, Bentley Dr, Porter RR (1984) A molecular map of the human major histocompatability complex class III region linking complement genes C4, C2 and factor B. Nature 307: 237–241

Churchill WH, Weintraub RM, Borsos T, Rapp HJ (1967) Mouse complement: the effect of sex hormones and castration on two of the late acting components. J Exp Med 125: 657–673

Cole FS, Colten HR (1988) Complement biosynthesis. In: Till GO, Rother KO (eds) The complement system. Springer, Berlin Heidelberg New York, pp 44–70

Colten HR (1972) Ontogeny of the human complement system: in vitro biosynthesis of individual complement components by fetal tissues. J Clin Invest 51: 725–750

Colten HR (1976) Biosynthesis of complement. Adv Immunol 22: 67–118

Colten HR (1985) Biology of disease. Molecular basis of complement deficiency syndromes. Lab Invest 52: 468–474

Daoudaki ME, Becherer JD, Lambris JD (1988) A 34-amino acid peptide of the third component of complement mediates properdin binding. J Immunol 140: 1577–1580

Darlington GJ, Wilson Dr, Lachmann LB (1986) Monocyte-conditioned medium, interleukin 1, and tumor necrosis factor stimulate the acute phase response in human hepatoma cells in vitro. J Cell Biol 103: 787–793

Davies KE, Jackson J, Williamson R, Harper PS, Sarfarazi M, Meredith L, Fey GH (1983) Linkage analysis of myotonic dystrophy and sequences on chromosome 19 using a cloned complement 3 gene probe. J Med Genet 20: 259–263

deBruijn MHL, Fey GH (1985) Human complement component C3: cDNA coding sequence and derived primary structure. Proc Natl Acad Sci USA 82: 708–712

Dodgson JB, Engel JD (1983) The nucleotide sequence of the adult chicken α-globin genes. J Biol Chem 258: 4623–4629

Domdey H, Wierbauer K, Kazmaier M, Muller V, Odink K, Fey G (1982) Characterization of the mRNA and cloned cDNA specifying the third component of mouse complement. Proc Natl Acad Sci USA 79: 7619–7623

Eggertsen G, Lundwall A, Hellman U, Sjoquist J (1983) Antigenic relationships between human and cobra complement factors C3 and cobra venom factor (CVF) from the Indian cobra (*Naja naja*). J Immunol 131: 1920–1923

Einstein LP. Hansen P, Ballow M, Davis E, Davis JS, Alper CA, Rosen FS, Colten HR (1977) Biosynthesis of the third component of complement (C3) in vitro by monocyte from both normal and homozygous C3-deficient humans. J Clin Invest 60: 963–969

Felber BK, Orkin SH, Hamer DH (1982) Abnormal RNA splicing causes one form of α thalassemia. Cell 29: 895–902

Fey G, Gauldie J (1989) The acute phase response of the liver in inflammation. In: Popper H, Schaffner F (eds) Progress in liver disease. Saunders, Philadelphia (to be published)

Fey G, Odink K, Chapuis RM (1980) Synthesis of the mouse complement component C4 (Ss-protein) by peritoneal macrophages: kinetics of secretion and glycosylation of the subunits. Eur J Immunol 10: 75–82

Fey G, Domdey H, Wiebauer K, Whitehead AS, Odink K (1983) Structure and expression of the C3 gene. Springer Semin Immunopathol 6: 119–147

Fey G, Wiebauer K, Domdey H, Kazmaier M, Southgate C, Muller V (1984) Structural analysis of cloned mouse and human DNA sequences specifying C3, the third component of complement. In: Muftuoglu AU, Barlas N (eds) Recent advances in immunology. Plenum, New York, pp 101–106

Goldberger G, Thomas ML, Tack BF, Williams J, Colten HR, Abraham GN (1981) NH2-terminal structure and cleavage of guinea pig pro-C3, the precursor of the third complement component. J Biol Chem 256: 12617–12619

Goldberger G, Arnaout MA, Aden D, Kay R, Rits M, Colten HR (1984) Biosynthesis and postsynthetic processing of human C3b/C4b inactivator (factor I) in three hepatoma cell lines. J Biol Chem 259: 6492–6497

Green MR (1986) Pre-mRNA splicing. Annu Rev Genet 20: 671–708

Guiguet M, Exilie Frigere M-F, Dethieux M-C, Bidan Y, Mack G (1987) Biosynthesis of the third component of complement in rat liver epithelial cell lines and its stimulation by effector molecules from cultured human mononuclear cells. In Vitro Cell Dev Biol 23: 821–829

Hamilton AO, Jones L, Morrison L, Whaley K (1987) Modulation of monocyte complement synthesis by interferons. Biochem J 242: 809–815

Hase S, Kikuchi N, Ikenaka T, Inoue K (1985) Structure of sugar chains of the third component of human complement. J Biochem 98: 963–874

Hirani S, Lambris JD, Müller-Eberhard HJ (1986) Structural analysis of the asparagine-linked oligosaccharides of human complement component C3. Biochem J 233: 613–616

Hostetter MK, Krueger RA, Schmeling DJ (1984) The biochemistry of opsonization: central role of the reactive thiol ester of the third component of complement. J Infect Dis 150: 653–661

Hubbard SC, Ivatt RJ (1981) Synthesis and processing of asparagine-linked oligosaccharides. Annu Rev Biochem 50: 555–583

Iijima M, Tobe T, Sakamoto T, Tomita M (1984) Biosynthesis of the internal thiolester bond of the third component of complement. J Biochem 96: 1539–1546

Johnson E, Hetland G (1988) Mononuclear phagocytes have the potential to synthesize the complete functional complement system. Scand J Immunol 27: 489–493

Kai C, Yoshikawa Y, Yamamuchi K, Okada H (1983) Isolation and identification of the third component of complement of Japanese quails. J Immunol 130: 2814–2820

Kan CC, Solomon E, Belt KT, Chain AC, Hiorns LR, Fey G (1985) Nucleotide sequence of cDNA encoding human a2-macroglobulin and assignment of the chromosomal locus. Proc Natl Acad Sci USA 82: 2282–2286

King CR, Piatigorsky J (1983) Alternative RNA splicing of the murine α A-crystallin gene: protein-coding information within an intron. Cell 32: 707–712

Knowles BB, Howe CC, Aden DP (1980) Human hepatocellular carcinoma cell lines secrete the major plasma proteins and hepatitis B surface antigen. Science 209: 497–499

Kohler PF, Müller-Eberhard HJ (1967) Immunochemical quantitation of the third, fourth and fifth components of human complement: concentrations in the serum of healthy adults. J Immunol 99: 1211–1216

Koj A (1985) Definition and classification of acute phase proteins. In: Gordon AH, Koj A (eds) The acute phase response to injury and infection. Elsevier, Amsterdam, pp 139–144

Koj A (1986) Biological functions of acute-phase proteins and the cytokines involved in their induced synthesis. In: Reutter W, Popper H, Arias IM, Heinrich PC, Keppler D, Landmann L (eds) Modulation of liver cell expression. MTP Press, Lancaster, pp 331–341

Kusano M, Choi N, Tomita M, Yamamoto K, Migita S, Sekiya T, Nishimura S (1986) Nucleotide sequence of cDNA and derived amino acid sequence of rabbit complement component C3 α chain. Immunol Invest 15: 365–378

Kushner I (1982) The phenomenon of the acute phase response. Ann NY Acad Sci 389: 39–48

Lambris JD (1988) The multifunctional role of C3, the third component of complement. Immunol Today 9: 387–393

Lambrid JD, Müller-Eberhard HJ (1986) The multifunctional role of C3: structural analysis of its interactions with physiological ligands. Mol Immunol 23: 1237–1242

Lambris JD, Ganu VS, Hirani S, Müller-Eberhardt HJ (1985) Mapping of the C3d receptor (CR2)-binding site and a neoantigenic site in the C3d of the third component of complement. Proc Natl Acad Sci USA 82: 4235–4239

Lambris JD, Avila D, Becherer JD, Müller-Eberhard HJ (1988) A discontinuous factor H binding site in the third component of complement as delineated by synthetic peptides. J Biol Chem 263: 2147–2150

Lappin D, Hamilton AD, Morrison L, Aref M, Whaley K (1986) Synthesis of complement components (C3, C2, B and C1-inhibitor) and lysozyme by human monocytes and macrophages. J Clin Lab Immunol 20: 101–105

Lernhardt W, Rascke Wc, Melchers F (1986) Alpha-type B cell growth factor and complement component C3: their possible structural relationship. In: Melchers F, Potter M (eds) Mechanisms in B-cell neoplasia. Springer, Berlin Heidelberg New York, pp 98–104 (Current topics in microbiology and immunology, vol 132)

Lévi-Strauss M, Mallat M (1987) Primary cultures of murine astrocytes produce C3 and factor B, two components of the alternative pathway of complement activation. J Immunol 139: 2361–2366

Lundwall A, Wetsel RA, Domdey H, Tack BF, Fey GH (1984) Structure of murine complement component C3. 1. Nucleotide sequence of cloned complementary and genomic DNA coding for the β chain. J Biol Chem 259: 13851–13856

Lundwall AB, Wetsel RA, Kristensen TK, Whitehead AS, Woods E, Ogden RC, Colten HR, Tack BF (1985) Isolation and sequence of a cDNA clone encoding the fifth complement component. J Biol Chem 260: 2108–2112

Lusis AJ, Heinzmann C, Sparkes RS, Scott J, Knott TJ, Geller R, Sparkes MC, Mohandas T (1986) Regional mapping of human chromosome 19: organization of the genes for plasma lipid transport (APOC1, -C2, and -E and LDLR) and the genes C3, PEPD, and GPI. Proc Natl Acad Sci USA 83: 3929–3933

Maniatis T, Goodbourn S, Fischer JA (1987) Regulation of inducible and tissue-specific gene expression. Science 236: 1237–1245

Minta JO, Isenman DE (1987) Biosynthesis of the third component of complement by the human monocyte-like cell line, U937. Mol Immunol 24: 1105–1111

Morris KM, Aden DP, Knowles BB, Colten HR (1982b) Complement biosynthesis by the human hepatoma-derived cell line HepG2. J Clin Invest 70: 906–913

Morris KM, Goldberger G, Colten HR, Aden DP, Knowles BB (1982a) Biosynthesis and processing of a human precursor complement, pro-C3, in a hepatoma-derived cell line. Science 215: 399–400

Mount SM (1982) A catalogue of splice junction sequences Nucleic Acids Res 10: 459–472

Müller-Eberhard HJ, Schreiber RD (1980) Molecular biology and chemistry of the alternative pathway of complement. Adv Immunol 29: 1–53

Natsumme-Sakai S, Moriwaki K, Amano S, Hayakawa JI, Kaidoh T, Takahashi M (1979) Genetic mapping of the locus controlling structural variations of murine C3 in the chromosome 17. J Immunol 123: 216–221

Nichols WK (1984) LPS stimulation of complement (C3) synthesis by a human monocyte cell line. Complement 1: 108–115

Nonaka M, Iwaki M, Nakai C, Nozaki M, Kaidoh T, Nonaka M, Natsuume-Sakai S, Takahashi M (1984a) Purification of a major serum protein of rainbow trout (*Salmo gairdneri*) homologous to the third component of mammalian complement. J Biol Chem 259: 6327–6333

Nonaka M, Fujii T, Kaidoh T, Natsumme-Sakai S, Nonaka M, Yamaguchi N, Takahashi M (1984b) Purification of a lamprey complement protein homologous to the third component of the mammalian complement system. J Immunol 133: 3242–3249

Ogata RT, Schreffler DC, Sepich DS, Lilly SP (1983) cDNA clone spanning the α-γ subunit in the precursor of the fourth complement component. Proc Natl Acad Sci USA 80: 5061–5065

Padgett RA, Grabowski PJ, Konarska MM, Seiler S, Sharp PA (1986) Splicing of messenger RNA precursors. Annu Rev Biochem 55: 119–1150

Penalva DaSilva F, Hoecker GF, Day NK, Vienne K, Rubinstein P (1978) Murine complement component 3: genetic variation and linkage to H-2. Proc Natl Acad Sci USA 75: 963–965

Propp RP, Alper CA (1968) C'3 synthesis in the human fetus and the lack of transplacental passage. Science 162: 672–674

Pussell BA, Bourke E, Nayef M, Morris S, Peters K (1980) Complement deficiency and nephritis: a report of a family. Lancet 1: 675–677

Rittner C, Schneider PM (1988) Genetics and polymorphism of the complement components. In: Rother K, Till GO (eds) The complement system. Springer, Berlin Heidelberg New York, pp 80–135

Ross SC, Denson P (1984) Complement deficiency states and infection: epidemiology, pathogenesis and consequences of neisserial and other infections in an immune deficiency. Medicine 63: 243–273

Ruddy S, Carpenter CB, Chin KW, Knostman JN, Soter NA, Gotze O, Müller-Eberhard HJ, Austen KF (1975) Human complement metabolism: an analysis of 144 studies. Medicine 54: 165–178

St John Sutton M, Strunk RC, Cole FS (1986) Regulation of the synthesis of the third component of complement and factor B in cord blood monocytes by lipopolysaccharide. J Immunol 136: 1366–1372

Sekizawa A, Fujii T, Katagiri C (1984) Isolation and characterization of the third component of complement of the serum of the clawed frog, *Xenopus laevis*. J Immunol 133: 1436–1443

Shaw DJ, Meredith AL, Brook JD, Sarfarazi M, Harley HG, Huson SM, Bell GI, Harper PS (1986) Linkage relationships of the insulin receptor gene with the complement component 3, LDL receptor, apolipoprotein C2 and myotonic dystrophy loci on chromosome 19. Hum Genet 74: 267–269

Sottrup-Jensen L (1987) Alpha$_2$-macroglobulin and related thiol ester plasma proteins. In: Putman FW (ed) The plasma proteins: structure, function, and genetic control, vol 5. Academic, Orlando, pp 192–291

Steiner DF, Docherty K, Carroll R (1984) Golgi/Granule processing of peptide hormone and neuropeptide precursors: a mini review. J Cell Biochem 24: 121–130

Strunk RC, Tashjian AH, Colten HR (1975) Complement biosynthesis in vitro by rat hepatoma cell strains. J Immunol 114: 331–335

Strunk RC, Kunke KS, Giclas PC (1983) Human peripheral blood monocyte-derived macrophages produced haemolytically active C3 in vitro. Immunology 49: 483–488

Strunk RC, Whitehead AS, Cole FS (1985) Pretranslational regulation of the synthesis of the third component of complement in human mononuclear phagocytes by the lipid A portion of lipopolysaccharide. J Clin Invest 76: 985–990

Strunk RC, Eilden DM, Mason RJ (1988) Pulmonary aveolar type II epithelial cells synthesize and secrete proteins of the classical and alternative complement pathways. J Clin Invest 81: 1419–1426

Sundstrom C, Nilsson K (1976) Establishment and characterization of a human histiocytic lymphoma cell line (U937). Int J Cancer 17: 565–577

Szpirer J, Islam MQ, Cooke NE, Szpirer C, Evan G (1988) Assignment of three rat genes coding for plasma proteins, transferrin, third component of complement, and beta-fibrinogen to rat chromosomes 8, 9, and 2. Cytogenet Cell Genet 47: 42–45

Tack BF (1983) The β-Cys-γ-Glu thiolester bond in human C3, C4 and $\alpha_2$macroglobulin. Springer Semin Immunopathol 6: 259–282

Tack BF, Prahl JW (1976) Third component of human complement: purification from plasma and physicochemical characterization. Biochemistry 15: 4513–4521

Tartakoff AM (1983) Perturbation of vesicular traffic with the carboxylic ionophore monensin. Cell 32: 1026–1028

Thomas ML, Tack BF (1983) Identication and alignment of a thiol ester site in the third component of guinea pig complement. Biochemistry 22: 942–947

Tomana M, Niemann M, Gardner C, Volanakis JE (1985) Carbohydrate composition of the second, third and fifth components and factors B and D of human complement. Mol Immunol 22: 107–111

Traut TW (1988) Do exons code for structural or functional units in proteins? Proc Natl Acad Sci USA 85: 2944–2948

Treisman R, Proundfoot NJ, Shander M, Maniatis T (1982) A single-base change at a splice site in a β-thalassemic gene causes abnormal RNA splicing. Cell 29: 903–911

Ueki A, Sai T, Oka H, Tabata M, Hosokawa K, Mochizuki Y (1987) Biosynthesis and secretion of the third component of complement by human endothelial cells in vitro. Immunology 61: 11–14

Warren HB, Pantazis P, Davies PF (1987) The third component of complement is transcribed and secreted by cultured endothelial cells. Am J Pathol 129: 9–13

Wetsel RA, Lundwall A, Davidson F, Gibson T, Tack BF, Fey GH (1984) Structure of murine complement component C3. II. Nucleotide sequence of cloned complementary DNA coding for the α chain. J Biol Chem 259: 13857–13862

Wetsel RA, Ogata RT, Tack BF (1987) Primary structure of the fifth component of murine complement. Biochemistry 26: 737–743

Wetsel RA, Lemons RS, LeBeau MM, Barnum SR, Noack D, Tack BF (1988) Molecular analysis of human complement component C5: localization of the structural gene to chromosome 9. Biochemistry 27: 1474–1482

Whaley K (1980) Biosynthesis of the complement components and the regulatory proteins of the alternative complement pathway by human peripheral blood monocytes. J Exp Med 151: 501–516

Wheat WH, Wetsel RA, Falus A, Tack BF, Strunk RC (1987) The fifth component of complement (C5) in the mouse. Analysis of the molecular basis for deficiency. J Exp Med 165: 1442–1447

Whitehead AS, Goldberger G, Woods DE, Markham AF, Colten HR (1983) Use of a cDNA clone for the fourth component of human complement (C4) for analysis of a genetic deficiency of C4 in guinea pig. Proc Natl Acad Sci USA 80: 5387–5391

Whitehead AS, Sim RB, Bodner WF (1981) A monoclonal antibody against human complement component C3: the production of C3 by human cells in vitro. Eur J Immunol 11: 140–146

Whitehead AS, Solomon E, Chambers S, Bodmer WF, Povey S, Fey G (1982) Assignment of the structural gene for the third component of human complement to chromosome 19. Proc Natl Acad Sci USA 79: 5021–5025

Wieacker P, Fey G, Vorculescu I, Roper HH (1983) Exclusion of the C3 gene from the 19q133 to 19qter region by Southern analysis of human rodent somatic cell hybrids employing a cloned genomic C3 fragment. Acta Anthropogenet 7: 107–112

Wiebauer K, Domdey H, Diggelmann H, Fey G (1982) Isolation and analysis of genomic DNA clones encoding the third component of mouse complement. Proc Natl Acad Sci USA 79: 7077–7081

Wieme RJ, Demulenaare B (1967) Genetically determined electrophoretic variant of the human complement component C'3. Nature 214: 1042–1043

Wieringa B, Meyer F, Reiser J, Weissman C (1983) Unusual splice sites revealed by mutagenic inactivation of an authentic splice site of the rabbit β-globulin gene. Nature 301: 38–43

Wieringa B, Hofer E, Weissman C (1984) A minimal intron length but no specific internal sequence is required for splicing the large rabbit β-globin intron. Cell 37: 915–925

Wright SS, Reddy PA, Jong MTC, Erickson BW (1987) C3bi receptor (complement receptor type 3) recognizes a region of complement protein C3 containing the sequence Arg-Gly-Asp. Proc Natl Acad Sci USA 84: 1965–1968

Yu CY, Belt TK, Giles CM, Campbell DR, Porter RR (1986) Structural basis of the polymorphism of human complement components C4A and C4B: gene size, reactivity and antigenicty. EMBO J 5: 2873–2881

Zimmer B, Hartung HP, Scharfenberger G, Bitter-Suermann D, Hadding U (1982) Quantitative studies of the secretion of complement component C3 by resident, elicited and activated macrophages. Comparison with C2, C4 and lysosomal enzyme release. Eur J Immunol 12: 426–430

Molecular Aspects of C3 Interactions and Structural/Functional Analysis of C3 from Different Species*

J. D. BECHERER, J. ALSENZ, and J. D. LAMBRIS

1 Introduction 45
2 C3: General Features 46
3 Degradation of C3 During Complement Activation 46
4 C3 Interactions 50
4.1 Interaction of C3 Fragments with Complement Receptor Type 1 50
4.2 Interaction of C3 Fragments with Complement Receptor Type 2 52
4.3 Interaction of C3 Fragments with Complement Receptor Type 3 55
4.4 Interaction of C3 Fragments with Factor H 56
4.5 Interaction of C3 Fragments with Factor B 57
4.6 Interaction of C3 Fragments with Properdin 57
4.7 Interaction of C3 Fragments with Conglutinin 58
5 Structure/Function Analysis of C3 in Different Species 59
5.1 Mammalian C3 59
5.2 Avian C3 61
5.3 Reptilian C3 61
5.4 Amphibian C3 62
5.5 Osteichthian C3 63
5.6 Cyclostome C3 64
6 Concluding Remarks 64
References 65

1 Introduction

C3 plays a critical role in both pathways of complement activation due to its ability to bind to numerous other complement proteins. In addition, its interactions with several cell surface receptors make it a key participant in phagocytic and immunoregulatory processes. It is the purpose of this chapter to review the characteristics and unique structural features of human C3 which permit it to bind to various ligands and receptors. (For the purpose of this review, "ligands" of C3 are taken as those serum proteins which bind C3 and are distinguished from C3 receptors, which are cell surface proteins.) Here we also enunciate, from a structural

---

Basel Institute for Immunology, Grenzacherstr. 487, 4005 Basel, Switzerland
*J.A. has an EMBO long-term fellowship (ALTF 298 – 1987). The Basel Institute for Immunology was founded by and is supported entirely by F. Hoffman–La Roche Ltd. Co., Basel, Switzerland.

viewpoint, our current knowledge of C3 from other species and discuss how their similarities, along with those of other homologous proteins, are used to further our understanding of the structure/function relationship of C3.

2 C3: General Features

Human C3 is a 190-kDa glycoprotein with a serum concentration of 1–2 mg/ml. It is synthesized predominantly in the liver (ALPER et al. 1969) as a single chain pre-pro molecule with β–α-chain sequence (DOMDEY et al. 1982; GOLDBERGER et al. 1981) and assumes its native two-chain form after enzymatic processing (see BARNUM et al., this volume). The α-chain (115-kDa) is linked to the β-chain (75-kDa) by a single disulfide bond and noncovalent forces (MATSUDA et al. 1985; JANATOVA 1986). The complete primary structures for human (DE BRUIJN FEY 1985) and mouse (WETSEL et al. 1984; LUNDWALL et al. 1984a) C3 and the partial structures for rabbit (KUSANO et al. 1986) and *Xenopus* (GROSSBERGER et al. 1989) C3 have been elucidated. Carbohydrate analysis revealed that human C3 possesses two N-linked carbohydrate moieties positioned at residue 63 of the β-chain ($Man_5GlcNac_2 + Man_6GlcNac_2$) and at residue 917 of the α-chain ($Man_8GlcNaC_2 + Man_9GlcNaC_2$) which together account for 1.5% of the molecular weight of C3 (HASE et al. 1985; HIRANI et al. 1986). [The numbering of residues of C3 throughout this review has been ascertained from the report of DE BRUIJN and FEY (1985) after subtracting the signal peptide.]

One of the distinguishing characteristics of C3 is its ability to bind covalently to acceptor molecules on cell surfaces (MÜLLER-EBERHARD et al. 1966) via ester or amide linkages (LAW el al. 1979). This feature has been attributed to the thiolester bond which is present within the C3d region of C3, and which is sensitive to nucleophilic attack. The thiolester bond is the product of an intramolecular transacylation between the thiol group of cysteine and the γ-amide group of the glutamine within the C3 sequence Gly-Cys^{988}-Gly-Glu-Gln^{991}-Asn (for review see TACK 1983 and LEVINE et al., this volume). This thiolester moiety is also found in C4 and α_2-macroglobulin, two plasma proteins homologous to C3. In native C3, the thiolester group appears to be protected within a hydrophobic pocket and is exposed in the C3b fragment upon cleavage of C3 by the C3 convertase. Thus, the transiently expressed thiolester group can now participate in a transacylation reaction with nucleophilic groups present on the acceptor molecules. This attachment of C3b to surface structures is important for deposition of additional C3b molecules on the activating surface, for initiation of the membrane attack complex (MAC), for phagocytosis of foreign particles, and for enhancement of effector cell-target cell contact.

3 Degradation of C3 During Complement Activation

Activation of C3 is essential for the continuation and amplification of the enzymatic cascade of the complement system. Native C3 expresses no biological activity in its native form, but cleavage to C3a and C3b by either the classical or alternative pathway convertases results in profound conformational changes within the molecule (ISENMAN et al. 1981; ISENMAN and COOPER 1981; NILSSON and NILSSON 1986; HACK et al. 1988). The nascent C3b molecule is metastable with respect to its newly exposed thiolester group which is now susceptible to nucleophilic attack by acceptor molecules on target membranes. Metastable C3b that fails to bind to the target membrane is hydrolyzed by H_2O and can no longer bind to target surfaces. It is estimated that the half-life of the metastable binding site with respect to its inactivation by H_2O is 60 ms (SIM et al. 1981). The generation of C3b leads to the expression of binding sites for factors B, H, and I, C5, C4 binding protein (C4bp), properdin, CR1, and membrane cofactor protein (MCP; Fig. 1). Which of these proteins bind to C3b determines whether the C3 convertase is amplified and initiation of the MAC occurs, or whether C3b is inactivated by cleavage to iC3b. In the case of surface-bound C3b, characteristics of the acceptor surface have been shown to play a critical role in determining the fate of further complement activation (reviewed in PANGBURN 1986b; ATKINSON and FARRIES 1987).

Control of the amplification of the classical or alternative pathway C3 convertase occurs by the inactivation of C3b (also C4b) by the serine protease factor I (see Fig. 1). This inactivation is dependent on the presence of one of several cofactor molecules (factor H, C4bp, CR1, MCP) which allow factor I to cleave the α' chain of C3b between the Arg-Ser bonds at residues 1281–1282 and 1298–1299 of the C3 sequence. A small peptide, C3f (DAVIS and HARRISON 1982), is released, and the resulting molecule, termed iC3b, expresses additional binding sites for CR2, CR3, and CR4 (p150,95). A third factor I cleavage site, with CR1, CR2, or factor H serving as cofactors (ROSS et al. 1982; MEDOF et al. 1982; MEDICUS et al. 1983; MITOMO et al. 1987) has been reported to occur at residues 932–933 (Arg-Glu) of the α'-chain of C3, generating the C3c and C3dg fragments (DAVIS et al. 1984). From the above cleavages and those of C4 (PRESS and GAGNON 1981; CAMPBELL et al. 1981) it appears that factor I preferentially cleaves Arg-X bonds, but that this specificity is restricted by the required presence of one of several cofactor molecules (factor H, C4bp, CR1, CR2, and MCP). Finally, the C3d fragment can be generated by trypsin, plasmin, or elastase cleavage of C3dg (LACHMANN et al. 1982; ROSS et al. 1982), but whether or not this fragment is generated under physiological conditions is uncertain. Also, a fragment termed C3dK, which extends nine amino acids from the NH_2-terminus of C3dg, is generated by kallikrein digestion of iC3b, and this fragment has been shown to induce leukocytosis (MEUTH et al. 1983). A similar leukocytosis-inducing activity is also observed in the C3e fragment of C3 (GHEBREHIWET and MÜLLER-EBERHARD 1979), and it is therefore thought that C3e and C3dK contain overlapping sequences. However, nothing more is known about the structural features of the C3e fragment or its receptor.

The third cleavage by factor I has been questioned based on experiments in which inhibitors specific for serine proteases were included, and no third cleavage was observed (SIM et al. 1981; MALHOTRA and SIM 1984). Since factor I has been reported to be insensitive to the serine protease inhibitors DFP, PMSF, and benzamidine (CROSSLEY and PORTER 1980), it was concluded that the third cleavage is due to a contaminating enzyme which is sensitive to the inhibitors used in those studies. Further support of this argument is that, in mouse and rabbit C3, the sequence Gln-Gly is found at the proposed third factor I cleavage site (WETSEL et al. 1984;

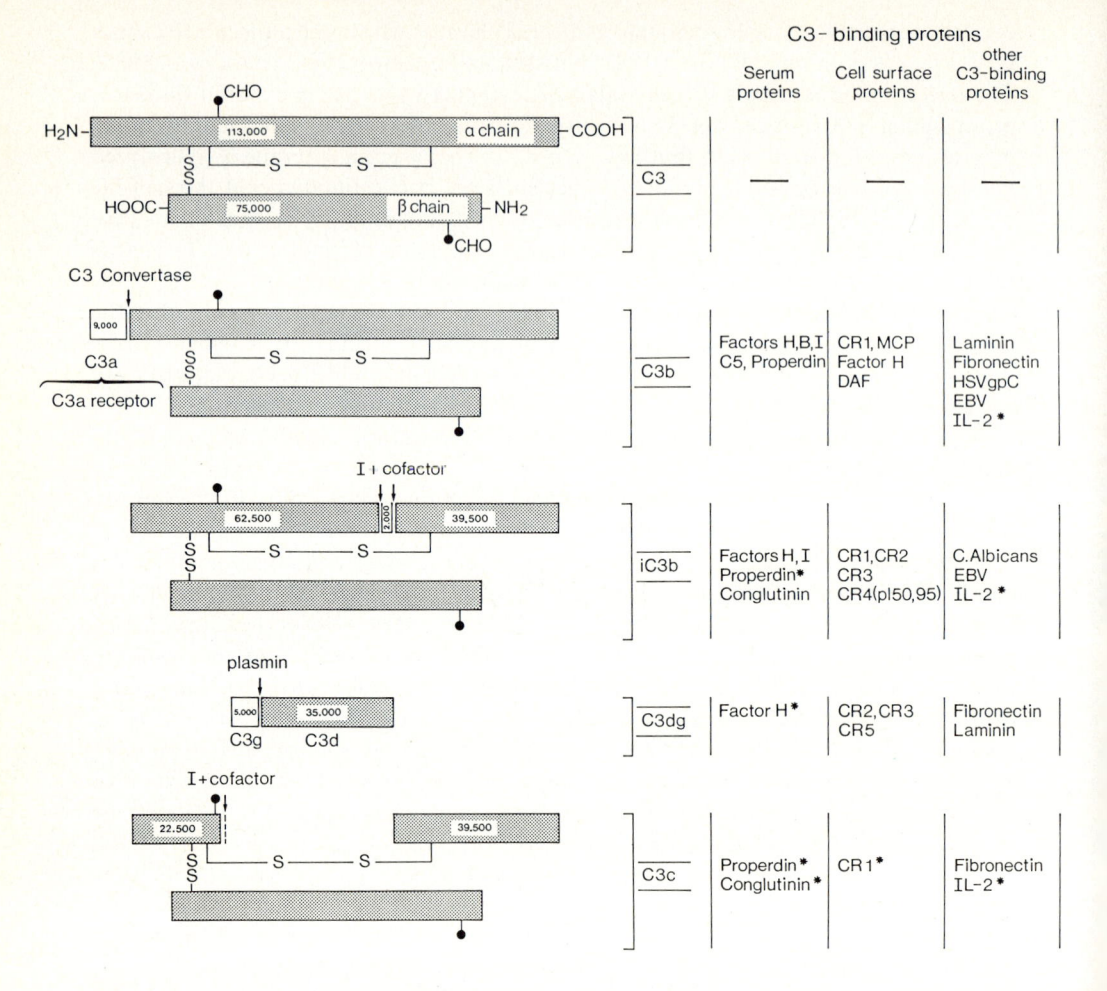

Fig. 1. Degradation of C3 and reactivity of its fragments with other complement proteins, cellular receptors, and proteins on foreign particles. *Asterisks*, interactions observed when the C3 fragments are coated on microtiter plates, but whether these occur under physiological conditions is unknown. These interactions, however, have been essential in localizing the ligand and receptor binding domains in C3

KUSANO et al. 1986), which would suggest that factor I must possess a broader specificity than previously proposed (see Fig. 2).

Thus, two questions are obviously raised: a) Why is factor I such a novel serine protease that it cannot be inactivated by DFP, yet its primary structure (CATTERALL et al. 1987) correlates well with that of other serine proteases? b) Does the third cleavage site actually occur at the N-terminus of C3dg (between the Arg-Ser bond, residues 932–933), or could the initial cleavage occur NH_2-terminal to this sequence followed by another cleavage yielding the C3dg fragment? The latter question is especially curious since SEYA and NAGASAWA (1985) reported the isolation of a C3dg fragment which resembles the C3dK fragment characterized by MUETH and collegues

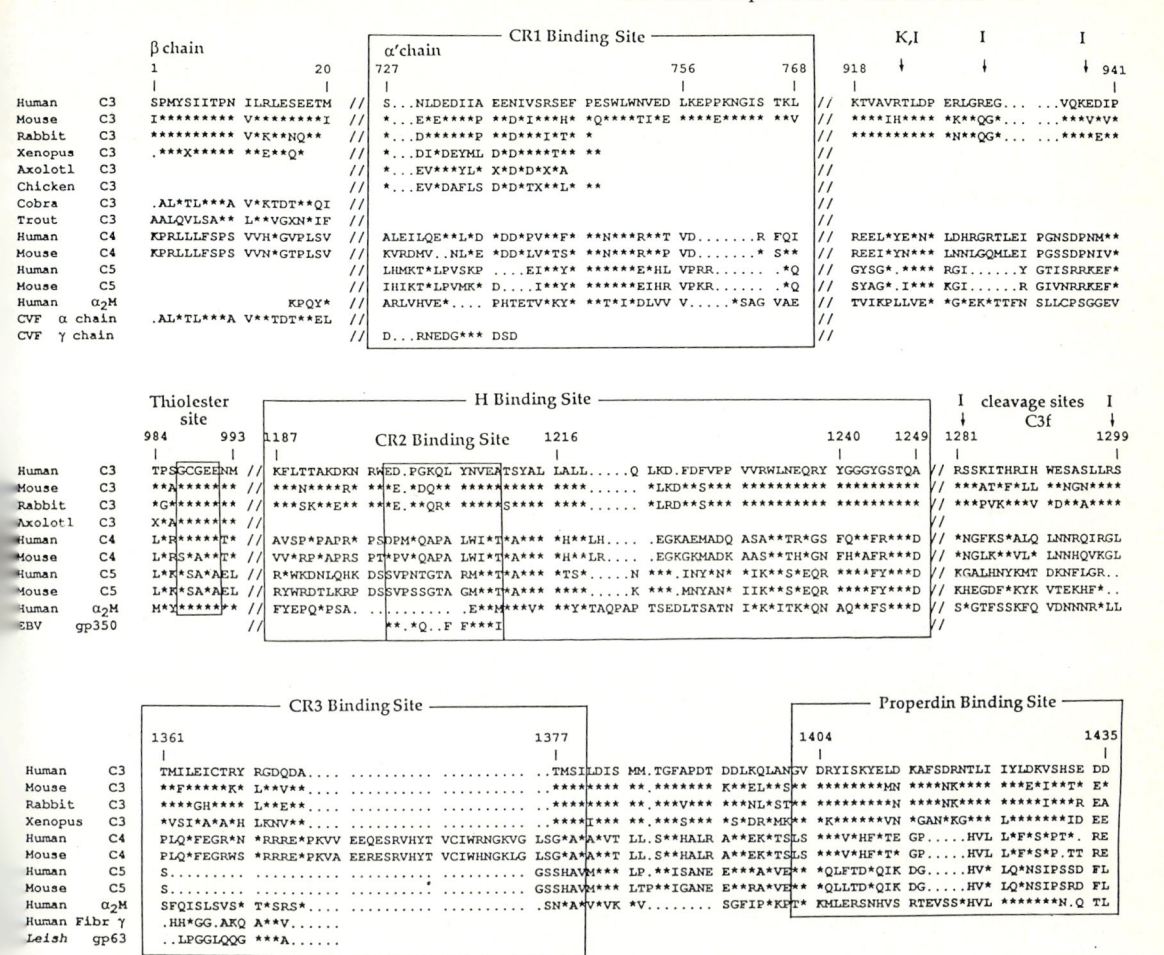

		β chain		//	α'chain	┌─── CR1 Binding Site ───┐		//	K,I ↓	I ↓	I ↓
		1	20		727		756 768		918		941
Human	C3	SPMYSIITPN	ILRLESEETM	//	S...NLDEDIIA	EENIVSRSEF PESWLWNVED	LKEPPKNGIS TKL	//	KTVAVRTLDP	ERLGREG...	...VQKEDIP
Mouse	C3	I*********	V*********I	//	*...E*E****P	**D*I****H* *Q****TI*E	****E****** **V	//	****IH****	*K**QG*...	...***V*V*
Rabbit	C3	**********	V*K**NQ**	//	*...D******P	*D****I*T* *		//	**********	*N**QG*...	...****E**
Xenopus	C3	.***X*****	**E**Q*	//	*...DI*DEYML	D*D****T** **		//			
Axolotl	C3			//	*...EV****YL*	X*D*D*X*A		//			
Chicken	C3			//	*...EV*DAFLS	D*D*TX**L* **		//			
Cobra	C3	.AL*TL***A	V*KTDT**QI	//				//			
Trout	C3	AALQVLSA**	L**VGXN*IF	//				//			
Human	C4	KPRLLLFSPS	VVH*GVPLSV	//	ALEILQE**L*D	*DD*PV**F* **N***R**T	VD.......R FQI	//	REEL*YE*N*	LDHRGRTLEI	PGNSDPNM**
Mouse	C4	KPRLLLFSPS	VVN*GTPLSV	//	KVRDMV..NL*E	*DD*LV*TS* **N****R**P	VD.......* S**	//	REEI*YN***	LNNLGQMLEI	PGSSDPNIV*
Human	C5			//	LHMKT*LPVSKP	....EI**Y* ******E*HL	VPRR...... .*Q	//	GYSG*.****	RGI......Y	GTISRRKEF*
Mouse	C5			//	IHIKT*LPVMK*	D....I**Y* ******EIHR	VPKR...... .*Q	//	SYAG*.I***	KGI......R	GIVNRRKEF*
Human	α₂M		KPQY*	//	ARLVHVE*....	PHTETV*KY* **T*I*DLVV	V.....*SAG VAE	//	TVIKPLLVE*	*G*EK*TTFN	SLLCPSGGEV
CVF	α chain	.AL*TL***A	V**TDT**EL	//				//			
CVF	γ chain			//	D...RNEDG*** DSD			//			

		Thiolester site		//	┌──────── H Binding Site ────────┐				//	I cleavage sites I	
		984	993		1187	┌── CR2 Binding Site ──┐ 1216		1240 1249		1281 ↓ C3f ↓	1299
Human	C3	TPSGCGEENM		//	KFLTTAKDKN	RWED.PGKQL YNVEATSYAL	LALL.....Q	LKD.FDFVPP VVRWLNEQRY YGGGYGSTQA	//	RSSKITHRIH	WESASLLRS
Mouse	C3	**A******		//	***N****R*	**E..DQ** **********	****.....	*LKD**S*** ********** **********	//	***AT*F*LL	**NGN****
Rabbit	C3	*G*******		//	***SK**E**	**E..*QR* *****S****	****.....	*LRD**S*** ********** **********	//	***PVK***V	*D**A****
Axolotl	C3	X*A******		//					//		
Human	C4	L*R*****T*		//	AVSP*PAPR*	PSDPM*QAPA LWI*T*A***	*H**LH....	.EGKAEMADQ ASA**TR*GS FQ**FR***D	//	*NGFKS*ALQ	LNNRQIRGL
Mouse	C4	L*RS*A**T*		//	VV**RP*APRS	PT*PV*QAPA LWI*T*A***	*H**LR....	.EGKGKMADK AAS***TH*GN FH*AFR***D	//	*NGLK**VL*	LNNHQVKGL
Human	C5	L*K*SA*AEL		//	R**WKDNLQHK	DSSVPNTGTA RM**T*A***	*TS*....N	***.INY*A** *IK**S*EQR ****FY***D	//	KGALHNYKMT	DKNFLGR..
Mouse	C5	L*K*SA*AEL		//	RYWRDTLKRP	DSSVPSSGTA GM**T*A***	****.....K	***.MNYAN* IIK**S*EQR ****FY***D	//	KHEGDF*KYK	VTEKHF*..
Human	α₂M	M*Y*****P*		//	FYEPQ*PSA.	....E**M	***V* **Y*TAQPAP	TSEDLTSATN I*K**ITK*QN AQ***FS***D	//	S*GTFSSKFQ	VDNNNR*LL
EBV	gp350			//		**.*Q..F F****I			//		

		┌───────── CR3 Binding Site ─────────┐				┌───── Properdin Binding Site ─────┐	
		1361		1377		1404	1435
Human	C3	TMILEICTRY	RGDQDA.........	............TMSI	LDIS MM.TGFAPDT DDLKQLANGV	DRYISKYELD KAFSDRNTLI	IYLDKVSHSE DD
Mouse	C3	**F******K*	L**V**.........	.........***A***	** ******* K**EL**S**	********MN ****NK****	***E*I**T* E*
Rabbit	C3	****GH****	L**E**.........	.........****I***	**.***V** ***NL*ST**	**********N ****NK****	*****I****R EA
Xenopus	C3	*VSI*A*A*H	LKNV**..........	............*T***	**.***S*** *S*DR*MK**	*K*******VN *GAN*KG***	L*******ID EE
Human	C4	PLQ*FEGR*N	*RRRE*PKVV EEQESRVHYT	VCIWRNGKVG LSG*A*A*VT	LL.S**HALR A**EK*TSLS	***V*HF*TE GP.....HVL	L*F*S*PT*. RE
Mouse	C4	PLQ*FEGRWS	*RRRE*PKVA EERESRVHYT	VCIWHNGKLG LSG*A**T	LL.S**HALR A**EK*TSLS	***V*HF*T* GP.....HVL	L*F*S*P.TT RE
Human	C5	S.........	....	.........GSSHAVM**	LP.**ISANE E***A*VE**	*QLFTD*QIK DG.....HV*	LQ*NSIPSSD FL
Mouse	C5	S.........	.................*	.........GSSHAVM**	LTP**IGANE E***RA*VE**	*QLLTD*QIK DG.....HV*	LQ*NSIPSRD FL
Human	α₂M	SFQISLSVS*	T*SRS*.....	....SN*A*V*VK *V....	SGFIP*KFT**	KMLERSNHVS RTEVSS*HVL	*******N.Q TL
Human	Fibr γ	.HH*GG.AKQ	A**V......				
Leish	gp63	..LPGGLQQG	***A......				

Fig. 2. Amino acid sequence comparison of functionally important sites within human C3 and C3 from other species and other homologous proteins (e.g., C4, C5, α₂-macroglobulin). In addition, the regions of similarity between human C3 and fibrinogen γ-chain, gp350/220 of EBV, and gp63 of *Leishmania* promastigotes are shown. Numbering of the human C3 residues is adapted from DE BRUIN and FEY (1985) after subtracting the signal peptide. *Boxed sequences,* areas of functional importance; *asterisks,* identical residues; *periods,* gaps, introduced for maximal sequence alignment. The sequence of guinea pig C3 surrounding the thiolester site has been reported by Thomas and Tack (1983). Partial amino acid sequences for rabbit, *Xenopus,* axolotl, chicken, and cobra C3 were obtained from the purified proteins (see Fig. 4) and are reported here for the first time. Briefly, NH₂-terminal amino acid sequence was obtained by electroblotting the C3 chains separated by sodium dodecyl sulfate polycrylamide gel electrophoresis onto polyvinylidene difluoride membranes and sequencing the excised bands directly from the membrane (see BECHERER and LAMBRIS 1988 for details)

(1983) in that it possessed leukocytosis-inducing activity. This activity was not observed in the C3dg characterized by LACHMAN and collegues (1982). Since the C3dg and C3dK fragments differ by nine amino acids in their NH₂-termini, and these nine amino acids were responsible for the leukocytosis activity (HOEPRICH et al. 1985), it seems possible that SEYA's C3dg was actually C3dK (no sequence data was reported). This would imply that factor I can cleave further up in the sequence, possibly between the Arg-Thr or Arg-Leu at residues 923–924 or 929–930 respectively (see Fig. 2). Recent results from our laboratory suggest that three different C3dg-like fragments are generated in serum. This is based on NH_2-terminal sequence data and on the reactivity of several monoclonal antibodies (NILSSON et al. 1988) which recognize different epitopes within residues 929–946 of C3. The three C3dg-like fragments had their NH_2-termini starting at residues 924 (cleavage between Arg-Thr), 933 (cleavage between Arg-Glu), and 938 (cleavage between Lys-Glu). That factor I is the enzyme responsible for "inactivating" C3b was confirmed by the ability of a monoclonal anti-factor I antibody to inhibit all three cleavages of C3b (NILSSON-EKDAHL et al. 1989). Another interesting observation is that all the factor I mediated cleavages were inhibited when factor I was treated with the protease inhibitor DFP in the presence of C3b. This suggests that factor I binds to C3b, and that this binding exposes the active site of the serine protease to DFP. These results are in agreement with an earlier report showing that factor I binds to Sepharose-coupled C3b (NAGASAWA et al. 1980).

4 C3 Interactions

The binding of numerous plasma and membrane proteins to the degraded products of C3 account for the molecule's ability to mediate a variety of biological responses (LAMBRIS 1988; Becherer et al. 1989a). Fluid-phase and surface-bound fragments of C3 are generated during complement activation, both of which can bind to other complement components and the numerous regulatory molecules of the complement system. Upon binding of these fragments to their respective complement receptors a variety of responses are elicited, with the fluid-phase fragments capable of migrating within their local environment and the surface bound fragments capable of forming a bridge between the receptor-bearing cell and the target. To understand the numerous responses mediated by the degraded products of C3, a great deal of effort has been placed on understanding the structural features of C3 responsible for its multifunctionality. Below, we review the available information concerning the binding sites for CR1, CR2, CR3, factor H, factor B, properdin, and conglutinin in C3 (see Fig. 3). The interactions of the surface proteins on foreign pathogens (i.e., Epstein-Barr virus, or EBV, herpes simplex virus, etc.) which bind C3, as well as the anaphylatoxin C3a with its receptor, will not be mentioned since comprehensive reviews by DIERICH et al. and HUGLI are presented elsewhere in this volume.

4.1 Interaction of C3 Fragments with Complement Receptor Type 1

Complement receptor type 1 (CR1, CD35) is a polymorphic membrane glycoprotein (160–280 kDa) that binds the C3b and C4b fragments of C3 and C4, and its structure is comprised of the short consensus repeats common to other C3b/C4b proteins (reviewed by FEARON and AHEARN, this volume). During opsonization of foreign particles by complement, numerous molecules of C3b are fixed on the activating surface, and it appears that the multivalency of surface-bound C3b may be physiologically important in CR1-mediated functions. This is supported by the fact that dimeric C3b (K_a approximately $5 \times 10^7 \, M^{-1}$) binds CR1 with a much higher affinity than monomeric C3b (K_a approximately $0.5 \times 10^6 - 2 \times 10^6 \, M^{-1}$). CR1 can also bind the iC3b and C3c fragments of C3 but with lower affinity (Ross et al. 1983; BECHERER and LAMBRIS 1988). The binding site on C3c for CR1 has been localized (BECHERER and LAMBRIS 1988; see Figs. 2, 3). Elastase cleavage of C3 yielded four fragments of C3c which varied in their reactivity toward CR1. NH_2-terminal sequence analysis revealed that the C3c fragments which did not bind CR1 were missing eight and nine amino acids from the NH_2-termini of their 27-kDa α'-chain. The COOH-termini of this chain, which was approximated by the presence or absence of the carbohydrate moiety at position 917 of the C3 sequence (DE BRUIJN and FEY 1985; HIRANI et al. 1985), played no role in CR1 binding.

Both a synthetic peptide ($C3^{727-768}$), which spans residues 727–768 of the C3 sequence, and the generated antipeptide antibody (anti-$C3^{727-768}$) inhibit the binding of C3b and C3c to CR1. The finding that $C3^{727-768}$ binds the receptor with a lower affinity than either C3b or C3c suggests that secondary sites of contact and/or conformational constraints inherent to C3c and C3b are important in their interaction with CR1. This region of C3 has also been shown to be involved in the interaction of C3b with factors H and B, a finding which supports the idea that, within C3b, there are sites of interaction common to several of the C3b/C4b binding proteins whose structures are characterized by short consensus repeats (LAMBRIS 1988; LAMBRIS et al. 1988; see following section). This supposition is reflected by the weak affinity of C4 binding protein for C3b (FUJITA and NUSSENZWEIG 1979), by the observations that CR1 and factors H and B compete for binding to C3b (DISCIPIO 1981; PANGBURN 1986a) and by the ability of a monoclonal antibody to inhibit the binding of all of these proteins to C3b (BECHERER and LAMBRIS, unpublished observation). Furthermore, the recent work by KLICKSTEIN using deletion mutants of recombinant CR1 showed that two C3b binding sites exist in CR1 (KLICKSTEIN et al. 1988). Whether CR1 recognizes two distinct sites within C3b, or whether the receptor can interact multivalently with the C3b-C3b complexes generated during complement activation remains to be determined.

Several interesting features were observed when anti-$C3^{727-768}$ antibodies (affinity purified on a Sepharose C3b column) were used to study the conformational changes that occur within the CR1 binding domain of C3 during complement activation. These antibodies, which inhibit C3b and C3c binding to CR1, bound to C3b, iC3b, and C3c but not to native C3. Furthermore, the binding of anti-$C3^{727-768}$ to these C3 fragments was inhibited by peptides $C3^{727-768}$ and $C3^{727-745}$ but not by $C3^{741-768}$. Thus, anti-$C3^{727-768}$ recognized a neoepitope expressed after cleavage of C3 to C3b that is confined within the 14 NH_2-terminal amino acids (residues 727–740) of the α'-chain of C3 (BECHERER et al. 1989b). This antibody, or a monoclonal antibody generated against this region, could therefore be used to detect C3 degradation fragments in biological fluids.

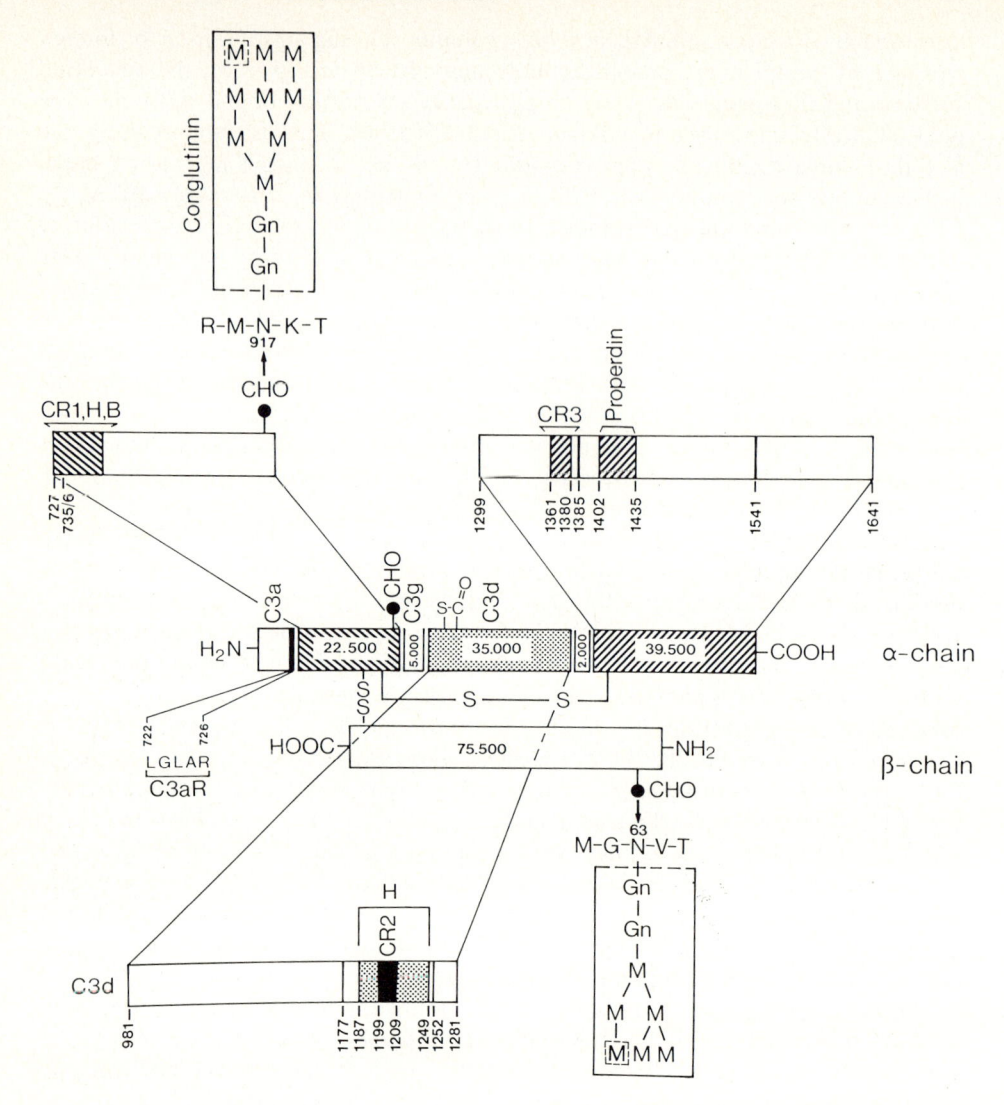

Fig. 3. A schematic representation of the C3 molecule showing the carbohydrate structures and the regions of C3 which have been identified to bind factors H and B, CR1, CR2, CR3, C3aR, properdin, and conglutinin (see text for details). Residues 1385/1252 of the α-chain denote the endpoints of the CNBr fragments originally generated which aided in the localization of the properdin and CR2 binding sites, respectively. Residues 735 and 736 of the α-chain represent the NH$_2$ termini of the C3c fragments which led to the localization of the CR1 binding site. (For numbering of the C3 residues, see Fig. 2)

4.2 Interaction of C3 Fragments with Complement Receptor Type 2

Complement receptor type 2 (CR2/EBVR, CD21) is a 140-kDa membrane glycoprotein originally identified on B cells, but recent reports have indicated a wider distribution, including endothelial cells, follicular dendritic cells, some T-cell lines, and immature thymocytes (for review see COOPER et al. 1988; BECHERER et al. 1989a and FEARON and AHEARN, this volume). CR2 binds the iC3b, C3dg, and C3d fragments of C3 (Ross et al. 1983) while a low-affinity interaction with C3b has also been observed (Ross and MEDOF 1985). In addition, CR2 serves as the receptor for EBV, an interaction mediated by the surface glycoprotein gp350/220 of EBV (NEMEROW et al. 1987; TANNER et al. 1987).

The binding site on C3d for CR2 was initially identified within an 8.6-kDa CNBr fragment of C3d, and subsequent studies employing synthetic peptides from this region identified residues 1199–1210 of the C3 sequence as comprising the CR2 binding site (LAMBRIS et al. 1985). This site is sandwiched within the discontinuous factor H binding site (LAMBRIS et al. 1988) and may yield a plausible explanation for the H-like cofactor activity of CR2 (MITOMO et al. 1987). Also, this region of C3d (see Fig. 2) was found to be similar to a region of gp350/220 (residues 21–31), the surface glycoprotein of EBV involved in binding CR2 (NEMEROW et al. 1987; TANNER et al. 1987). This suggested that EBV and C3d possess similar structural features which mediate binding to CR2. The proposal is supported by the binding of a synthetic peptide, corresponding to the aforementioned region of gp350/220, to CR2 and by its ability to inhibit: (a) the binding of gp350/220 and C3dg to B cells (NEMEROW et al. 1989), (b) EBV-induced B-cell proliferation (NEMEROW et al. 1989), and (c) the growth-supporting effects on Raji cells of a multivalent, CR2 binding synthetic peptide from C3d (SERVIS and LAMBRIS 1989). Furthermore, this agrees with the observations that the CR2 specific monoclonal antibody OKB-7 inhibited both C3d and EBV binding to CR2 (NEMEROW et al. 1985a), and that a deletion mutant of gp350/220, missing two amino acids from the region of similarity with C3d, did not bind CR2 (TANNER et al. 1988). Scatchard analysis of gp350/220 binding to CR2/EBVR, however, demonstrated both high- and low-affinity receptor binding sites (TANNER et al. 1988); thus one of these sites appears to be specific for the domain of gp350/220 that shares amino acid similarity with the CR2 binding site in C3d. The recent findings using CR2-specific monoclonal antibodies and an anti-idiotypic anti-CR2 antibody indicate different binding sites on CR2/EBVR for C3d and gp350/220 (BAREL et al. 1988), and, when taken together with the observed inhibition by OKB-7, these observations are not inconsistent with the above Scatchard analysis indicating the presence of two binding sites for EBV on CR2. Therefore, further refinement of both the ligand and receptor binding sites involved in the CR2 interaction with gp350/220 and C3d is necessary to clarify the relationship between these two interactions. One possibility is that CR2 possesses multiple sites of interactions for fragments of C3 and for gp350/220; such a model would be analogous to the CR1 interaction with C3b and C4b. Supporting the idea that multiple binding sites exist in CR2 for EBV is the recent finding that mouse CR2, despite binding human C3d, cannot bind EBV (FINGEROTH et al. 1989).

In addition to its dual ligand binding properties, CR2 has long been implicated in modulating the growth of B cells, although the precise manner by which CR2 mediates these effects is still unclear. However, the nature of the ligand interacting with the receptor appears important since monovalent fluid-phase C3 fragments generally inhibit in vitro T- and B-cell responses while multivalent CR2 ligands have been shown to induce growth and differentiation of B cells. The latter regulatory function of CR2 has been demonstrated by the cross-linking of CR2 on the cell surface by polyclonal or monoclonal anti-CR2 antibodies (FRADE et al. 1985; NEMEROW et al. 1985b; WILSON et al. 1985; PETZER et al. 1988) or by particle-bound or aggregated C3d/C3dg (MELCHERS et al. 1985; BOHNSACK and COOPER 1988). Furthermore, C3 (HATZFELD et al. 1988) and anti-CR2 monoclonal antibodies (PERNEGGER et al. 1988) have been shown to support the growth of the B-lymphoblastoid cell line Raji in serum-free conditions. Thus, in view of the findings that CR2 can be phosphorylated and translocated to the nucleus (CHANGELIAN and FEARON 1986; BAREL et al. 1986; DELCAYRE et al. 1987), that CR2 synergistically enhances the anti-IgM induced increase of cytoplasmic free Ca^{2+} (CARTER et al. 1988), and that CR2 cocaps with membrane IgM in the presence of anti-CR2 and anti-IgM antibodies (TSOKOS et al. 1988), it appears that engagement of the receptor by multivalent ligands delivers inductive signals to the cells. To further explore the mechanisms that lead to B-cell activation and growth, especially in view of the fact that C3d reacts with several serum and cell surface proteins, we have used both monovalent and multivalent CR2-binding synthetic peptides in an attempt to mimic the effects of CR2 ligation by C3d. Under serum-free conditions, a multivalent CR2-binding peptide, $C3^{(1202-1214)4}$, supported the growth of Raji cells, and this effect could be inhibited by the monovalent CR2-binding peptide $C3^{1201-1214}$ (SERVIS and LAMBRIS 1989). A similar inhibition was observed by a synthetic peptide which spans residues 19–30 of the NH_2-terminus of gp350/220. In addition, a multimeric form of the same gp350/220 peptide stimulated Raji cell growth (ESPARZA and LAMBRIS, unpublished observation), supporting the idea that cross-linking of the receptor is essential for proliferation of these cells. These findings are substantiated by the recent report of a synthetic peptide from the same region of gp350/220 which blocks EBV-induced B-cell proliferation (NEMEROW et al. 1989).

The recent observation that the amino acid sequence of the CR2 binding site in C3d is similar to the sequence between residues 295 and 306 of C3 (Table 1) and the finding that a synthetic peptide representing this sequence competes with the synthetic peptide $C3^{1201-1214}$ for binding to CR2 (ESPARZA and LAMBRIS 1989) suggests that, within C3, there are multiple sites which are involved in its interaction with CR2. Although it is not known whether this site contributes to the binding of iC3b to CR2, one possible scenario is that the fluid-phase iC3b fragment delivers a stimulatory signal to the CR2-bearing cell as a result of its multivalent nature; subsequent to this binding, cleavage of iC3b by factor I and CR2 (acting now as a cofactor) produces the monovalent C3c and C3dg fragments which would transmit an inhibitory signal to the cell. This multiple-site model for the interaction of C3 fragments with CR2 is similar to that demonstrated for C3b binding to factor H, another member of the family of C3b/C4b binding proteins (see below).

Another region of C3, similar in amino acid sequence to the CR2 binding site, was found between residues 744 and 755 of the α-chain (Table 1). These residues are within the identified CR1 and H binding domain located in the NH_2 terminus of the α'-chain (Figs. 2, 3). The contribution of these amino acids in this region of C3 to the H and CR1 binding remains to be determined. Nonetheless, since factor H has been shown to bind to a region of C3d incorporating the CR2 binding site, one could speculate that these molecules bind to common regions within C3 due to the structural and functional similarities between CR1, CR2, and factor H.

Table 1. Amino acid sequence similarity between the identified CR2 binding site in C3d and other segments of C3 and the gp$^{350/220}$ protein of EBV

	Residues	Sequence
CR2 binding site in C3d	1199–1210	E D . P G K Q L Y N V E A
β-Chain of human C3	295–306	* * L V * * S * * . * S *
α-Chain of human C3	744–755	* F . * E S W * W * * * D
CR2 binding site in gp350	21–30	* * . * * . . F F * * * I

Identical residues and inserted gaps are denoted as in Fig. 2

4.3 Interaction of C3 Fragments with Complement Receptor Type 3

CR3 (Mac-1, CD11b/18) is membrane glycoprotein that participates in the phagocytosis of iC3b-opsonized particles and in cell adhesion (see ROSEN and LAW, this volume). It is a heterodimer consisiting of noncovalently linked α- (165-kDa) and β-chains (95-kDa) and is a member of a superfamily of cell surface receptors termed integrins (HYNES 1987).

CR3 binds to the iC3b fragment of C3, and this binding is dependent on divalent cations (BELLER et al. 1982; ROSS and LAMBRIS 1982). One binding site in iC3b for CR3 has been localized to residues 1361–1380 of C3. This region of C3 contains an RGD sequence, and WRIGHT et al. demonstrated that erythrocytes coated with a synthetic peptide spanning these residues bind to macrophages, and that this binding is inhibited by monoclonal antibodies against CR3 (WRIGHT et al. 1987). In addition, this peptide could also inhibit CR3-mediated binding of polymorphonuclear leukocytes to endothelial cells (LO and WRIGHT 1988), suggesting that adhesion of polymorphonuclear leukocytes to endothelial cells is mediated via the same site on CR3. Tertiary conformation and amino acids adjacent to the RGD sequence, as is true for the other integrin ligands (RUOSLAHTI and PIERSCHBACHER 1987), must also be important for CR3 binding since in mouse C3, which binds well to MAC-1, and rabbit C3 the Arg is replaced by Leu (WETSEL et al. 1984; KUSANO et al. 1986), and since CR3 failed to recognize the hexapeptide GRGDSP (WRIGHT et al. 1987). The binding of neutrophil and macrophage CR3 to fibrinogen (WRIGHT et al. 1988) and *Leishmania* gp63 (RUSSELL and WRIGHT 1988), respectively, and analysis of their sequences surrounding the CR3 binding site led to the hypothesis that the positively charged Arg is important, and that, in mouse C3 and fibrinogen, this charge requirement is maintained by the Lys situated two residues NH$_2$-terminal to the position of the substituted Arg (Fig. 2). Cloning of *Xenopus* C3 has revealed that it lacks the RGD sequence but shows high amino acid similarity with a segment of the human and rabbit sequences immediately COOH-terminal to the RGD (GROSSBERGER et al. 1989; see Fig. 2). Since *Xenopus* macrophages bind *Xenopus* iC3b (SEKIZAWA et al. 1984a), it would be interesting to determine whether *Xenopus* iC3b binds to human CR3.

Recently, fibrinogen has been shown to bind to CR3 on ADP-stimulated monocytes, and this binding is not inhibited by the RGD-containing synthetic peptides derived from the iC3b and fibrinogen sequences (ALTIERI et al. 1988). This, coupled with the observation that iC3b, fibrinogen, and coagulation factor X, which also binds CR3 (ALTIERI and EDGINGTON 1988), mutually inhibit one another in binding CR3 (ALTIERI et al. 1988), suggests that CR3 possesses additional recognition specificities for these molecules that are different from its RGD recognition site. For a more comprehensive review on the binding of these ligands to CR3 see ROSEN and LAW (this volume).

In addition to its specificity for iC3b, CR3 has been shown to bind the C3dg fragment of C3 (GAITHER et al. 1987), suggesting that CR3 can interact with at least two different sites in C3. Also, CR3 binds to lipopolysaccharide (WRIGHT and JONG 1986) and zymosan (Ross et al. 1985), apparently via a polysaccharide binding site, since the latter binding is inhibited by N-acetyl-D-glucosamine. Recently, it has been shown that the lipopolysaccharide binding site in CR3 is distinct from the iC3b binding site (WRIGHT et al. 1989). This ligand multispecificity expressed by CR3 is a common characteristic of the integrin superfamily of cell receptors (RUOSLAHTI and PIERSCHBACHER 1986) and underscores the ability of CR3 to mediate a variety of processes.

4.4 Interaction of C3 Fragments with Factor H

Factor H is a serum glycoprotein which down-regulates the amplification of the alternative pathway by binding to either surface-bound or fluid-phase C3b and accelerating the decay of the C3bBb convertase or by acting as a cofactor in the factor I mediated breakdown of C3b (for details see VIK et al., this volume). Several reports have identified different regions of the C3b molecule to be involved in H binding, thus suggesting that H and C3b interact via multiple sites. One of the H binding sites has been localized within the C3d fragment of C3 using anti-idiotypic antibodies and synthetic peptides (LAMBRIS et al. 1988). The results have shown that H binds to a discontinuous site surrounding the CR2 binding site, spanning residues 1187–1249 of the C3 sequence (see Fig. 2). From these studies it has been proposed that the conformation of this segment in C3b is such that the H site is preferentially exposed in C3b, whereas upon cleavage to iC3b this segment undergoes a "conformational inversion" with the concomitant appearance and disappearance of the CR2 and H binding sites, respectively. This would explain why CR2 is most efficient as a cofactor for the cleavage of iC3b to C3c and C3dg. Another possibility is that factor H can bind simultaneously to additional sites within C3b, thereby promoting the third factor I mediated cleavage.

Another site in C3b which has been reported to mediate factor H binding has been identified to the N-terminal 40 amino acids of the α'-chain based on the inhibition of factor H binding to EC3b by synthetic peptides and antipeptide antibodies (GANU and MÜLLER-EBERHARD 1985). This domain of C3 has also been identified as a site of interaction for CR1 (BECHERER and LAMBRIS 1988) and factor B (GANU and MÜLLER-EBERHARD 1985). This would, on the one hand, explain the ability of factor H to compete with factor B and CR1 for binding to C3b (PANGBURN and MÜLLER-EBERHARD 1978; PANGBURN 1986a) and, on the other, the ability of factor H to participate in the third factor I cleavage. Since factor H, CR1, CR2, and C4bp are all cofactors for the I mediated cleavages of C3b, and since they are all structurally similar, one possibility is that multiple recognition sites within C3b and iC3b exist that are common, albeit with different affinities, to the different cofactor molecules.

A third site of interaction in C3 for factor H has been localized to the COOH-terminal 40-kDa α'-chain fragment of Cec based on studies in which an anti-C3 antibody, affinity purified on an anti-factor H affinity column, could inhibit factor H binding to C3b and serve as a cofactor for factor I mediated cleavage of C3b (NILSSON and NILSSON 1987). For sites of interaction on H for C3b see VIK et al. (this volume).

4.5 Interaction of C3 Fragments with Factor B

Factor B is a 93-kDa serine protease which participates in the initiation and feedback amplification of the alternative pathway by serving as the catalytic subunit of the C3 and C5 convertases (PANGBURN 1986b; VOLANAKIS, this volume). Binding of factor B to C3b appears to involve multiple sites within both molecules. One site was postulated to reside in the Ba fragment due to the presence of the consensus repeating units found in other C3b binding proteins within this fragment of factor B. This was confirmed using monoclonal antibodies against the Ba fragment which inhibited C3b binding to factor B (UEDA et al. 1987) and by direct interaction of the Ba fragment itself (PRYZDIAL and ISENMAN 1987). The NH_2-terminus of the α'-chain of C3b has been shown to be involved in B binding (GANU and MÜLLER-EBERHARD 1985). Since this site has also been shown to mediate the binding of CR1 (BECHERER and LAMBRIS 1988) and factor H (GANU and MÜLLER-EBERHARD 1985), two C3b binding proteins which also contain the consensus repeating units, this region of C3b may be involved in the binding of the Ba fragment.

A second C3b binding site in factor B was localized within the 33-kDa COOH-terminal end of the Bb fragment, and its binding, like that of factor B, is Mg^{2+} dependent (LAMBRIS and MÜLLER-EBERHARD 1984). A domain in C3, spanning residues 933–942, has been predicted to be involved in factor B binding (O'KEEFE et al. 1988). This is based on the differential binding of factor B to C3c and a C3c-like fragment generated by a protease from cobra venom glands. The latter fragment has been proposed to be structurally similar to cobra venom factor (CVF); however, it is not known whether this region of C3b is involved in binding the 33-kDa Bb fragment. That multiple binding sites exist in C3b for factor B is supported by data showing that monoclonal antibodies from both the C3c and C3d domains inhibit factor B binding to C3b (BURGER et al. 1982; KOISTINEN et al. 1989).

4.6 Interaction of C3 Fragments with Properdin

Properdin is a plasma glycoprotein whose initial isolation led to the elucidation of the alternative (properdin) pathway (PILLEMER et al. 1954). It allows a rapid amplification of the surface-bound C3b by stabilizing the C3bBb convertase. The precise molecular mechanisms by which properdin stabilizes the C3 convertase are unknown, although its presence does not seem to be essential for initiation (FARRIES et al. 1988).

To further understand the mechanisms by which properdin stabilizes the C3 convertase, the binding site in C3b for properdin has been identified. Originally, properdin was shown to bind both C3b and C3c (CHAPITIS and LEPOW 1976), and subsequent studies placed the properdin binding site within the 40-kDa COOH-terminal α-chain of C3c (LAMBRIS et al. 1984). Cleavage of this 40-kDa fragment by CNBr yielded a 17-kDa fragment (residues 1385–1541 of C3) that bound to properdin (DAOUDAKI et al. 1988). Further localization of the properdin binding site was achieved using synthetic peptides. Comparison of the amino acid sequences of human, mouse, and rabbit C3 (properdin binding proteins) with those sequences from human and mouse C4, C5, and α_2-macroglobulin (homologous but not properdin binding proteins) identified a region, residues 1402–1435 of the human C3 sequence, that was conserved and therefore a possible candidate for the properdin binding site (see Fig. 2). A synthetic peptide ($C3^{1402-1435}$) corresponding to this segment of C3 was shown to bind properdin, inhibit properdin binding to C3, and inhibit the activation of the alternative pathway by rabbit erythrocytes (DAOUDAKI et al. 1988). These results show conclusively that properdin stabilization of the C3 convertase is necessary for efficient amplification of the enzyme cascade during complement activation and is in agreement with the studies on patients with properdin deficiency that indicated properdin to be essential for optimal complement activation (BRACONIER et al. 1983; NIELSEN and KOCH 1987). Further studies using overlapping synthetic peptides localized the properdin binding site to residues 1424–1432, and synthetic analogues indicate that His^{1431} and Ser^{1432} are important for the binding activity of this peptide (ALSENZ et al. 1989). The binding site in properdin for C3b is unknown, but since properdin exists as cyclic oligomers constructed from the association of single monomers (SMITH et al. 1984; FARRIES et al. 1987), it is possible that it binds more than one molecule of C3b.

4.7 Interaction of C3 Fragments with Conglutinin

Conglutinin is a 300-kDa collagen-like plasma protein that binds to yeast cell walls and to surface-bound iC3b (LACHMANN and MÜLLER-EBERHARD 1968). Originally identified in bovine plasma (LACHMANN and MÜLLER-EBERHARD 1968) and later in human plasma (THIEL et al. 1987), conglutinin binding to iC3b is Ca^{2+} dependent and can be inhibited by N-acetyl-D-glucosamine. Recently, LOVELESS et al. have shown conglutinin to be a lectin which recognizes terminal N-acetylglucosamine, mannose, and fucose residues (LOVELESS et al. 1989). By using C3 fragments fixed to microtiter plates, conglutinin was also shown to bind C3b, C3c, and the α-chain of C3 but not the β-chain or to C3d. Molecular dissection of the C3c molecule identified the conglutinin binding site within the NH_2-terminal α'-chain fragment of C3c. Since endo-H removal of the α-chain carbohydrate moiety abolished conglutinin binding to C3c, it was concluded that this carbohydrate, attached to the Asn residue at position 917 of the C3 sequence, is the conglutinin binding site (HIRANI et al. 1985). Little is known about the physiological role of conglutinin, but it may be a member of the family of Ca^{2+}-dependent endogenous lectins recently reviewed by DRICKAMER (1988).

5 Structure/Function Analysis of C3 from Different Species

Although complement and other nonspecific lytic systems have been described in lower vertebrates and invertebrates, relatively little is known about the individual components that make up these systems. The best phylogenetically characterized component of complement is C3, no doubt due to its prominent role in both pathways of activation. C3 has been purified and characterized from numerous vertebrates including monkeys, mice, rats, guinea pigs, cats, rabbits, birds, frogs, snakes, and lampreys, and the complete or partial amino acid sequences of human, mouse, rabbit, and *Xenopus* C3 have been deduced from the cloning of their cDNA. Comparison of the C3 amino acid sequences between species and correlation of this information with the ability of these forms of C3 to bind the different ligands and receptors is instrumental in identifying structural features of C3 that are important for its functionality. Homologous proteins such as C4, C5, and α_2-macroglobulin serve as "natural analogues" and yield additional information regarding which C3 residues are critical for ligand binding. Below we review briefly the structural and functional characteristics of C3 from other species and present some recent work comparing the primary sequences of C3 from various species with regard to the identified binding sequences in human C3. Although a number of studies on the phylogenetic compatability of the complement system have been reported (GIGLI and AUSTEN 1971; KAIDOH and GIGLI 1987; VON ZABERN 1988), our focus here is on the C3 molecule, and inclusion of the other complement components is beyond the scope of this review.

5.1 Mammalian C3

Mouse C3. Murine C3 was first purified to homogeneity in 1977 by GYONGYOSSY (GYONGYOSSY and ASSIMEH 1977) who showed that it consisted of a two-chain structure with physiochemical properties similar to those of human C3. The major difference between mouse and human C3 is that the murine β-chain lacks carbohydrates. Cloning of the murine C3 (WETSEL et al. 1984; LUNDWALL et al. 1984a) and comparison of its sequence with the human C3 sequence indicated that the murine C3 α-chain is one amino acid longer than the human α-chain, and that this extra amino acid is located within the C3a domain, while the murine β-chain is three amino acids shorter than its human counterpart. Human and mouse C3 share 77% and 79% identity on the amino acid and nucleotide levels, respectively. The observations that murine EAC3b is cleaved by human factor I and rosettes with human erythrocytes, and that murine EC3bi/C3dg rosettes with Raji cells suggest that the binding sites for human factor I, CR1, and CR2 are conserved in mouse C3 (KINOSHITA and NUSSENZWEIG 1984; DIERICH et al. 1974; ROSS et al. 1976). The latter cross-reactivities are supported by the sequence conservation between the two species within the identified CR1 and CR2 binding domains (Fig. 2).

Rabbit C3. Rabbit C3 has been purified to homogeneity (GICLAS et al. 1981; HORSTMANN and MÜLLER-EBERHARD 1985a) and was found to be composed of glycosylated α- (123-kDa) and β-chains (70 kDa; see Figs. 2, 4). Its amino acid composition resembles that of human C3 while its carbohydrate content and its concentration in serum are significantly lower than in the corresponding human component (Horstmann and Müller-Eberhard 1985a). Identification of the nucleotide sequence for a region of approximately 70% of the α-chain of rabbit C3 has revealed that it shares 78% amino acid similarity with the human and mouse sequences with conservation of the cysteinyl residues between species. Also, this region of the α-chain contains the binding sites for factor H, properdin, CR2, CR3, as well as the thiolester site and the factor I cleavage sites, and allows the primary sequences of these sites to be compared between species (see Fig. 2). For example, a peptide from the rabbit C3 sequence was synthesized that corresponded to the identified properdin binding site in human C3. This peptide inhibited properdin binding to human C3b (Alsenz et al. 1989) and provided additional information regarding which residues are important in the C3b-properdin interaction. Furthermore, rabbit C3 has been shown to bind human factor H, CR1, CR2, and MCP, and this suggests that the binding sites for these human components are conserved in rabbit C3 (Dierich et al. 1974; Horstmann et al. 1985b; Becherer et al. 1987; Manthei et al. 1988). This observation is supported by the sequence similarity between the two species within these regions of C3 (Fig. 2).

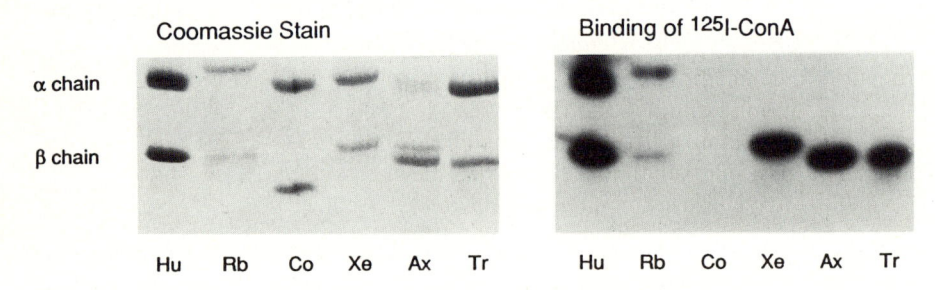

Fig. 4. Sodium dodecyl sulfate polyacrylamide gel electrophoresis and binding of ^{125}I-labeled concanavalin A (^{125}I-ConA) to C3 from different species. Human, rabbit, cobra, *Xenopus*, axolotl, and trout C3 were electrophoresed on a 7.5% sodium dodecyl sulfate polyacrylamide gel under reducing conditions, and proteins were stained with Coomassie blue. A duplicate gel procedure was carried out, and the proteins were electroblotted to nitrocellulose, incubated with ^{125}I-labeled concanavalin A, washed, and subjected to autoradiography. The purification of C3 from different species was achieved as follows: plasma, after fractionation with polyethylene glycol, was subjected successively to anion exchange and gel filtration chromatography (for details see Avila and Lambris 1989) and the C3-containing fraction was then adsorbed, after treatment with methylamine, to activated thiol-Sepharose. After extensive washing the purified C3 was eluted from the activated thiol-Sepharose with L-cysteine. Rabbit C3 was purified as previously described (Horstman and Müller-Eberhard 1985a)

C3 from Other Mammals. The third component of rat (Daha et al. 1979), guinea pig (Shin and Mayer 1968), porcine (Paques 1980), and feline (Jacobse-Geels et al. 1980) complement has been purified and found to be similar to that of human C3 in its chain structure and its participation in complement activation. The carbohydrate content of rat, porcine, and feline C3 has not been determined, but guinea pig C3, unlike human C3 but similar to mouse C3, appears to lack carbohydrate on its β-chain (THOMAS and TACK 1983). Additional studies have shown that a C3 convertase could be formed using purified rat C3 and human components B, D, and P, indicating that the binding sites for factor B and properdin are conserved between rat and human C3. Likewise, feline C3 could form a classical pathway C5 convertase which resulted in lysis of the target erythrocyte upon addition of human C5–C9. This suggests that the C5 binding site in C3 is conserved between felines and humans. In contrast to the compatibilities observed between these species, guinea pig C3 is incompatible with both human factor B (BRADE et al. 1976) and C5 (VON ZABERN et al. 1979).

5.2 Avian C3

Quail C3. Among avian species C3 has been purified and characterized in the Japanese quail and in the chicken. Using an antiserum prepared against quail serum-treated zymosan, quail C3 was found to be a two-chain structure ($M_r = 184000$; KAI et al. 1983), and during ontogeny synthesis of quail C3 appears to be associated with cells belonging to the reticuloendothelial system (KAI et al. 1985). Quail complement resembles that of mammalian in that: (a) inulin or zymosan led to the activation of quail complement, (b) hydrazine or methylamine eluted quail C3 from zymosan, and (c) quail C3 was deposited on the membrane of cells activating quail complement, all of which suggest the presence of an internal thiolester bond in C3. However, one discriminating feature of quail complement system is its inability to be activated by CVF, suggesting that CVF does not make a C3 convertase with quail factors B and D (KAI et al. 1985).

Chicken C3. The C3 of chicken has been characterized as a two-chain structure (α-chain, 118 kDa; β-chain, 68 kDa) that participates in both the classical and alternative pathways of complement activation in chickens. It has a pI between 6.4 and 6.6, a concentration in plasma of 0.4–0.5 mg/ml and is sensitive to methylamine treatment; in contrast to quail complement, the chicken complement system can be activated by CVF. Also, immunoblotting experiments reveal that chicken C3 exists in multiple forms, but as yet no genetic polymorphism has been demonstrated (KOCH 1988).

5.3 Reptilian C3

Interest in the complement system of cobras began upon the identification of CVF which was able to activate the alternative pathway of complement. CVF forms a stable enzyme complex when exposed to factors B and D that mimics the C3/C5 convertase of the alternative pathway in its ability to cleave C3 and C5. The CVF,Bb complex has a half-life approximately 280 times that of C3b,Bb (VOGEL and MÜLLER-EBERHARD 1982), cannot be disassembled by factor H (LACHMANN and HALBWACHS 1975; NAGAKI et al. 1978), and can continuously cleave C3 and C5, leading to severe complement depletion in animals injected with CVF (COCHRANE et al. 1970; VOGEL and MÜLLER-EBERHARD 1984; RYAN et al. 1986).

The functional similarities between CVF and C3b led to the elucidation of structural similarities as well. CVF resembles the C3c fragment of C3 in its chain structure, amino acid composition, NH_2-terminal sequence of the individual chains, pI value, and secondary structure while common antigenic determinants exist berween the two molecules (EGGERTSEN et al. 1983; VOGEL et al. 1984; GRIER et al. 1987). Furthermore, the cross-reactivity of an anti-CVF antibody with cobra C3 led to its purification from cobra plasma (EGGERTSEN et al. 1983). This molecule resembled human C3 in its chain structure, and common antigenic determinants were found between the β-chains of human and cobra C3 and in the α-chain of CVF. NH_2-terminal sequence (see Fig. 2) analysis of the α-, β-, and γ-chains of CVF revealed homology with the β- and α-chains of human C3 (VOGEL et al. 1984; LUNDWALL et al. 1984b; O'KEEFE et al. 1988). Based on the structural and functional similarities between human C3, CVF, and cobra C3, it was suggested that CVF is a physiologically derived fragment of cobra C3 (EGGERTSEN et al. 1983). This was supported by the recent report of O'KEEFE which describes a protease in cobra venom that can cleave human C3 into C3c-like and C3d-like fragments (O'KEEFE et al. 1988). This C3c-like fragment, unlike the C3c generated by trypsin or elastase, retains the ability to bind factor B and has led to the identification of a region in C3 thought to be involved in the factor B interaction (see Sect. 4.5). Further structural analysis of CVF and cobra C3 will determine unequivocally whether CVF is a proteolytic fragment of cobra C3. The inability of ^{125}I-labeled concanavalin A to bind cobra C3 suggests that, if cobra C3 is indeed glycosylated, then its carbohydrate structure differs from that of human C3 (Fig. 4).

5.4 Amphibian C3

Xenopus C3. Identification of a hemolytic activity against sheep red blood cells in the serum of the clawed frog, *Xenopus laevis,* allowed the purification of C3 from this species (SEKIZAWA et al. 1984b). *Xenopus* C3 resembled human C3 in its chain structure (α-chain, 125 kDa; β-chain, 85 kDa), its electrophoretic mobility (β-globulin), its amino acid composition, and its susceptibility to methylamine treatment. The recently described cDNA and amino acid sequence corresponding to a COOH-terminal 34-kDa α-chain fragment of *Xenopus* C3 revealed 57% and 49% similarity with human C3 on the nucleic and amino acid levels, respectively, although three additional cysteine residues were found within this segment of *Xenopus* C3 (GROSSBERGER et al. 1989). Also, regions of high and low similarity were identified when the *Xenopus* sequence was compared with the other mammalian C3 sequences. For example, the properdin binding domain of human C3b (DAOUDAKI et al. 1988) appears to be highly conserved in *Xenopus* C3, but the RGD sequence predicted to be involved in iC3b binding is replaced by LKN (see Fig. 2). Since *Xenopus* C3b/iC3b-coated sheep erythrocytes bind to *Xenopus* macrophages (SEKIZAWA et al. 1984a), either this binding is mediated through a different receptor than *Xenopus* CR3 (possibly *Xenopus* CR1 or its equivalent) or *Xenopus* iC3b binding to *Xenopus* CR3 does not involve an RGD sequence. It is interesting to note that the 20 amino acids COOH-terminal to the RGD sequence are highly conserved in the human, mouse, rabbit, and *Xenopus* C3 sequences (see Fig. 2). In addition, a synthetic peptide, derived from the *Xenopus* C3 sequence that is homologous to the properdin binding domain in human C3, was found to inhibit binding of human properdin to human C3b (DAOUDAKI and LAMBRIS, unpublished observation) and aided in the sublocalization of the properdin binding site (For details see Sect. 4.6). These results suggest that the binding site in C3 for properdin is conserved among the species so far tested.

Axolotl C3. Axolotl C3 was purified from plasma based on the ability of an anti-human C3 monoclonal antibody and a polyclonal anti-CVF antibody to cross-react with axolotl C3 (AVILA and LAMBRIS, 1989). It consists of a disulfide-linked α- (110-kDa) and β-chain (72-kDa; Fig. 4) and it contains, within the α-chain, an internal thiolester bond as deduced from its amino acid sequence and the incorporation of [^{14}C]-methylamine within the C3d fragment of axolotl C3. NH$_2$-terminal sequence from tryptic fragments of axolotl C3 (see Fig. 2) and amino acid composition indicated that this molecule is highly homologous to C3 from other vertebrates. Furthermore, axolotl C3 was found to bind human C5 and factor H, suggesting that these binding sites have been conserved throughout evolution. However, in contrast to human C3 which has carbohydrate structures on both its chains, axolotl C3 lacks concanavalin A binding carbohydrate on its α-chain (AVILA and LAMBRIS 1989; and Fig. 4), a moiety which, in human C3, is responsible for binding conglutinin.

5.5 Osteichthian C3

The complement system of rainbow trout was originally identified by the existence of a serum lytic system homologous to mammalian complement (NONAKA et al. 1981). Both a classical and alternative pathway could be demonstrated that were dependent on divalent cations, and lipopolysaccharide, zymosan, and inulin depleted the lytic activity, suggesting that a C3-like molecule is essential. Consequently, trout C3 was isolated and, like human C3, was found to possess an α-β-chain structure and a methylamine-sensitive thiolester site, in addition to its integral role in complement-mediated hemolysis (NONAKA et al. 1984b). Similar to *Xenopus* and axolotl C3, only the β-chain of trout C3 contains concanavalin A binding carbohydrate (Fig. 4). Additionally, the amino acid composition and NH$_2$-terminal amino acid sequences of the α- and β-chains of trout C3 show homology to human C3 and CVF (NONAKA et al. 1984b; Fig. 2). A variant of trout C3 which lacks hemolytic activity has been identified (NONAKA et al. 1985). Termed C3-2, it differs from the hemolytically active C3 (C3-1) in its antigenicity and tryptic peptide map, while, from the available amino acid sequences of C3-1 and C3-2, (26 amino acids of the α-chain, 21 of the β-chain) only two substitutions (both in the α-chain) were observed (NONAKA et al. 1985). The reason for the hemolytic inactivity of C3-2 is unknown, but since its internal thiolester allows it to bind to activating surfaces, one could speculate that its ability to form an effective C5 convertase is impaired. The genetic origin of these variants is unknown, but isotypic variants have been identified at both the protein and DNA levels for human and mouse C4 (SHREFFLER 1976; O'NEILL et al. 1978; CHAPLIN et al. 1983; CARROLL et al. 1984) although no isotypic variants of mammalian forms of C3 have been identified (WHITEHEAD et al. 1982; WIEBAUER et al. 1982). Further analysis is required to resolve the origin of these two C3 molecules in trout.

5.6 Cyclostome C3

One of the most interesting findings to emerge from these studies of C3 in different species is a lamprey complement protein homologous to mammalian C3 (NONAKA et al. 1984a). Lamprey C3 resembles C4 in its chain structure, being comprised of α- (84-kDa), β- (74-kDa), and γ-chains (32-kDa). However, it is believed to be the lamprey equivalent of C3 based on its essential role in the phagocytosis of rabbit erythrocytes by lamprey phagocytes. It contains an internal thiolester bond that is susceptible to methylamine treatment, and the location of this bond is within a 35-kDa fragment of the α-chain. This segment shows homology with the sequences of the thiolester region of mammalian C3 (NONAKA et al. 1984a). Other characteristic features are its ability to bind covalently to activating surfaces and its ability to mediate phagocytosis. This latter characteristic is the major reason that this molecule has been described as lamprey C3 and not C4. Somewhat surprising, however, is the finding that lamprey C3 is not involved in the naturally occuring hemolytic activity in lamprey serum. The factor responsible for this lytic activity has been identified and shown not to be related to complement (GEWURZ et al. 1966). Thus lamprey C3 appears to function primarily in opsonization of foreign particles. As a result of the structural and functional similarities between lamprey C3 and the C3 and C4 from other species, as well as the primordial nature of the cyclostomes, it has been speculated that lamprey C3 typifies a common ancestor of both C3 and C4, and that lampreys possess a relatively simplified complement system (NONAKA et al. 1984a).

6 Concluding Remarks

The biological functions of proteins almost invariably depend upon their direct, physical interaction with other molecules. This is no more evident than with C3, whose interactions with at least 16 different serum and cell surface proteins, some of which have been reviewed in detail here, make it the central molecule of the complement system. Understanding this multifunctionality of C3 on a molecular level is essential if one is to control complement activation in such disadvantageous situations as transplantation rejection and autoimmune disorders. Therefore, elucidation of the molecular features involved in the C3-ligand and -receptor interactions should facilitate the development and design of specific inhibitors of the complement cascade. Still, much remains unknown about the reactions of C3 at the cellular level and about the biological responses mediated by C3 receptors, particularly in the immune response and phagocytosis. The further analysis of C3 interactions with the homologous C3b/C4b regulator proteins, which have overlapping functions, should determine whether these molecules share binding domains within C3, and how these shared domains are differentially recognized by the regulators, leading to inactivation of the cascade. This latter regulatory function is particularly interesting since certain pathogens possess surface proteins which allow them to escape complement neutralization. Finally, the chemical synthesis and genetic engineering of peptides and proteins with tailor-made conformations and reactivities that mimic C3 will greatly facilitate our understanding of the multifunctionality of this molecule.

Acknowledgements. We thank Drs. D. GOUNDIS and I. ESPARZA for their critical reading of the manuscript.

References

Alper CA, Johnson AM, Birtch AG, Moore FD (1969) Human C'3: evidence for the liver as the primary site of synthesis. Science 163: 286–288
Alsenz J, Becherer JD, Esparza I, Daoudaki ME, Avita D, Oppermann S, Lambris JD (1989) Structure and function analysis of C3 from different species. Complement Inflamm 6: 307
Altieri DC, Edgington TS (1988) The saturable high affinity association of factor X to ADP-stimulated monocytes defines a novel function of the Mac-1 receptor. J Biol Chem 263: 7007–7015
Altieri DC, Bader R, Mannucci PM, Edgington TS (1988) Oligospecificity of the cellular adhesion receptor MAC-1 encompasses an inducible recognition specificity for fibrinogen. J Cell Biol 107: 1893–1900
Atkinson JP, Farries T (1987) Separation of self from non-self in the complement system. Immunol Today 8: 212–215
Avila D, Lambris JL (1989) Isolation and characterization of the third complement component of axolotl. (manuscript submitted)
Barel M, Vazquez A, Charriaut C, Aufredou MT, Galanaud P, Frade R (1986) gp 140, the C3d/EBV receptor (CR2), is phosphorylated upon in vitro activation of human peripheral B lymphocytes. FEBS Lett 197: 353–356
Barel M, Fiandino A, Delcayre AX, Lyamani F, Frade R (1988) Monoclonal and anti-idiotypic anti-EBV/C3d receptor antibodies detect two binding sites, one for EBV and one for C3d on glycoprotein 140, the EBV/C3dR, expressed on human B lymphocytes. J Immunol 141: 1590–1595
Becherer JD, Lambris JD (1988) Identification of the C3b-receptor binding domain in the third component of complement. J Biol Chem 263: 14586–14591
Becherer JD, Daoudaki ME, Lambris JD (1987) Conservation of the C3 ligand binding sites within different species. Fed Proc 46: 771
Becherer JD, Alsenz J, Servis C, Myones BL, Lambris JD (1989a) Cell surface proteins reacting with activated complement components. Complement Inflammation 6: 142–165
Becherer JD, Alsenz J, Hack E, Drakopoulov E, Lambris JD (1989b) Identification of common binding domains in C3b for members of the complement family of C3b-binding proteins. Complement Inflamm 6: 313
Beller DI, Springer TA, Schreiber RD (1982) Anti-Mac-1 selectively inhibits the mouse and human type three complement receptor. J Exp Med 156: 1000–1009
Bohnsack JF, Cooper NR (1988) CR2 ligands modulate human B cell activation. J Immunol 141: 2569–2576
Braconier JH, Sjoholm AG, Soderstrom C (1983) Fulminant meningococcal infections in a family with inherited deficiency of properdin. Scand J Infect Dis 15: 339–344
Brade V, Dieminger L, Schmidt G, Vogt W (1976) Incompatibility between C3b and B of guinea pig and man and its influence on the alternative pathway factors D and B in these two species. Immunology 30: 171–179
Burger R, Deubel U, Hadding U, Bitter-Suermann D (1982) Identification of functionally relevant determinants on the complement component C3 with monoclonal antibodies. J Immunol 129: 2042–2050
Campbell RD, Gagnon J, Porter RR (1981) Amino acid sequence around the thiol and reactive acyl groups of human complement component C4. Biochem J 199: 359–370

Carroll MC, Campbell RD, Bentley DR, Porter RR (1984) A molecular map of the human major histocompatibility complex class III region linking complement genes C4, C2 and factor B. Nature 307: 237–241

Carter RH, Spycher MO, Ng YC, Hoffman R, Fearon DT (1988) Synergistic interaction between complement receptor type 2 and membrane IgM on B lymphocytes. J Immunol 141:457–463

Catterall CF, Lyons A, Sim RB, Day AJ, Harris TJR (1987) Characterization of the primary amino acid sequence of human complement control protein factor I from an analysis of cDNA clones. Biochem J 242: 849–856

Changelian PS, Fearon DT (1986) Tissue-specific phosphorylation of complement receptors CR1 and CR2. J Exp Med 163: 101–115

Chapitis J, Lepow IH (1976) Multiple sedimenting species of properdin in human and interaction of purified properdin with the third component of complement. J Exp Med 143: 241–257

Chaplin DD, Woods DE, Whitehead AS, Goldberger G, Colten HR, Seidman JG (1983) Molecular map of the murine S region. Proc Natl Acad Sci USA 80: 6947–6951

Cochrane CG, Müller-Eberhard HJ, Aikin BS (1970) Depletion of plasma complement in vivo by a protein of cobra venom: its effect on various immunologic reactions. J Immunol 105: 55–62

Cooper R, Moore MD, Nemerow GR (1988) Immunobiology of CR2, the B lymphocyte receptor for Epstein-Barr virus and the C3d complement fragment. Annu Rev Immunol 6: 85–113

Crossley LG, Porter RR (1980) Purification of the human complement control protein C3b inactivator. Biochem J 191: 173–182

Daha MR, Stuffers-Heiman M, Kijlstra A, van Es LA (1979) Isolation and characterization of the third component of rat complement. Immunology 36: 63–69

Daoudaki ME, Becherer JD, Lambris JD (1988) A 34-amino acid peptide of the third component of complement mediates properdin binding. J Immunol 140: 1577–1580

Davis AE III, Harrison RA (1982) Structural characterization of factor I mediated cleavage of the third component of complement. Biochemistry 21: 5745–5749

Davis AE III, Harrison RA, Lachmann PJ (1984) Physiologic inactivation of fluid phase C3b: isolation and structural analysis of C3c, C3d,g, (α2D), and C3g. J Immunol 132: 1960–1966

De Bruijn MHL, Fey GH (1985) Human complement component C3: cDNA coding sequence and derived primary structure. Proc Natl Acad Sci USA 82: 708–712

Delcayre AX, Fiandino A, Barel M, Frade R (1987) gp 140, the EBV/C3d receptor (CR2) of human B lymphocytes, is involved in cell-free phosphorylation of p120, a nuclear ribonucleoprotein. Eur J Immunol 174: 1827–1833

Dierich MP, Pellegrino MA, Ferrone S, Reisfeld RA (1974) Evaluation of C3 receptors on lymphoid cells with different complement sources. J Immunol 112: 1766–1773

DiScipio RG (1981) The binding of human complement proteins C5, factor B, β1H and properdin to complement fragment C3b on zymosan. Biochem J 199: 485–496

Domdey H, Wiebauer K, Kazmaier M, Müller V, Odink K, Fey G (1982) Characterization of the mRNA and cloned cDNA specifying the third component of mouse complement. Proc Natl Acad Sci USA 79: 7619–7623

Drickamer K (1988) Two distinct classes of carbohydrate-recognition domains in animal lectins. J Biol Chem 263: 9557–9560

Eggertsen G, Lundwall A, Hellman U, Sjöquist J (1983) Antigenic relationships between human and cobra complement factors C3 and cobra venom factor (CVF) from the Indian cobra (*Naja naja*). J Immunol 131: 1920–1923

Esparza I, Lambris JD (1989) Identification of a second binding site in iC3b for CR2. Complement (in press)

Farries TC, Finch JT, Lachmann PJ, Harrison RA (1987) Resolution and analysis of "native" and "activated" properdin. Biochem J 243: 507–517

Farries TC, Lachmann PJ, Harrison RA (1988) Analysis of the interactions between properdin, the third component of complement (C3), and its physiological activation products. Biochem J 252: 47–54

Fingeroth JD, Benedict MA, Levy DN, Strominger JL (1989) Identification of murine complement receptor type 2. Proc Natl Acad Sci USA 86: 242–246

Frade R, Crevon MC, Barel M, Vazquez A, Krikorian L, Charriaut C, Galanaud P (1985) Enhancement of human B cell proliferation by an antibody to the C3d receptor, the gp140 molecule. Eur J Immunol 15: 73–76

Fujita T, Nussenzweig V (1979) The role of C4-binding protein and β1H in proteolysis of C4b and C3b. J Exp Med 150: 267–276

Gaither TA, Vargas I, Inada S, Frank MM (1987) The complement fragment C3d facilitates phagocytosis by monocytes. Immunology 62: 405–411

Ganu VS, Müller-Eberhard HJ (1985) Inhibition of factor B and factor H binding to C3b by synthetic peptide corresponding to residues 749–789 of human C3. Complement 2: 27

Gewurz H, Finstad J, Muschel LH, Good RA (1966) Phylogenetic inquiry into the origins of the complement system. In: Smith RT, Miescher PA, Good RA (eds) Phylogeny of immunity. University of Florida Press, Gainsville, pp 105

Ghebrehiwet B, Müller-Eberhard HJ (1979) C3e: an acidic fragment of human C3 with leukocytosis-inducing activity. J Immunol 123: 616–621

Giclas PC, Keeling PJ, Henson PM (1981) Isolation and characterization of the third and fifth components of rabbit complement. Mol Immunol 18: 133–123

Gigli I, Austen KF (1971) Phylogeny and function of the complement system. Annu Rev Microbiol 25: 309–332

Goldberger G, Thomas ML, Tack BF, Williams J, Colten HR, Abraham GN (1981) NH2-terminal structure and cleavage of guinea pig pro-C3 the precursor of the third complement component. J Biol Chem 256: 12617–12619

Grier AH, Schultz M, Vogel CW (1987) Cobra venom factor and human C3 share carbohydrate antigenic determinants. J Immunol 139: 1245–1252

Grossberger D, Marcuz A, Du Pasquier L, Lambris JD (1989) Conservation of structural and functional domains in complement component C3 of *Xenopus* and mammals. Proc Natl Acad Sci USA 86: 1323–1327

Gyongyossy MIC, Assimeh SN (1977) Isolation of the third component of mouse complement. J Immunol 118: 1032–1035

Hack CE, Paardekooper J, Smeenk RJT, Abbink J, Eerenberg AJM, Nuijens JH (1988) Disruption of the internal thioester bond in the third component of complement, (C3) results in the exposure of neodeterminants also present on activation products of C3. An analysis with monoclonal antibodies. J Immunol 141: 1602–1609

Hase S, Kikuchi N, Ikenaka T, Inoue K (1985) Structures of sugar chains of the third component of human complement. J Biochem (Tokyo) 98: 863–874

Hatzfeld A, Fischer E, Levesque JP, Perrin R, Hatzfeld J, Kazatchkine MD (1988) Binding of C3 and C3dg to the CR2 complement receptor induces growth of an Epstein-Barr virus-positive human B cell line. J Immunol 140: 170–175

Hirani S, Lambris JD, Müller-Eberhard HJ (1985) Localization of the conglutinin binding site on the third component of human complement. J Immunol 134: 1105–1109

Hirani S, Lambris JD, Müller-Eberhard HJ (1986) Structural analysis of the asparagine-linked oligosaccharides of human complement component C3. Biochem J 233: 613–616

Hoeprich PDJr, Dahinden CA, Lachmann PJ, Davis AE III, Hugli TE (1985) A synthetic nonapeptide corresponding to the NH2-terminal sequence of C3d-K causes leukocytosis in rabbits. J Biol Chem 260: 2597–2600

Horstmann RD, Müller-Eberhard HJ (1985a) Isolation of rabbit C3, factor B, and Factor H and comparison of their properties with those of the human analog. J Immunol 134: 1094–1100

Horstmann RD, Pangburn MK, Müller-Eberhard HJ (1985b) Species specificity of recognition by the alternative pathway of complement. J Immunol 134: 1101–1104

Hynes RO (1987) Integrins: a family of cell surface receptors. Cell 48: 549–554

Isenman DE, Cooper NR (1981) The structure and function of the third component of human complement 1. The nature and extent of conformational changes accompanying C3 activation. Mol Immunol 18: 331–339

Isenman DE, Kells DIC, Cooper NR, Müller-Eberhard HJ, Pangburn MK (1981) Nucleophilic modification of human complement protein C3: correlation of conformational changes with acquisition of C3b-like functional properties. Biochemistry 20: 4458–4467

Jacobse-Geels HE, Daha MR, Horzinek MC (1980) Isolation and characterization of feline C3 and evidence for the immune complex pathogenesis of feline infectious peritonitis. J Immunol 125: 1606–1610

Janatova J (1986) Detection of disulphide bonds and localization of interchain linkages in the third (C3) and the fourth (C4) components of human complement. Biochem J 233: 819–825

Kai C, Yoshikawa Y, Yamanouchi K, Okada H (1983) Isolation and identification of the third component of complement of Japanese quails. J Immunol 130: 2814–2820

Kai C, Yoshikawa Y, Yamanouchi K, Okada H, Morikawa S (1985) Ontogeny of the third component of complement of Japanese quails. Immunology 54: 463–470

Kaidoh T, Gigli I (1987) Phylogeny of C4b–C3b cleaving activity: similar fragmentation patterns of human C4b and C3b produced by lower animals. J Immunol 139: 194–201

Kinoshita T, Nussenzweig V (1984) Regulatory proteins for the activated third and fourth components of complement (C3b and C4b) in mice. I. Isolation and characterization of factor H: the serum cofactor for the C3b/C4b inactivator (factor I). J Immunol Methods 71: 247–257

Klickstein LB, Bartow TJ, Miletic V, Rabson LD, Smith JA, Fearon DT (1988) Identification of distinct C3b and C4b recognition sites in the human C3b/C4b receptor (CR1, CD35) by deletion mutagenesis. J Exp Med 168: 1699–1717

Koch C (1988) Complement system in avian species. Avian Immunol 2: 43–55

Koistinen V, Wessberg S, Leikola J (1989) Common binding region of complement factors B, H, and CRI on C3b revealed by monoclonal anti-C3d. Complement Inflammation 6: 270–280

Kusano M, Choi NH, Tomita M, Yamamoto K, Migita S, Sekiya T, Nishimura S (1986) Nucleotide sequence of cDNA and derived amino acid sequence of rabbit complement component C3 alpha-chain. Immunol Invest 15: 365–378

Lachmann PJ, Halbwachs L (1975) The influence of C3b inactivator (KAF) concentration on the ability of serum to support complement activation. Clin Exp Immunol 21: 109–114

Lachmann PJ. Müller-Eberhard HJ (1968) The demonstration in human serum of "conglutinogen-activating factor" and its effect on the third component of complement. J Immunol 100: 691–698

Lachmann PJ, Pangburn MK, Oldroyd RG (1982) Breakdown of C3 after complement activation. Identification of a new fragment, C3g, using monoclonal antibodies. J Exp Med 156: 205–216

Lambris JD (1988) The multifunctional role of C3, the third component of complement. Immunol Today 9: 387–393

Lambris JD, Müller-Eberhard HJ (1984) Isolation and characterization of a 33000-dalton fragment of complement factor B with catalytic and C3b binding activity. J Biol Chem 259: 12685–12690

Lambris JD, Alsenz J, Schulz TF, Dierich MP (1984) Mapping of the properdin-binding site in the third component of complement. Biochem J 217: 323–326

Lambris JD, Ganu VS, Hirani S, Müller-Eberhard HJ (1985) Mapping of the C3d receptor (CR2)-binding site and a neoantigenic site in the Ced domain of the third component of complement. Proc Natl Acad Sci USA 82: 4235–4239

Lambris JD, Avilla D, Becherer JD, Müller-Eberhard HJ (1988) A discontinuous factor H binding site in the third component of complement as delineated by synthetic peptides. J Biol Chem 263: 12147–12150

Law SK, Lichtenberg NA, Levine RP (1979) Evidence for an ester linkage between the labile binding site of C3b and receptive surfaces. J Immunol 123: 1388

Lo SK, Wright SD (1988) CR3 mediates binding of PMN to endothelial cells (EC) via its RGD binding site, not the LPS binding site. FASEB J 2: A1236

Loveless RW, Feizi T, Childs RA, Mizuochi T, Stoll MS, Oldroyd RG, Lachmann PJ (1989) Bovine serum conglutinin is a lectin which binds non-reducing terminal N-acetylglucosamine, mannose and fucose residues. Biochem J 258: 109–113

Lundwall A, Wetsel RA, Domdey H, Tack BF, Fey GH (1984a) Structure of murine complement component C3. I. Nucleotide sequence of cloned complementary and genomic DNA coding for the β chain. J Biol Chem 259: 13851–13862

Lundwall A, Hellman U, Eggertsen G, Sjöquist J (1984b) Chemical characterization of cyanogen bromide fragments from the β-chain of human complement factor C3. FEBS Lett 169: 57–62

Malhotra V, Sim RB (1984) Role of complement receptor CR1 in the breakdown of soluble and zymosan-bound C3b. Biochem Soc Trans 12: 781–782

Manthei U, Nickells MW, Barnes SH, Ballard LL, Cui W, Atkinson JP (1988) Identification of a C3b/iC3 binding protein of rabbit platelets and leukocytes. J Immunol 140: 1228–1235

Matsuda T, Seya T, Nagasawa S (1985) Location of the inter-chain disulfide bonds of the third component of human complement. Biochem Biophys Res Commun 127: 264–269

Medicus RG, Melamed J, Arnaout MA (1983) Role of human factor I and C3b receptor in the cleavage of surface-bound C3b. Eur J Immunol 13: 465–470

Medof ME, Iida K, Mold C, Nussenzweig V (1982) Unique role of the complement receptor CR1 in the degradation of C3b associated with immune complexes. J Exp Med 156: 1739–1754

Melchers F, Erdei A, Schulz T, Dierich MP (1985) Growth control of activated, synchronized murine B cells by the C3d fragment of human complement. Nature 317: 264–267

Meuth JL, Morgan EL, DiScipio RG, Hugli TE (1983) Suppression of T lymphocyte function by human C3 fragments. I. Inhibition of human T cell proliferative responses by a kallikrein cleavage fragment of human iC3b. J Immunol 130: 2605–2611

Mitomo K, Fujita T, Iida K (1987) Functional and antigenic properties of complement receptor type 2, CR2. J Exp Med 165: 1424–1429

Müller-Eberhard HJ, Dalmasso AP, Calcott MA (1966) The reaction mechanism of β1c-globulin (C'3) in immune hemolysis. J Exp Med 123: 33–54

Nagaki K, Iida K, Okubo M, Inai S (1978) Reaction mechanism of β1H globulin. Int Arch Allergy Appl Immunol 57: 221–232

Nagasawa S, Ichihara C, Stroud RM (1980) Cleavage of C4b by C3b inactivator: production of a nicked form of C4b, C4b', as an intermediate cleavage product of C4b by C3b inactivator. J Immunol 125: 578–582

Nemerow GR, Wolfert R, McNaughton ME, Cooper NR (1985a) Identification and characterization of the Epstein-Barr virus receptor on human B lymphocytes and its relationship to the C3d complement receptor (CR2). J Virol 55: 347–351

Nemerow GR, McNaughton ME, Cooper NR (1985b) Binding of monoclonal antibody to the Epstein Barr virus (EBV)/CR2 receptor induces activation and differentiation of human B lymphocytes. J Immunol 135: 3068–3073

Nemerow GR, Mold C, Keivens Schwed V, Tollefson V, Cooper NR (1987) Identification of gp350 as the viral glycoprotein mediating attachment of Epstein-Barr virus (EBV) to the EBV/C3d receptor of B cells: sequence homology of gp350 and C3 complement fragment C3d. J Virol 61: 1416–1420

Nemerow GR, Houghten RA, Moore MD, Cooper NR (1989) Identification of an epitope in the major envelope protein of Epstein-Barr virus that mediates viral binding to the B lymphocyte EBV receptor (CR2). Cell 56: 369–377

Nielsen HE, Koch C (1987) Congenital properdin deficiency and meningococcal infection. Clin Immunol Immunopathol 44: 134–139

Nilsson B, Nilsson UR (1986) SDS denaturation of complement factor C3 as a model for allosteric modifications occuring during C3b binding: demonstration of a profound conformational change by means of circular dichroism and quantitative immunoprecipitation. Immunol Lett 13: 11–14

Nilsson B, Nilsson UR (1987) Anti-idiotypic antibodies in antisera against human C3 and factor H and their application in the enrichment of antibodies specific for H-binding domains of C3. J Immunol 138: 1858–1863

Nilsson B, Nilsson Ekdahl K, Avila D, Nilsson UR, Lambris JD (1988) Monoclonal anti-C3 antibodies specific for bound iC3b: mapping the recognized epitopes on C3 by synthetic peptides. Complement 5: 206

Nilsson-Ekdahl K, Nilsson B, Becherer JD, Nilsson UR, Lambris JD (1989) Further studies on the degradation of complement factor C3 by factor I. Inhibition of factor I with DFP. Seventh international congress of immunology, July 1989, Berlin Gustav Fischer Verlag Stuttgart p 131 (Abstract 22–24)

Nonaka M, Yamaguchi N, Natsuume-Sakai S, Takahashi M (1981) The complement system of rainbow trout (Salmo Gairdneri). J Immunol 126: 1489–1494

Nonaka M, Fujii T, Kaidoh T, Natsuume-Sakai S, Yamaguchi N, Takahashi M (1984a) Purification of a lamprey complement protein homologous to the third component of the mammalian complement system. J Immunol 133: 3242–3249

Nonaka M, Iwaki M, Nakai C, Nozaki M, Kaidoh T, Natsuume-Sakai S, Takahashi M (1984b) Purification of a major serum protein of rainbown trout (*Salmo gairdneri*) homologous to the third component of mammalian complement. J Biol Chem 259: 6327–6333

Nonaka M, Irie M, Tanabe K, Kaidoh T, Natsuume-Sakai S, Takahashi M (1985) Identification and characterization of a variant of the third component of complement (C3) in rainbow trout (*Salmo gairdneri*) serum. J Biol Chem 260: 809–814

O'Keefe MC, Caporale LH, Vogel CW (1988) A novel cleavage product of human complement components C3 with structural and functional properties of cobra venom factor. J Biol Chem 263: 12690–12697

O'Neill GJ, Yang SY, DuPont B (1978) Two HLA-linked loci controlling the fourth component of human complement. Proc Natl Acad Sci USA 75: 5165–5169

Pangburn MK (1986a) Differences between the binding sites of the complement regulatory proteins DAF, CR1, and factor H on C3 convertases. J Immunol 136: 2216–2221

Pangburn MK (1986b) The alternative pathway. In: Ross GD ed. Immunobiology of the complement system. An introduction for research and clinical medicine. Academic, Orlando, FL, pp 45–62

Pangburn MK, Müller-Eberhard HJ (1978) Complement C3 convertase: cell surface restriction of β1H control and generation of restriction on neuraminidase-treated cells. Proc Natl Acad Sci USA 75: 2416–2420

Paques EP (1980) Purification and partial characterization of the third component of the complement system from porcine serum (C3) and of a crystallizable degradation product of the fourth component of the complement system from human serum (C4). Hoppe Seylers Z Physiol Chem 361: 445–456

Pernegger G, Schulz TF, Hosp M, Myones BL, Petzer AL, Eigentler A, Bock G, Wick G, Dierich MP (1988) Cell cycle control of a Burkitt lymphoma cell line: responsiveness to growth signals engaging the C3d/EBV receptor. Immunology 65: 237–241

Petzer AL, Schulz TF, Stauder R, Eigentler A, Myones BL, Dierich MP (1988) Structural and functional analysis of CR2/EBV receptor by means of monoclonal antibodies and limited tryptic digestion. Immunology 63: 47–53

Pillemer L, Blum L, Lepow IH, Ross OA, Todd EW, Wardlaw AC (1954) The properdin system and immunity. I. Demonstration and isolation of a new serum protein, properdin, and its role in immune phenomena. Science 120: 280–285

Press EM, Gagnon J (1981) Human complement component C4. Biochem J 199: 351–357

Pryzdial ELG, Isenman DE (1987) Alternative complement pathway activation fragment Ba binds to C3b. J Biol Chem 262: 1519–1525

Ross GD, Lambris JD (1982) Identification of a C3bi-specific membrane complement receptor that is expressed on lymphocytes, monocytes, neutrophils, and erythrocytes. J Exp Med 155: 96–110

Ross GD, Medof ME (1985) Membrane complement receptors specific for bound fragments of C3. Adv Immunol 37: 217–267

Ross GD, Rabellino EM, Polley MJ (1976) Mouse leukocyte C3 receptors. Fed Proc 35: 254

Ross GD, Lambris JD, Cain JA, Newman SL (1982) Generation of three different fragments of bound C3 with purified factor I or serum. I. Requirements for factor H vs CR1 cofactor activity. J Immunol 129: 2051–2060

Ross GD, Newman SL, Lambris JD, Devery-Pocius JE, Cain JA, Lachmann PJ (1983) Generation of three different fragments of bound C3 with purified factor I or serum. II. Location of binding sites in the C3 fragments for factors B and H, complement receptors, and bovine conglutinin. J Exp Med 158: 334–352

Ross GD, Cain JA, Lachmann PJ (1985) Membrane complement receptor type three (CR3) has lectin-like properties analogous to bovine conglutinin and functions as a receptor for zymosan and rabbit erythrocytes as well as a receptor for iC3b. J Immunol 134: 3307–3315

Ruoslahti E, Pierschbacher MD (1986) Arg-Gly-Asp: a versatile cell recognition system. Cell 44: 517–518

Ruoslahti E, Pierschbacher MD (1987) New perspectives in cell adhesion: RGD and integrins. Science 238: 491–497

Russell DG, Wright SD (1988) Complement receptor type 3 (CR3) binds to an Arg-Gly-Asp-containing region of the major surface glycoprotein, gp63, of *Leishmania* promastigotes. J Exp Med 168: 279–292

Ryan AF, Catanzaro A, Wasserman SI, Harris JP, Vogel CW (1986) The effect of complement depletion on immunologically mediated middle ear effusion and inflammation. Clin Immunol Immunopathol 40: 420–421

Sekizawa A, Fujii T, Tochinai S (1984a) Membrane receptors on *Xenopus* macrophages for two classes of immunoglobulins (IgM and IgY) and the third complement component (C3). J Immunol 133: 1431–1435

Sekizawa A, Fujii T, Katagiri C (1984b) Isolation and characterization of the third component of complement in the serum of the clawed frog, *Xenopus laevis*. J Immunol 133: 1436–1443

Servis C, Lambris JD (1989) C3 synthetic peptides support growth of CR2 positive lymphoblastoid cell lines. J Immunol 142: 2207–2212

Seya T, Nagasawa S (1985) Limited proteolysis of complement protein C3b by regulatory enzyme C3b inactivator: isolation and characterization of a biologically active fragment, C3d,g. J Biochem (Tokyo) 97: 373–382

Shin HS, Mayer MM (1968) The third component of the guinea pig complement system. I. Purification and characterization. Biochemistry 7: 2991–2996

Shreffler DC (1976) The S region of the mouse major histocompatibility (H-2): genetic variation and functional role in complement system. Transplant Rev 32: 140–167

Sim E, Wood AB, Hsiung L, Sim RB (1981) Pattern of degradation of human complement fragment, C3b. FEBS Lett 132: 55–60

Sim RB, Twose TM, Paterson DS, Sim E (1981) The covalent-binding reaction of complement component C3. Biochem J 193: 115–127

Smith CA, Pangburn MK, Vogel CW, Müller Eberhard HJ (1984) Molecular architecture of human properdin, a positive regulator of the alternative pathway of complement. J Biol Chem 259: 4582–4588

Tack BF (1983) The β-Cys-γ-Glu thiolester bond in human C3, C4, and α_2-macroglobulin. Springer Semin Immunopathol 6: 259–282

Tanner J, Weis J, Fearon D, Whang Y, Kieff E (1987) Epstein-Barr virus gp350/220 binding to the B lymphocyte C3d receptor mediates adsorption, capping, and endocytosis. Cell 50: 203–213

Tanner J, Whang Y, Sample J, Sears A, Kieff E (1988) Soluble gp350/220 and deletion mutant glycoprotein block Epstein-Barr virus adsorption to lymphocytes. J Virol 62: 4452–4464

Thiel S, Baatrup G, Friis-Christiansen P, Svehag SE, Jensenius JC (1987) Characterization of a lecitin in human plasma analogous to bovine conglutinin. Scand J Immunol 26: 461–468

Thomas ML, Tack BF (1983) Identification and alignment of a thiol ester site in the third component of guinea pig complement. Biochemistry 22: 942–947

Tsokos GC, Thyphronitis G, Jack RM, Finkelman FD (1988) Ligand-loaded but not free complement receptors for C3b/C4b and C3d co-cap with cross-linked B cell surface IgM and IgD. J Immunol 141: 1261–1266

Ueda A, Kearney JF, Roux KH, Volanakis JE (1987) Probing functional sites on complement protein B with monoclonal antibodies. J Immunol 138: 1143–1149

Vogel CW, Müller-Eberhard HJ (1982) The cobra venom factor-dependent C3 convertase of human. J Biol Chem 257: 8292–8299

Vogel CW, Smith CA, Müller-Eberhard HJ (1984) Cobra venom factor: structural homology with the third component of human complement. J Immunol 133: 3235–3241

von Zabern I (1988) Species-dependent incompatabilities. In: Rother K, Till GO (eds) The complement system. Springer, Berlin Heidelberg New York, pp 196–202

von Zabern I, Nolte R, Vogt W (1979) Incompatability between complement components C3 and C5 of guinea pig and man, an indication of their interaction in C5 activation by classical and alternative C5 convertases. Scand J Immunol 9: 69–74

Wetsel RA, Lundwall A, Davidson F, Gibson T, Tack BF, Fey GH (1984) Structure of murine complement component C3. II. Nucleotide sequence of cloned complementary DNA coding for the alpha chain. J Biol Chem 259: 13857–13862

Whitehead AS, Solomon E, Chambers S, Bodmer WF, Povey S, Fey G (1982) Assignment of the structural gene for the third component of complement to chromosome 19. Proc Natl Acad Sci USA 79: 5021–5025

Wiebauer K, Domdey H, Diggelmann H, Fey G (1982) Isolation and analysis of genomic DNA clones encoding the third component of mouse complement. Proc Natl Acad Sci USA 79: 7077–7081

Wilson BS, Platt JL, Kay NE (1985) Monoclonal antibodies to the $140000 M_r$ glycoprotein of B lymphocyte membranes (CR2 receptor) initiates proliferation of B cells in vitro. Blood 66: 824–829

Wright SD, Jong MTC (1986) Adhesion-promoting receptors on human macrophages recognize *Escherichia coli* by binding to lipopolysaccharide. J Exp Med 164: 1876–1888

Wright SD, Reddy PA, Jong MT, Erickson BW (1987) C3bi receptor (complement receptor type 3) recognizes a region of complement protein C3 containing the sequence Arg-Gly-Asp. Proc Natl Acad Sci USA 84: 1965–1968

Wright SD, Weitz JI, Huang AJ, Levin SM, Silverstein SC, Loike JD (1988) Complement receptor type three (CD11b/CD18) of human polymorphonuclear leukocytes recognizes fibrinogen. Proc Natl Acad Sci USA 85: 7734–7738

Wright SD, Levin SM, Jong MTC, Chad Z, Kabbash LG (1989) CR3 (CD11b/Cd18) expresses one binding site for Arg-Gly-Asp-containing peptides and a second site for bacterial lipopolysaccharide. J Exp Med 169: 175–183

The Thioester Bond of C3

R. P. Levine[1,2] and A. W. Dodds[2]

1 Introduction   73
2 The Covalent Bond   74
3 Mechanism of Covalent Bond Formation and the Covalent Binding Specificities of C3 and C4   75
4 Environment of the Thioester Bond of C3 and the Covalent Bond Between C3b and Receptive Surfaces   78
5 Biological Role of the Thioester Bond   79
References   80

1 Introduction

Thioester bonds are not uncommon in biochemical systems, being formed as intermediates in the reaction mechanisms of a number of enzymes (Liu 1976; Douglas 1987). Thioester bonds are, however, important structural and functional features of certain proteins, including C3. C3 shares the thioester bond with several other proteins, principally the fourth complement protein, C4, and α_2-macroglobulin. It is this bond that confers upon C3 the ability to form covalent bonds with cell-surface macromolecules, immune complexes, and a variety of small molecules in solution (Law and Levine 1977; Law et al. 1979, 1980, 1981). The bond in C3 lies in the α-polypeptide linking the $-SH$ group of a cysteinyl residue to the γ-carboxyl group of a glutaminyl residue. The same obtains in C4 and α-macroglobulin, and the three proteins share extensive amino acid homology throughout their structure, particularly in the vicinity of the bond.

The early studies of covalent bond formation by C3 led to the speculation (Law et al. 1979) that "the formation of the bond involves a trans-esterification, a reaction that necessarily implies the existence of an ester within C3." Several investigations provided experimental evidence for the presence of the thioester bond within the α-polypeptide of C3 (Tack et al., 1980; Thomas et al. 1982), C4 (Janatova and Tack 1981; Harrison et al. 1981), and α_2-macroglobulin (Sottrup-Jensen et al. 1981; Howard 1981). The mode of formation of the internal thioester bond during the

---

[1] James S. McDonnell Department of Genetics, Washington University School of Medicine, St. Louis, Missouri, USA
[2] MRC Immunochemistry Unit, Department of Biochemistry, University of Oxford, Oxford, UK

biosynthesis of these proteins is not known. The liver is the major site for C3 synthesis, and it has been reported (IIJIMA et al. 1984) that homogenates of mouse liver contain a protein that catalyzes the formation of the internal thioester bond of C3.

2 The Covalent Bond

By virtue of the cleavage of the α-polypeptide the activation of C3 by either the classical or the alternative pathway yields C3a, a 9000 dalton polypeptide having anaphylatoxic activity, and C3b, 180000 daltons, that can bind covalently to a variety of molecules. That C3b might possibly form covalent bonds became apparent on examination of the results of BHAKDI et al. (1974). Following the classical pathway activation of ^{125}I-labeled C3 on the surface of antibody-coated sheep erythrocytes they observed high molecular weight radiolabeled complexes that persisted after the solubilization of the erythrocyte membranes in sodium dodecyl sulfate (SDS). Subsequent investigations (LAW and LEVINE 1977; LAW et al. 1979, 1980) focused on efforts to define the nature of the bond or bonds that hold C3b to the erythrocyte surface and the polypeptide chain of C3 that is involved in forming the bond. High molecular weight C3b-membrane protein complexes that appear following the SDS polyacrylamide gel electrophoresis of preparations in which either C3 or membrane proteins have been radiolabeled can be cleaved either by ammonolysis or NH_2OH at high pH (LAW and LEVINE 1977). When ^{125}I-labeled C3b-membrane protein complexes are separated in one dimension by SDS polyacrylamide gel electrophoresis under reducing conditions and then subjected to separation in a second dimension following incubation in NH_2OH, it can be seen that the covalent bond is between the C3b α-polypeptide and a membrane protein or proteins (LAW and LEVINE 1977). C3b is able to bind to several membrane proteins, and on erythrocytes the principal protein to which it binds appears to be glycophorin (PARKER et al. 1984).

The observation that C3b can be released from molecules to which it has become bound on either the erythrocyte cell surface or zymosan by NH_2OH suggested that it is bound via an ester bond. Support for this notion comes from experiments in which the kinetics of the release of covalently bound C3b are measured (LAW et al. 1979).

It was observed that the second-order rate constants for C3b release from zymosan at both neutral and alkaline pH resemble those obtained for the release of a variety of compounds that contain ester bonds. In addition, the product of the release of C3b by NH_2OH was found to be an hydroxamate derivative of C3b; this is expected if the bond formed between the C3b and a cell-surface macromolecule is an ester bond (LAW et al. 1979). It must be pointed out, however, that the maximum release of C3b from both erythrocyte membrane macromolecules and zymosan requires, in addition to NH_2OH, the presence of a detergent such as SDS (LAW and LEVINE 1977), suggesting that hydrophobic interactions are also important in the association between C3b and cell surface macromolecules. On the other hand, it may be that a denaturing detergent is required in order to expose the cell surface C3b bond to nucleophilic attack. In addition, a significant amount of C3b bound to the surface of both erythrocytes and zymosan is resistent to the combined effects of NH_2OH and SDS (Law and Levine 1977; Law et al. 1979). From this it is inferred that C3b can also bind via amide bonds.

Further evidence for the formation of covalent bonds by C3 comes from fluid-phase binding studies (Law et al. 1981, 1984b) in which C3 is activated by trypsin in the presence of small molecules possessing either $-OH$ or free $-NH_2$ groups. C3b binds with a range of binding efficiencies to a variety of molecules in solution (Law et al. 1981). It can form ester bonds at neutral pH and amide bonds at pH values greater than 9.0. The ester bonds are most likely formed through the $-CH_2OH$ groups of sugars and short-chain alcohols and through the $-OH$ groups of serine and threonine (Law et al. 1981). With the exeption of serine and threonine, C3b bound to $-OH$ group containing molecules is released on treatment of the complexes with NH_2OH. In addition, it is not released from lysine, from which it is inferred that an amide bond has been formed between the two molecules. Since the bond between C3b and either serine or threonine is also NH_2OH-resistant, it too appears to be amide in nature. On the other hand, it is not possible to exclude that initially these molecules form ester bonds with C3b, and that subsequently amide bonds form as a consequence of an intramolecular O to N shift from the β-OH to the α-NH_2 group of either serine or threonine. An experiment to test this possibility as an explanation for the apparent formation of amide bonds between C3b and serine was carried out by taking advantage of the fact that the efficiency of covalent binding to C3b of a variety of small molecules in the fluid phase can be inferred from measurements of their ability to inhibit ester bond formation between C3b and [^{3}H]glycerol (Law et al. 1984b). The ability of serine, O-methylserine, and N-acetylserine to inhibit [^{3}H]glycerol binding was ascertained in this manner, and it was found (Law et al. 1984b) that serine and N-acetylserine, but not O-methylserine, inhibited glycerol binding. It was concluded from these results that the covalent binding reaction between C3 band serine is via the $-OH$ group of the latter, and that an intramolecular rearrangement accounts for the NH_2OH-resistant, presumably amide bonds between C3b and this amino acid and, by inference, threonine. These results, however, do not account for the fact that lysine and C3b can form NH_2OH-resistant bonds at neutral pH (Law et al. 1981).

Overall, from the observations and the results of the experiments described in brief here it was concluded (Law et al. 1981, 1984b) that C3b possesses a binding site whereby it can form covalent bonds with a variety of small molecules in solution and with cell-surface macromolecules; that ester bonds are of the form $R-O-CO-C3b$ rather than $R-CO-O-C3b$, where R refers to either a molecule on a receptive surface or a molecule in solution; and that, by inference, amide bonds, because of their resistance to cleavage by NH_2OH, are also formed.

3 Mechanism of Covalent Bond Formation and the Covalent Binding Specificities of C3 and C4

Fluid-phase binding experiments in which C3 is converted to C3b by trypsin in the presence of either glycerol or glycine revealed that whereas binding to glycerol is pH independent, significant binding to glycine occurs only at pH values greater than 9, values at which the α-amino group is deprotonated. It has been concluded from these observations that under physiological conditions C3 is essentially an ester bond forming molecule. On the other hand, as pointed out earlier, a significant number of NH_2OH-resistant bonds are formed under physiological conditions between C3b and both erythrocytes and zymosan (LAW and LEVINE 1977) and under fluid-phase binding conditions between C3b and lysine (LAW et al. 1981) and immunoglobulins (TAKATA et al. 1984). If indeed these bonds are truly amide in nature and not merely highly protected ester bonds, it must be assumed that they are formed under conditions in which the local environment provides for the deprotonation of free amino groups at neutral pH.

The covalent binding specificities of human C4 differ from those of C3 and are instructive with respect to efforts to understand the mechanism of covalent bond formation. The initial observation of C4 covalent binding revealed that C4, purified from pooled human serum, forms either ester or amide bonds and the latter at relatively high efficiency at neutral pH (LAW et al. 1984b). Further studies revealed, however, that the efficiency with which covalent bonds are formed depends on the allotype of C4 (LAW et al. 1984a).

Human C4 is specified by a pair of tandem genes (A and B) that lie within the major histocompatibility complex (O'NEILL et al. 1978), and it is possible to isolate and purify C4 of either the A or B allotypic forms. A comparison of the reaction rates of human C4A and C4B, with glycine and glycerol at pH 7.0 (DODDS et al. 1986) is shown in Table 1, where it can be seen that C4A forms amide bonds with high efficiency, but not ester bonds, and that C4B can form both amide and ester bonds. Human C4A and C4B differ from each other by only four amino acid substitutions (BELT et al. 1985; YU et al. 1986), and these substitutions appear to account for the difference in the reactivity of the two C4 types (DODDS et al. 1986; CARROLL et al. 1989). The mouse expresses a single C4 type, the relevant sequence of which is a hybrid of the two human C4 types (NONAKA et al. 1985; SEPICH et al. 1985). The binding reactivity of mouse C4 is very similar to that of human C4B. The amino acids that confer the difference in reactivity appear to be Ile His in human C4B and mouse C4 and Leu Asp in human C4A. It is unlikely that an Ile/Leu substitution could alter the reactivity of the proteins. It seems more likely that the His/Asp substitution at position 1106 in the human pro-C4 molecule is the specificity-defining residue in these proteins (DODDS and LAW 1988; VENKATESH and LEVINE 1988). An examination of the amino acid sequences reveals that at the homologous position C3 also has a His residue, and that the binding properties of C3 are more similar to those of human C4B and mouse C4 than to human C4A.

It has been hypothesized (DODDS and LAW 1988; VENKETESH and LEVINE 1988) that the imidazole ring of the His residue can participate in a general base catalysis that would increase the nucleophilicity of $-CH_2OH$ groups thus facilitating their attack on the internal thioester bond whereas the carboxylate ion of the Asp residue can participate in the deprotonation of free amino groups at physiological pH. Support for this hypothesis comes from the report of the results of experiments in which the His of C4B1 was replaced with the Asp of C4A3 by site-specific mutagenesis. The covalent binding capacities of the altered forms of C4 were inferred from measurements of their contribution to the hemolytic activity, taking advantage of the fact that C4B allotypes are hemolytically more active than the C4A allotypes

Table 1. The reaction rates of [^{3}H]glycine and [^{3}H]glycerol binding with C4 allotypes. (After DODDS et al. 1986)

C4type	k'/k_0 (M^{-1})	
	Glycine	Glycerol
C4A		
RP A3	17020	1.52
CA A4	13250	1.52
HP A6	11270	0.50
AT A6	13800	ND
CB A2/A3	17020	1.02
AD A1	18570	2.56
Average	15150 ± 2800	1.42 ± 0.75
C4B		
RP B1	114	16.2
CA B2	93	14.9
AW B3	106	12.4
KC B3	106	12.9
TA B5	59	12.4
Average	95.6 ± 21.8	13.8 ± 1.7
JM A3/A4/B5	4285	7.5

Binding efficiency is defined as the fraction of active C4 bound to radioactive glycine or glycerol on activation by C1s. Binding efficiency (BE) is converted to reaction rate using the equation: BE = $k'[G]/(k_0 + k'[G])$ (LAW et al. 1984b), where k_0 is the first-order reaction rate of C4b with water, k' is the second-order reaction rate with substrate molecules, and [G] is the concentration of the substrate molecule under study. The k'/k_0 values for the reaction of glycine and glycerol with the various allotypes are shown

because of the their greater binding efficiency to the erythrocyte cell surface. The hemolytic assays showed that substituting Asp for His reduced the hemolytic activity of C4B1.

Quite obviously, because of their covalent binding capacity, the thioester bond containing proteins of the complement system are unique, and further study of both the mechanism of thioester bond formation and bond cleavage and subsequent acyl ester and amide bond formation to cell surface macromolecules and to immune complexes is of special interest.

Relevant to the covalent bond forming capacity of C3b is the extensive polymorphism of human C4; there is an array of alleles (at least 35) at both loci, including null alleles (SIM and DODDS 1987).

Since the A allotypes bind more efficiently than C4B allotypes to immunoglobulins, either free in solution or aggregated (LAW et al. 1984a; KISHORE et al. 1988), they might be considered to have evolved in response to natural selection favoring a molecule that can bind to proteins in which free $-NH_2$ groups are available. The selective advantage of the ester-forming B isotypes, on the other hand, may be in defense against bacterial infections since these forms of C4 might be expected to bind more effectively to bacterial polysaccharide capsules that are relatively rich in −OH groups. This appears to be the case for C4B1, which binds more efficiently than C4A3 to the bacterium *Hemophilus influenzae* type b (A. W. DODDS, unpublished material).

Associations between either C4A-null or C4b-null phenotypes with immune complex diseases, on the one hand, and bacterial infections, on the other, might give clues as to the relative significance of amide versus acyl ester bond formation. To data there is a good evidence for a significant association between the C4A-null phenotype and systemic lupus erythematosis (FIELDER et al. 1983; REVEILLE et al. 1985; RANFORD et al. 1987; DUNKLER et al. 1987). Clearly, the polymorphism of human C4 and the distinct functional differences that characterize the A and B variants need to be investigated from the point of view of how these proteins participate in either the pathogenesis or pathophysiology of immune complex diseases and bacterial infections, as well as in the variety of phenomena of the immune system with which the complement system is involved.

4 Environment of the Thioester Bond of C3 and the Covalent Bond Between C3b and Receptive Surfaces

The proteolytic cleavage of the α-polypeptide chain of C3 either by the classical or the alternative pathway C3 convertases or by trypsin induces conformational changes in the molecule (ISENMAN et al. 1981). The changes, detected spectroscopically, are believed to expose the thioester bond to either hydrolysis or nucleophilic attack and are a prerequisite for the covalent binding reaction to occur.

The reaction itself does not require the proteolytic cleavage of C3, however, for ester bond formation between C3 and glycerol, for example, can occur in the presence of chaotropic agents or certain protein denaturants (LAW 1983), all of which presumably alter the secondary and tertiary structure of the protein in a way that exposes the thioester bond to either hydrolysis or nucleophilic attack. When the thioester bond containing proteins are exposed to severely denaturing conditions, especially at elevated temperature and at high pH, another reaction occurs. A peptide bond is cleaved between the two glutamic acid residues in the sequence: Cys Gly Glu Glu. In the reaction the new N-terminal Glu residue becomes cyclized to give pyroglutamic acid. It is unlikely that this reaction has any physiological significance, however (SIM and SIM, 1983).

The region encompassing the thioester bond in C3, C4, and α_2-macroglobulin is relatively hydrophobic, and there is evidence to suggest that the thioester sequence of C3 and α_2-macroglobulin lies within a pocket some 12–20 Å deep (LEVINE et al. 1983; ZHAO et al. 1988).

It seems reasonable to assume that the sequences of relatively hydrophobic amino acid residues that bracket the thioester bond serve to shield it from the aqueous environment and from nucleophilic attack, and that further protection of the bond is gained by its location within a pocket or groove. Presumably, the conformational changes that occur upon the cleavage of C3 or upon its incubation in chaotropic agents or denaturants exposes the thioester bond. The outcome of this exposure depends upon the conditions under which it occurs. Proteolytic cleavage of C3 either by trypsin or by the C3 convertases of the classical and alternative pathways results in either the hydrolysis of the thioester or the covalent binding reaction with a receptive molecule or cell-surface macromolecule (R—OH). Strong nucleophiles such as CH_3NH_2 attack the thioester directly resulting in the production of an inactive C3 molecule. Chaotropic agents do not cleave C3 but allow either the hydrolysis of the thioester bond or the covalent binding reaction to occur with R—OH.

5 Biological Role of the Thioester Bond

The thioester bond of C3 appears to have two major roles in complement activation. The first is to localize complement-mediated damage to the immediate area of activation. The second is in the mechanism of activation of the alternative pathway.

It has been estimated that the half-life of the intact thioester in nascent C3b is in the region of 10^{-3}–10^{-4} s (GÖTZE and MÜLLER-EBERHARD 1970). This short time in which the thioester is reactive severely limits the distance that the molecules can diffuse from the site of activation. The deposition of active C3b and C4b is highly localized, therefore, avoiding potential damage to the host own cells.

As discussed earlier, denaturants can bring about the exposure of the thioester bond. However, even under nondenaturing conditions the thioester bond of C3 is subject to slow hydrolysis to a form in which the α-chain is intact but the thioester is broken. This form of C3 termed C3 (H_2O) is able to bind factor B which can then be activated by factor D to form an alternative pathway C3 convertase C3 (H_2O) Bb. It has been proposed that the low level generation of C3 (H_2O) Bb in plasma is the mechanism by which the alternative pathway becomes activated (PANGBURN et al. 1981). During its short lifetime each C3 (H_2O) Bb complex formed could cleave a number of C3 molecules to C3b. Under normal circumstances the thioester of the nascent C3b would react with water or other plasma constituants and would be rapidly inactivated by factors H and I. If an alternative pathway activator, such as zymosan or a bacterial surface, were present, some of the nascent C3b could become bound and protected from inactivation. A longer lived C3 convertase site would thus be formed, and alternative pathway activation could occur as more C3b was deposited around the initial site of deposition.

Recently, it has been shown that the deposition of C3b on a surface by C3 convertase is not entirely random. Of the initial C3 molecules deposited, a high percentage bind covalently to C4b in the classical pathway convertase (TAKATA et al. 1987) and to C3b in the alternative pathway convertase (KINOSHITA et al. 1988). In both cases the covalent attachment of the C3b to the convertase results in the generation of a high-affinity C5 binding site and of C5 convertase activity. C5 binds to sites in both the C3 and C4 in the classical pathway C5 convertase (KINOSHITA et al. 1989) and presumably also to sites in C3 molecules in the alterative pathway convertase. In order to ensure that the two sites are in the correct configuration for C5 binding, the binding of C3b other C3 convertase must be to a single specific site.

A number of functions can be ascribed to C3. (a) The covalent binding of C3b to a cell can lead to the cell's death either via cytolysis or phagocytosis. (b) When C3b binds covalently to antigen-antibody complexes, it prevents the complexes from forming large and ultimately precipitable lattices, and, in addition, it opsonizes the complexes for their clearance and ultimate elimination by phagocytosis. (c) C3b, covalently bound to certain molecules, is essential in an immune response to that molecule. Clearly, the biological significance of C3 lies in its potential to form ester bonds with $-OH$ groups that are of ubiquitous occurrence, and although there are certain exceptions, for example C3b does not bind to inositol (Law et al. 1981) or to double-stranded DNA (Edberg et al. 1988; Y. Kosono and R. P. Levine, unpublished material), the broad specificity (or more properly, the lack of specificity) that characterizes the binding capacity of C3 emphasizes its importance as a key component in the evolution and activities of the immune system as a whole. Indeed, along with the alternative complement pathway it can be considered to be part of a biologically ancient system whereby foreign bodies are recognized and eliminated.

The course of evolution of C3 and the other proteins containing internal thioester bonds is of interest. By taking advantage of the fact that proteins possessing an internal thioester bond can be labeled by virtue of the ability of CH_3OH to cleave the bond, C3 or α_2-macroglobulin-like proteins have been detected in the sera of animals of the major vertebrate phyla as well as in the hemolymph of certain invertebrates (Day et al. 1970; Gigli and Austen 1971; Grossberger et al. 1989; Koppenheffer 1987). The wide occurrence of these proteins among apparently disparate groups of animals attests to both their ancient origin and significance of their function. It remains to be determined, however, how in the course of the history of these proteins a mechanism has evolved for the biosynthesis of the thioester bond; how, once formed, the bond is protected from hydrolysis; and precisely how the covalent binding reaction occurs between these proteins and the variety of molecules with which they can form covalent bonds.

References

Belt KT, Carroll MC, Porter RR (1985) Polymorphism of human complement component C4. Immunogenetics 21: 173–180

Bhakdi S, Knüffermann H, Schmidt-Ullrich R, Fischer H, Wallach DFH (1974) Interaction between erythrocyte membrane proteins and complement components. I. The role of $-s-s$ linkages as revealed by two dimensional sodium dodecyl sulfate-polyacryamide gel electrophoresis. Biochim Biophys Acta 363: 39–53

Carroll MC, Fathallah DM, Bergamaschini L, Alicott EM, Isenman DE (1989) The chemical basis for the functional hemolytic difference between isotypes of human C4(Abstract). FASEB J 3: A367

Day NK, Gewurz H, Johannsen R, Finstedd J, Good RA (1970) Complement and complement-like activity in lower vertebrates and invertebrates. J Exp Med 132: 941–950

Dodds AW, Law SKA (1988) Structural basis for the binding specificity of the thioester containing proteins, C4, C3 and alpha-2-macroglobulin. Complement 5: 89–98

Dodds AW, Law SKA, Porter RR (1986) The purification and properties of some less common allotypes of the fourth component of human complement. Immunogenetics 24: 279–285

Douglas KT (1987) Mechanism of action of glutathione-dependent enzymes. Adv Enzymol 59: 103–107

Dunckler H, Gatenby PA, Dawkins B, Naito S, Serjeantson SW (1987) Deficiency of C4A is a genetic determinant of systemic lupus erythematosus in three ethnic groups. J Immunogenet 14: 209–218

Edberg JC, Tosic L, Wright EL, Sutherland WM, Taylor RP (1988) Quantitative analyses of the relationship between C3 consumption, C3b capture, and immune adherence of complement-fixing antibody/DNA immune complexes. J Immunol 141: 4258–4265

Fielder AHL, Walport MJ, Batchelor JR, Rynes RI, Black CM, Dodi RL, Hughes GRV (1983) Family study of the major histocompatibility complex in patients with systemic lupus erythematosus: importance of null alleles of C4A and C4B in determining disease susciptibility. Br Med J 286: 425–428

Gigli I, Austen KF (1971) Phylogeny and function of the complement system. Annu Rev Microbiol 25: 309–313

Götze O, Müller-Eberhard HJ (1970) Lysis of erythrocytes by complement in the absence of antibody. J Exp Med 132: 898–915

Grossberger D, Marcuz A, DuPasquier L, Lambris JD (1987) Conservation of structural and functional domains in complement component C3 of Xenopus and mammals. Proc Natl Acad Sci USA 86: 1323–1327

Harrison RA, Thomas ML, Tack BL (1981) Sequence determination of the thiolester site of the fourth component of human complement. Proc Natl Acad Sci USA 78: 7388–7392

Howard JB (1981) Reactive site in α_2-macroglobulin: circumstantial evidence for a thioester. Proc Natl Acad Sci USA 78: 2235–2239

Iijima M, Takashi T, Sakamoto T, Tomita M (1984) Biosynthesis of the internal thioester bond of the third component of complement. J Biochem 96: 1539–1546

Isenman DE, Kells DIC, Cooper NR, Müller-Eberhard HJ, Pangburn MK (1981) Nucleophilic modification of human complement protein C3: correlation of conformational changes with acquisition of C3b-like functional properties. Biochemistry 20: 4458–4467

Janatova J, Tack BF (1981) Fourth component of human complement: studies of an amine-sensitive site comprised of a thiol component. Biochemistry 20: 2394–2402

Kinoshita T, Takata Y, Kozono H, Takeda J, Hong K, Inoue K (1988) C5 convertase of the alternative complement pathway: covalent linkage between two C3b molecules within the trimolecular complex enzymes. J Immunol 141: 3895–3901

Kinoshita T, Dodds AW, Law SKA, Inoue K (1989) Biochem J 261: 743–748

Kishore N, Shah D, Skanes VM, Levine RP (1988) The fluid-phase binding of human C4 and its genetic variants, C4A3 and C4B1, to immunoglobulins. Mol Immunol 25: 811–819

Koppenheffer TL (1987) Serum complement systems of ectothermic vertebrates. Dev Comp Immunol 11: 279–286

Law SKA (1983) Non-enzymic activation of the covalent binding reaction of the complement protein C3. Biochem J 211: 381–389

Law SK, Levine RP (1977) Interaction between the third complement protein and cell surface macromolecules. Proc Natl Acad Sci 74: 2701–2705

Law SK, Lichtenberg NA, Levine RP (1979) Evidence for an ester linkage between the labile binding site of C3b and receptive surfaces. J Immunol 123: 1388–1394

Law SK, Lichtenberg NA, Levine RP (1980) Covalent binding and hemolytic activity of complement proteins. Proc Natl Acad Sci USA 77: 7194–7198

Law SK, Minich TM, Levine RP (1981) Binding reaction between the third human complement protein and small molecules. Biochemistry 20: 7457–7463

Law SKA, Dodds Aw, Porter RR (1984a) A comparison of the properties of two class C4A and C4B, of the human complement component C4. EMBO J 3: 1819–1823

Law SKA, Minich TM, Levine RP (1984b) The covalent binding efficiency of the third and fourth complement proteins in relation to pH, nucleophilicity and availability of hydroxyl groups. Biochemistry 23: 3267–3272

Levine RP, Finn R, Gross R (1983) Interactions between C3b and cell surface macromolecules. Ann NY Acad Sci 421: 235–245

Liu T (1976) The role of sulphur in proteins. In: Nevra TLH (ed) The proteins, vol 3. Academic, New York, p 239

Nonaka M, Nakayama K, Yeul YD, Takahashi M (1985) Complete nucleotide and derived amino acid sequence of the fourth component of mouse complement (C4). J Biol Chem 260: 10936–10943

O'Neill GJ, Yang SY, Dupont B (1978) Two HLA-linked loci controlling the fourth component of human complement. Proc Natl Acad Sci USA 75: 5165–5169

Parker CJ, Soldata CM, Ross WF (1984) Abnormality of glycophorin-α on paroxysmal nocturnal hemoglobinemia erythrocytes. J Clin Invest 73: 1130–1143

Pangburn MK, Schreiber RD, Müller-Eberhard HJ (1981) Formation of the initial C3 convertase of the alternative complement pathway: acquisition of C3b-like activities by spontaneous hydrolysis of the putative thioester in native C3. J Exp Med 154: 856–867

Ranford P, Serjeantson SW, Hay J, Dunckler H (1987) A high frequency of inherited deficiency of complement component C4 in Darwin aborigines. Aust NZ J Med 17: 420–423

Revielle JD, Arnett FC, Wilson RW, Bias WB, McLean RN (1985) Null alleles of the fourth component of complement and HLA haplotypes in familial systematic lupus erythematosus. Immunogenetics 21: 299–311

Sepich DS, Noonan DJ, Ogata RT (1985) Complement cDNA sequence of the fourth component of murine complement. Proc Natl Acad Sci USA 82: 5895–5899

Sim E, Dodds AW (1987) The fourth component of human complement – towards understanding an enigma of variations. In Whaley K (ed) Complement in health and disease. MTP Press, Lancaster pp 99–124

Sim RB, Sim E (1983) Autolytic fragmentation of complement components C3 and C4 and its relationship to covalent binding specificity. Ann NY Acad Sci 421: 259–276

Sottrup-Jensen L, Hansen HF, Mortensen SB, Petersen TE, Magnusson S (1981) Sequence and location of the reactive thiol ester in human α_2-macroglobulin. FEBS Lett 123: 145–148

Tack BF, Harrison RA, Janatova J, Thomas ML, Prahl JW (1980) Evidence for the presence of an internal thiolester bond in the third component of human complement. Proc Natl Acad Sci USA 77: 5764–5768

Takata Y, Tamura N, Fujita T (1984) Interaction of C3 with antigen-antibody complexes in the process of solubilization of immune complexes. J Immunol 132: 2531–2537

Takata Y, Kinoshita T, Kozono H, Takeda J, Tanaka E, Hong K, Inoue K (1987) Covalent association of C3b with C4b within C5 convertase of the classical complement pathway. J Exp Med 165: 1494–1507

Thomas ML, Janatova J, Gray WR, Tack BF (1982) Third component of human complement: lokalization of the internal thiolester bond. Proc Natl Acad Sci USA 79: 1054–1058

Venketesh YP, Levine RP (1988) The esterase-like activity of covalently bound human third complement protein. Mol Immunol 25: 821–828

Yu CY, Belt KT, Cortes CM, Campbell RD, Porter RR (1986) Structural basis of the polymorphism of human complement components C4A and C4B: gene size reactivity and antigenicity. EMBO J 5: 2873–2881

Zhao B, Musci G, Susawara Y, Berliner LJ (1988) Spin label and fluorescence studies of the thioester bonds in human α_2-macroglobulin. Biochemistry 27: 5304–5308

Complement Receptor Type 1 (C3b/C4b Receptor; CD35) and Complement Receptor Type 2 (C3d/Epstein-Barr Virus Receptor; CD21)

D. T. Fearon and J. M. Ahearn

1 Complement Receptor Type 1 (CR1; CD35)   83
1.1 Structure of CR1   83
1.2 Allotypes of CR1   85
1.3 Biosynthesis of CR1   87

2 Complement Receptor Type 2 (CR2; CD21)   88
2.1 Structure of CR2   88
2.2 Biosynthesis of CR2   89

3 Immunoregulatory Functions of CR1 and CR2   90

4 Determination of Epstein-Barr Virus Tissue Tropism by CR2   92

References   93

1 Complement Receptor Type 1 (CR1; CD35)

1.1 Structure of CR1

The primary sequence of the A allotype of CR1 has been deduced from the cDNA sequence and includes a 41-residue signal peptide, an extracellular domain of 1930 residues, a 25 amino acid transmembrane domain, and a 43 amino acid cytoplasmic region (Wong et al. 1985; Klickstein et al. 1987, 1988; Hourcade et al. 1988) (Fig. 1). Thirty short consensus repeats (SCRs) similar to those present in other C3/C4 binding complement proteins comprise the extracellular domain. Each SCR has four cysteines which may be disulfide bonded in the pattern, cysteine-1 to cysteine-3 and cysteine-2 to cysteine-4 (Lozier et al. 1984; Janatova et al. 1988). These disulfide bridges create three loops of 26–29 amino acids, 12–19 amino acids, and 12–16 amino acids, respectively, in CR1. The dimensions of each SCR in CR1, 2.9×2.8 nm, have been estimated by electron microscopy (Bartow et al. 1989) and are comparable to those of the SCRs of C4 binding protein (C4-bp; Perkins et al. 1986). The regions linking SCRs are short, ranging from three to five amino acids, except for that connecting SCR-28 to SCR-29 which is seven amino acids in length. This characteristic may limit the range of motion between SCRs, perhaps accounting for the filamentous appearance of CR1.

Department of Medicine and The Graduate Program in Immunology, The Johns Hopkins University School of Medicine, 725 N. Wolfe Street, Baltimore, MD 21205, USA

Complement Receptor Type 1 (CR1)

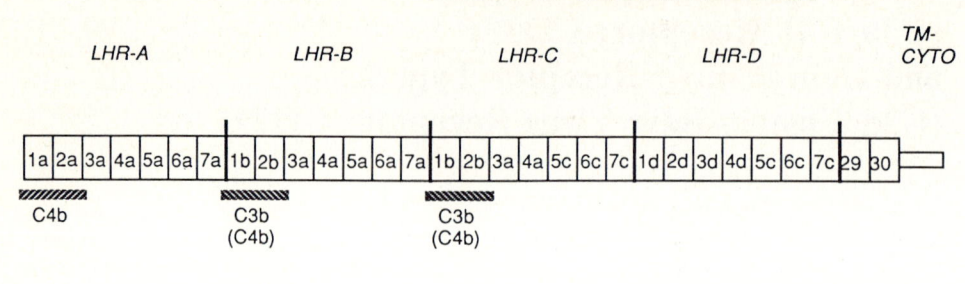

```
   LHR-A              LHR-B              LHR-C              LHR-D              TM-
                                                                               CYTO
|1a|2a|3a|4a|5a|6a|7a|1b|2b|3a|4a|5a|6a|7a|1b|2b|3a|4a|5c|6c|7c|1d|2d|3d|4d|5c|6c|7c|29|30|
 ▓▓▓▓▓▓▓▓            ▓▓▓▓▓▓▓▓            ▓▓▓▓▓▓▓▓
   C4b                 C3b                 C3b
                      (C4b)               (C4b)
```

Complement Receptor Type 2 (CR2)

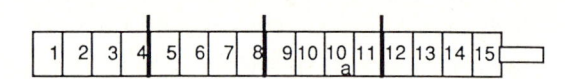

```
                                                                   TM-
                                                                   CYTO
                        |1|2|3|4|5|6|7|8|9|10|10a|11|12|13|14|15|
```

Fig. 1. Schematic diagrams demonstrating the structures of human CR1 and CR2. *Boxes*, short consensus repeats (SCRs). In CR1 28 of the 30 SCRs are divided into four long homologous repeats (*LHRs*). Identical numbers indicate highly homologous repeats and identical lower case letters indicate greater than 95% sequence identity. In CR2, 15 or 16 SCRs are organized into four homology groups which are not as conserved as are the LHRs of CR1. SCR-10a is an alternatively spliced repeat that is not present in all CR2 molecules. The transmembrane and cytoplasmic (*TM-CYTO*) domains are at the carboxy-termini of both receptors

Long homologous repeats (LHRs), termed LHR-A, -B, -C and -D, are formed by four groups of seven SCRs (KLICKSTEIN et al. 1987, 1988) (Fig. 1). SCRs in the same relative positions within LHRs have amino acid sequence identities ranging from 60% to 100%. Distinct binding sites for C3b and C4b in LHR-A, -B and -C have been demonstrated, indicating that the LHR represents a functional domain. Transient transfection of COS cells with a full-length cDNA construct caused the expression of a recombinant receptor having the same size as that of the A allotype of erythrocytes. This recombinant receptor mediated the formation of rosettes between the COS cells and sheep erythrocytes bearing C3b or C4b and served as a cofactor for the cleavage of C3b to iC3b by factor I. By analysis of the C3b and C4b binding function of deletion mutants of recombinant CR1, two distinct ligand binding sites for C3b were found to reside in LHR-B and -C, respectively, and an additional site that preferentially bound C4b was demonstrated in LHR-A; LHR-D had no C3 or C4 binding function. Only the recombinant deletion mutants having LHR-B or -C promoted the conversion of C3b to iC3b by factor I, indicating that the sites in CR1 that mediate binding of C3b also are involved for the cofactor function of the receptor. The specificities for C3b and C4b required the two NH_2-terminal SCRs of these LHRs. These functional studies are especially interesting when the primary structure of CR1 is considered. The amino acid sequences of LHR-A and -B differ only in the two NH_2-terminal SCRs; this region of LHR-C resembles that of LHR-B rather than LHR-A, and the nonbinding LHR-D differs from LHR-A, -B, and -C.

The low affinity of CR1 for monomeric C3b indicates that multipoint attachment between ligand and receptor is necessary for cellular uptake of immune complexes. Multivalent interaction is facilitated by two characteristics of CR1: the clustering of CR1 at the plasma membrane (FEARON et al. 1981; ABRAHAMSON and FEARON 1983; CHEVALIER and KAZATCHKINE 1989) and the presence of three binding sites per receptor, two for C3b and one for C4b. The separation of the binding sites by five SCRs, or approximately 14 nm, may permit the folding of the receptor during the process of multivalent attachment. Folding or "kinking" of the molecule has been observed in electron micrographs of complexes between CR1 and a monoclonal antibody, Yz-1, bivalently interacting with two epitopes within a single receptor (BARTOW et al. 1989).

Comparison of CR1 with two structurally and functionally related plasma proteins, factor H and C4-bp, are informative. Previous analyses of proteolytic fragments of these plasma proteins had suggested that their ligand binding sites resided in the NH_2-termini (ALSENZ et al. 1985; KRISTENSEN et al. 1987); the CR1 studies extend these findings by localizing the C3b/C4b binding sites to the first two SCRs of the tandemly repeated LHRs. Second, like CR1, C4-bp is multivalent, although it achieved this function by a posttranslational aggregation at the COOH-termini of identical subunits. Thirdly, whereas the regulatory C4b and C3b specificities reside in two distinct plasma proteins, both functions are present in CR1, perhaps reflecting the need to bind immune complexes containing both C4b and C3b.

1.2 Allotypes of CR1

CR1 is unusual in having four allotypic forms that differ in molecular weight (Table 1; DYKMAN et al. 1983, 1984, 1985; WONG et al. 1983; VAN DYNE et al. 1987). In addition to the common A and B allotypes of M_r 250000 and 290000 that occur with frequencies of 0.82 and 0.18, there exist two rare allotypes, C and D. Several observations suggest that the allotypes differ by the presence of variable numbers of LHRs. First, the allotypes differ in the sizes of their respective polypeptides (WONG et al. 1983; LUBLIN et al. 1986). Second, they differ by increments of approximately 40000 M_r, the calculated size of an LHR. Third, the mRNAs for the different allotypes differ by increments of 1.3–1.4 kb, and an LHR is encoded by 1.35 kb (WONG et al. 1986; HOLERS et al. 1987). Fourth, the HPLC profile of tryptic peptides of ^{125}I-labeled CR1 allotypes are indistinguishable (NICKELLS et al. 1986). Fifth, the B allele has an additional BamHI restriction fragment (WONG et al. 1986, 1989), and Southern blot analysis with a noncoding probe derived from this BamHI fragment hybridizes not only to this fragment but to two other fragments that are also present in individuals homozygous for the A allotype. Thus, an intervening sequence in a B allele-specific region is highly homologous to sequences in two other regions of the CR1 gene, indicating that the CR1 gene is comprised of highly conserved repetitive regions and that the B allele arose from the A allele by the duplication of such a genomic segment.

Mapping studies of the CR1 gene from an individual homozygous for the B allele support this proposal (WONG et al. 1989). The gene encoding the B allele is divided into nine regions extending over 160 kb. The 5' region contains untranslated and

Table 1. Characteristics of CR1 Allotypes

Nomenclature	M_r		mRNA (kb)		Frequency
F[1] (A)[2]	250000[3]	(190000)[4]	7.9, 9.2[5]	(8.6, 10.0)[6]	.82
S (B)	290000	(220000)	9.2 10.7	(10.0, 11.6)	.18
F (C)	210000	(160000)	6.5, 7.9	(7.3, 8.6)	<.01
(D)		(250000)		(11.6, 12.8)	<.01

[1] Nomenclature of WONG, W. W., and FEARSON, D. T., 1983,
[2] Nomenclature of DYKMAN, T. R. et al. 1983,
[3] The calculated molecular weight of the unglycosylated polypeptide of this allotype is 220000,
[4] M_r of unreduced CR1,
[5] From WONG, W. W. et al. 1986,
[6] From HOLERS, V. M. et al. 1987

leader sequences and is followed downstream by five regions of 18–30 kb having exons encoding the 28 SCRs comprising the LHRs. A short region contains the two 3′ SCRs that lie outside the LHRs and segments having exons encoding the transmembrane and cytoplasmic regions and the 3′ untranslated sequences. Four of the five genomic regions corresponding to the LHRs are identified by hybridization with cDNA probes specific for LHR-A, -B, -C, and -D, respectively. The fifth region of 20 kb resides between the LHR-A and -B genomic segments and is absent from overlapping genomic clones from a library prepared with DNA from a homozygote for the A allotype. The restriction map of the 5′ half of this genomic region is the same as that of the 5′ half of the LHR-B genomic region, and the 3′ half is identical to the corresponding portion of the LHR-A genomic region, suggesting that this genomic segment encodes a chimeric LHR. This B allotype has the predicted structure, LHR-A, -B/A, -B, -C, -D, and the gene may have arisen through an homologous recombination event involving the LHR-A and -B regions. The reciprocal allele of this crossover would have only three LHRs with the composition, LHR-A, -B/A, -D, and may encode the short C allotype having an M_r of 200000. This hypothesis would predict the presence of three and one C3b binding sites, respectively, in the B and C allotypes.

At least 42 exons are present in the CR1-B allele, including those encoding the leader sequence, the transmembrane and cytoplasmic sequences, and most of the SCRs. The second and perhaps the sixth SCR of each LHR is encoded by two exons (HOURCADE et al. 1988; WONG et al. 1989), as are the second SCRs of factor H (VIK et al. 1987, 1988) and C4-bp (BARNUM et al. 1987).

The human gene for CR1 is linked to the genes for CR2, decay-accelerating factor (DAF), C4-bp, factor H, and membrane cofactor protein (MCP) in the regulator of complement activation (RCA) gene cluster at band q32 of chromosome 1 (RODRIGUEZ DE CORDOBA et al. 1984, 1985; WEIS et al. 1987; REY-CAMPOS et al. 1987; LUBLIN et al. 1987). Physical maps of the RCA region have placed the genes in the order, MCP, CR1, CR2, DAF, and C4-bp (REY-CAMPOS et al. 1988; CARROLL et al. 1988; BORA et al. 1989). The H gene maps to a locus 6.9 cM from the C4-bp/CR1 locus (RODRIGUEZ DE CORDOBA and RUBENSTEIN 1987).

1.3 Biosynthesis of CR1

In man the following cells express CR1: erythrocytes, B-lymphocytes, a subset of T-lymphocytes, monocytes, neutrophils, eosinophils, glomerular podocytes, and follicular dendritic cells (FEARON 1979, 1980; WILSON et al. 1983; REYNES et al. 1985; GELFAND et al. 1975; KAZATCHKINE et al. 1982). Low concentrations of soluble, apparently intact and functional CR1 are also present in plasma (YOON and FEARON 1985). The number of receptors varies among these cell types, with peripheral blood neutrophils, monocytes, and B-lymphocytes having 20000–40000 CR1/cell, T-lymphocytes having one-tenth this number and erythrocytes having only 100–1000 CR1/cell. The CR1 number on erythrocytes is genetically regulated (WILSON et al. 1982) and correlates with a restriction fragment length polymorphism involving allelic HindIII fragments of 7.4 and 6.9 kb (WILSON et al. 1986), the latter fragment being caused by a new restriction site generated by a single base change in an intron located in the LHR-D region of the gene (WONG et al. 1989). Individuals homozygous for the 7.4-kb HindIII fragment have an average of 661 sites per cell to which monoclonal anti-CR1 binds; persons homozygous for the 6.9-kb HindIII fragment have 156 sites per cell; and heterozygotes have an intermediate number. There is no comparable variation in CR1 number on neutrophils or B-lymphocytes, although unequal expression of CR1 structural allotypes in these cells does occur.

In HL-60 cells (ATKINSON and JONES 1984) and B lymphoblastoid cells (LUBLIN et al. 1986) each allotype of CR1 is initially synthesized as a precursor molecule having high-mannose, N-linked oligosaccharides that are converted to complex-type oligosaccharides. Approximately six to eight oligosaccharide moieties are present per receptor, although there are 21 potential N-glycosylation sites in the A allotype (KLICKSTEIN et al. 1988; WONG et al. 1983; LUBLIN et al. 1986). The receptor lacks O-linked oligosaccharides (LUBLIN et al. 1986). Nonglycosylated CR1 may have diminished ligand binding activity and apparently is inserted into the plasma membrane with decreased efficiency (LUBLIN et al. 1986). The small M_r difference between CR1 of neutrophils and other cell types (DYKMAN et al. 1983) has been attributed to unspecified differences in glycosylation. Biosynthesis of CR1 has been shown to occur also in mature peripheral neutrophils (JACK and FEARON 1988) and may indicate a capacity for an adaptive response of these cells to their microenvironment (NEUMANN and JACK 1989).

During ontogeny, the B-lymphocyte gradually acquires the ability to express CR1: 15% of large pre-B cells, 35% of small pre-B cells, 60%–80% of immature B cells having membrane IgM but not IgD, and essentially all mature B cells express this receptor (TEDDER et al. 1983). Differentiation into plasma cells is associated with loss of CR1. The uniform expression by mature B cells suggests some critical role that has not been fully defined as yet.

CR1 is expressed relatively late in development of myelomonocytic cells (TEDDER et al. 1983). All segmented neutrophilic leukocytes and 75% of band forms, but only 4% of metamyelocytes and less than 1% of myelocytes, have the receptor. Thus, CR1 expression does not occur until neutrophil development into a competent phagocyte.

During ontogeny of the glomerulus, CR1 appears as early as the late S-body stage of nephron differentiation on the basolateral part of primitive podocytes (APPAY et al.

1985). The function of CR1 in the kidney remains unknown, and the receptor may not be essential since it is not present in nonprimate glomeruli; thus, the consequences of its early appearance in the human kidney are not known.

2 Complement Receptor Type 2 (CR2; CD21)

2.1 Structure of CR2

The primary structure of two forms of CR2 has been inferred from the analysis of nucleotide sequences of cDNA clones isolated from human tonsillar (WEIS et al. 1986, 1988) and Raji lymphoblastoid (MOORE et al. 1987) cDNA libraries (Fig. 1). Four of the five tonsillar cDNA clones specified a peptide of 1032 amino acids having a M_r of 112716, which is similar to the estimated M_r of 111000 for nonglycosylated CR2 synthesized by the SB lymphoblastoid cell line in the presence of the inhibitor tunicamycin (WEIS and FEARON 1985). This peptide is comprised of a 20 amino acid signal peptide, a 954-residue extracellular domain, a 24 amino acid putative transmembrane region, and a 34 amino acid cytoplasmic domain. The extracellular domain is comprised entirely of 15 tandem SCR sequences homologous to those described in CR1 and other C3/C4 binding proteins. One of the five clones isolated from the tonsillar library and the single clone obtained from Raji cells encoded a 16th SCR, termed SCR-10a, between SCR-10 and -11 of the 15-SCR form of CR2. It is most likely that the presence or absence of SCR-10a is a consequence of alternative splicing. The fully processed mRNA is 5 kb (WEIS et al. 1986), and the CR2 gene spans 25 kb (WEIS et al. 1988; FUJISAKU et al. 1989) within the RCA gene cluster located on chromosome 1, band q32 (WEIS et al. 1987). The CR2 gene is unusual in having single exons encoding SCR-1 and -2, SCR-5 and -6, SCR-9 and -10, and SCR-12 and -13, respectively (HOLERS 1989). In contrast, in the genes for CR1, factor H and C4-bp, all SCRs are encoded either by single or two exons.

Posttranslational modifications of CR2 in B cells include glycosylation (WEIS and FEARON 1985) and probable serine/threonine phosphorylation after activation of protein kinase C (CHANGELIAN and FEARON 1986). Analysis of the sequence of CR2 has identified 11 Asn-X-Ser/Thr sites for potential glycosylation within the extracellular domain of the 15 SCR form and two additional sites within SCR 10a, consistent with the 8–11 N-linked oligosaccharides predicted from biosynthetic studies CR2 (WEIS and FEARON 1985). The cytoplasmic domain contains three serine, two threonine, and four tyrosine residues. The sequence TSQK in which a basic amino acid is carboxy-terminal to the serine/threonine, is the best condidate for a protein kinase C substrate (KISHIMOTO et al. 1985). Although tyrosine phosphorylation has not been demonstrated as yet for CR2, the presence of four tyrosines in the cytoplasmic domain, which are conserved in murine CR2 (FINGEROTH et al. 1989) (Fig. 2), compared to none in this region of CR1, which is 35% homologous to the cytoplasmic region of CR2, suggests strongly the possibility of some function for these residues. The presence of the sequence EAREVY in which a tyrosine is carboxy-terminal to two acidic residues is a pattern that resembles the consensus sequence for tyrosine kinase substrates (HUNTER and COOPER 1985).

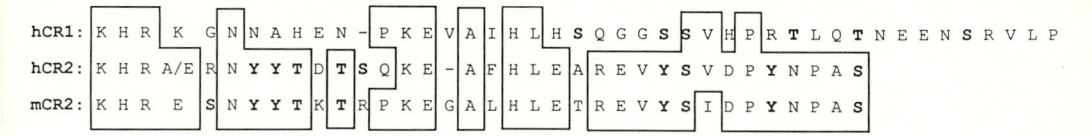

```
hCR1: K H R  K  G N N A H E N - P K E V A I H L H S Q G G S S V H P R T L Q T N E E N S R V L P
hCR2: K H R A/E R N Y Y T D T S Q K E - A F H L E A R E V Y S V D P Y N P A S
mCR2: K H R  E  S N Y Y T K T R P K E G A L H L E T R E V Y S I D P Y N P A S
```

Fig. 2. The amino acid sequences of human CR1 (*hCR1*), human (*hCR2*), and murine CR2 (*mCR2*) with identical residues enclosed within *boxes*. Serine, threonine, and tyrosine residues are shown in *bold type*. Different cDNA clones from tonsillar and Raji libraries have specified alanine and glutamic acid, respectively, at the fourth position

There are no LHRs within CR2 as there are in CR1, however dot matrix analysis reveals a less conserved repeating pattern of homology involving every fifth SCR such that 1, 5, 9, and 12 are homologous, as are 2, 6, 10, 13; 3, 7, 10a, 14; and 4, 8, 11. SCR 15 is unique and not part of this repeating pattern. Based upon the shared SCR motif, the tertiary structure of CR2 would be expected to consist of an extended array of triple loops maintained by disulfide bonds, similar to that predicted for CR1 (Fig. 1).

2.2 Biosynthesis of CR2

CR2 resides primarily on mature B-lymphocytes (Ross et al. 1973; Eden et al. 1973; Iida et al. 1983; Weis et al. 1984; Nadler et al. 1981; Tedder et al. 1984), although human thymocytes (Tsoukas and Lambris 1988), rare T lymphoblastoid cell lines such as Molt-4, Jurkat, and HPB-ALL (Menezes et al. 1977; Tatsumi et al. 1985; Fingeroth et al. 1988) pharyngeal (Young et al. 1986) and cervical (Sixbey et al. 1987) epithelium, and follicular dendritic cells (Reynes et al. 1985) have also been found to have CR2. B cells and Epstein-Barr virus (EBV) infected Burkitt B lymphoma cell lines express 24000–63000 receptors per cell as compared with the T-cell leukemia line Molt-4 which expresses only 8000 per cell (Fingeroth et al. 1984). Burkitt lymphoma cells not infected with EBV characteristically express low numbers of CR2, but following infection with the virus, cells increase receptor expression to the range of mature B cells. The single report of low CR2 expression on human platelets (Nunez et al. 1987) has not been consistently observed by other investigators (Vik and Fearon 1987, Yu et al. 1986).

Ontogenic analysis of CR2 expression during human B-cell differentiation demonstrated that the receptor became increasingly evident during development from the pre-B cell to the mature B cell (Tedder et al. 1984). Approximately half of the B cells in adult bone marrow expressed CR2, as did 76% and 95% of newborn and adult peripheral blood B cells, respectively, and 72%–97% of the B cells in adult spleen, tonsil, and lymph node. Differentiation into plasma cells was associated with the loss of CR2 expression. Thus, CR2 expression is most characteristic of the mature B-lymphocyte.

The expression of CR2 on human epithelium is also developmentally regulated (Young et al. 1986). Only 1%–9% of primary cultures of ectocervical epithalial cells were CR2 positive (Sixbey et al. 1986), and ultrastructural studies suggested that these cells were metabolically active and less differentiated. Similarly, stratified squamous epithelia from human tongue and extocervix expressed CR2 exclusively in the less differentiated basal and parabasal layers, with none being detectable in the most superficial, differentiated layers (Young et al. 1986; Sixbey et al. 1987). In contrast, epithelia from the soft palate, oropharynx, inner cheek, and Rosenmüller's fossa, from which EBV-infected nasopharyngeal carcinoma is thought to arise, express CR2 in all strata. Generally, CR2 appeared to be present in greater amounts in stratified squamous tissue than in the pseudostratified columnar epithelium of nasal and tracheal sites in which only basal cells were positive.

Biosynthetic studies of B lymphoblastoid CR2 demonstrated the presence of a precursor of M_r 134000 containing high-mannose n-linked oligosaccharides sensitive to endoglycosidase H that was processed to the mature M_r 145000 form of CR2 containing complex-type N-linked oligosaccharides (Weis and Fearon 1985). In the presence of tunicamycin, a nonglycosylated polypeptide of M_r 111000 was synthesized which was expressed at a low level at the plasma membrane. The nonglycosylated form of CR2 underwent accelerated catabolism, having a half-life of 2.8 h rather then 13.8 h as found for normally glycosylated CR2. Glycosylation was not required for receptor-ligand interaction since all three forms of CR2 were capable of binding to C3 Sepharose.

3 Immunoregulatory Functions of CR1 and CR2

Experiments performed in whole animals or in man suggest that the complement system may be able to alter the humoral immune response. For example, Klaus and Humphrey (1977) demonstrated that depletion of C3 from mice greatly impaired their capacity to generate memory B cells. Bottger and his colleagues (1985) have shown that guinea pigs genetically deficient in either C4 or C2 required higher doses of a T-dependent antigen for priming for a secondary immune response. A similar result has been reported for dogs lacking C3 (O'Neil et al. 1988) and a human lacking C4 (Jackson et al. 1979). The ability of IgM to enhance the primary immune response to its antigen is dependent on C3 and the complement-activating function of the IgM. The molecular and cellular mechanisms accounting for these findings are not known, although it seems likely that they involve C3 and C4, which can covalently attach to antigen and antibody and mediate their uptake lymphocytes and macrophages through interaction with the complement receptors, CR1, CR2, and CR3. Therefore, this section will discuss possible functions of CR1 and CR2 in regulating the functions of lymphocytes.

Cross-linking CR1 of human B cells does not induce a proliferative or differentiation response, nor does it cause the release of intracellular calcium. However, some signal may be presumed to occur based on the formation of caps; CR1 also co-caps with cross-linked membrane immunoglobulin when ligated with monoclonal anti-CR1 (Tsokos et al. 1988). Perhaps more relevant to the in vivo effects of the complement system on the immune response are the findings of enhanced antibody production in vitro. In a 6-day culture of human peripheral blood B and T cells stimulated with suboptimal concentrations of pokeweed mitogen, the presence of rabbit $F(ab')_2$

anti-CR1 augmented production of IgM, IgG, and IgA (DAHA et al. 1984). The augmenting effect was greater when the assay was performed with B cells having relatively higher as compared to lower numbers of CR1 sites, and when bivalent rather than monovalent Fab' anti-CR1 fragments were used. No effect was observed when optimal doses of pokeweed mitogen were employed in this study or in another in which monoclonal anti-CR1 was used (TEDDER et al. 1986). In an antigen-driven system, analogous results have been reported (WEISS et al. 1987). Production of anti-TNP by human peripheral blood B cells in the presence of T cells or T-cell derived soluble factors and suboptimal concentrations of trinitrophenyl-polyacrylamide beads was enhanced two- to fourfold by low concentrations of monoclonal anti-CR1. No augmentation was observed if optimal concentrations of antigen were present. The results of these studies resemble those of the in vivo studies of the C4- or C2-deficient guinea pigs in that the complement system had an accessory, but not obligatory or even sufficient, role which was most evident when antigen or mitogen was relatively limited.

Studies of the function of CR2 have been more extensive and have suggested that CR2 can modulate the growth of B-lymphocytes. Lipopolysaccharide-activated murine spleen cells enriched for B-cell blasts were stimulated to proliferate and mature into immunoglobulin-secreting cells by culture with Sepharose-bound or glutaraldehyde cross-linked human C3 or C3d (ERDEI et al. 1985; MELCHERS et al. 1985). Soluble monomeric C3 was inhibitory, and resting, as contrasted to lipopolysaccharide-activated, spleen cells were refractory to stimulation. These observations suggested that signaling required either that the cell be preactivated, or that CR2 itself must be in an "activated state " for it to be capable of driving B-cell proliferation. These possibilities also may apply to the more recent findings that C3dg and certain monoclonal anti-CR2 antibodies have growth-stimulating effects upon low-density cultures of Raji cells grown in a serum-free defined medium (HATZFELD et al. 1988; PERNEGGER et al. 1988), and that polymerized C3dg or monoclonal anti-CR2 induces proliferation of human B cells that are been activated with phorbol esters, but not of resting B cells (BOHNSACK and COOPER 1988). The phosphorylation of CR2 occurring with phorbol ester treatment of B cells (CHANGELIAN and FEARON 1986) may be related to the acquisition of mitogenic function by the receptor. In general, these findings that CR2 on resting B cells does not transmit a mitogenic signal are biologically reasonable since polyclonal B-cell expansion with each episode of complement activation would be deleterious. Earlier observations that several monoclonal anti-CR2 antibodies would drive B cells from G_0 through upper case S phase have not been reproducible (NEMEROW et al. 1985a; MITTLER et al. 1983; WILSON et al. 1985).

If CR2 alone is not capable of causing the progression of B cells through the cell cycle, a physiologic costimulating receptor would be membrane Ig (mIg) since this receptor confers specificity for antigen on the reaction. Furthermore, the capacity of C3 to attach covalently to immune complexes containing antigen would lead to the formation of complexes having ligands for both CR2 and mIg that would be capable of cross-linking these receptors. In support of the possibility that such complexes might have important functional properties is the finding that CR2 and mIgM synergistically increase the free intracellular calcium concentration when cross-linked with multivalent ligands (CARTER et al. 1988). Tonsillar B cells were loaded with the calcium-sensitive fluorescent probe indo-1; mIgM, CR1 or CR2 were cross-linked by monoclonal antibodies individually or in combination and fluorescence was monitored by flow cytometry. Although aggregation of CR1 or CR2 alone did not release intracullular calcium, cross-linking CR2 to mIgM lowered the concentration of anti-IgM necessary for an optimal B cell response by approximately 100-fold. CR1 was without effect. Thus, as was observed in the in vivo experiments assessing roles for complement in the immune response, an accessory, antigen-dependent function exists for CR2 in this in vitro system. The presence of CR1, CR2, and CR3 on follicular dendritic cells (REYNES et al. 1985) may indicate that immune complexes bearing C3 fragments is retained at these sites for interaction with antigen-specific memory B cells (KLAUS and HUMPHREY 1987) by a synergistic mechanism involving CR2 on the responding cells.

4 Determination of Epstein-Barr Virus Tissue Tropism by CR2

EBV, an oncogenic herpesvirus which causes acute infectious mononucleosis and is implicated in the pathogenesis of Burkitt's lymphoma and nasopharyngeal carcinoma, is capable of infecting and immortalizing human B-lymphocytes in vitro. Several observations have led to the conclusion that CR2 is the receptor for EBV. Treatment of peripheral blood B-lymphocytes with antibodies to CR2 blocks EBV binding and infection (FINGEROTH et al. 1984; NEMEROW et al. 1985b; FRADE 1986). Second, purified CR2 binds EBV (FINGEROTH et al. 1984; MOLD et al. 1986), and recombinant or purified gp350/220 membrane antigen of EBV binds CR2 (TANNER et al. 1987; NEMEROW et al. 1987). The tissue distribution of CR2 correlates with the tissue tropism of EBV: B-lymphocytes and certain epithelial cells (SIXBEY et al. 1987). Therefore, the role of CR2 in EBV infection clearly included the cellular uptake of the virus.

A comparison of the gp350/220 and C3d amino acid sequences has identified three regions of homology. One of these, which resides at the N-terminus of gp350/220 (EDPGFFNVE), corresponds to the region of C3d (EDPGKQLYNVE) that was predicted to bind CR2 (LAMBRIS et al. 1985). Soluble gp350/220 deletion mutants (TANNER et al. 1988) and synthetic peptides (NEMEROW et al. 1989) have recently been employed to demonstrate that this is the primary CR2 binding site for EBV. Despite the sequence homology between gp350/220 and C3d and the capacity of the monoclonal OKB-7 to block binding by B cells of C3d and EBV (RAO et al. 1985; NEMEROW et al. 1985a), BAREL et al. (1988) have recently described a panel of nine IgG1 monoclonal antibodies, of which four blocked 18%–64% of EBV binding to Raji cells but had no effect upon C3d binding and four others blocked 58%–90% of C3d binding but did not effect EBV binding. Analysis of the ligand binding sites of CR2 by site-directed and deletion mutagenesis, as has been accomplished for CR1, will provide more definitive evidence for the location of these binding sites.

These studies had established that CR2 binds EBV but did not determine whether cellular expression of CR2 was sufficient for infection. This question was addressed in part by stably transfecting murine L cells with pMT.CR2.neo.1, a eukaryotic expression construct containing a full-length human CR2 cDNA encoding the 15-SCR form of the receptor downstream from a mouse metallothionein I promoter (AHEARN et al. 1988). Transfectants expressed high levels of CR2, having an M_r identical to that of wild-type CR2 and specifically bound both C3d and EBV. Coculture of these cells with EBV led to infection of approximately 0.5% of the cells, as assessed by staining for Epstein-Barr nuclear antigen (EBNA). The infection appeared to be latent, as no fluorescent staining of cells was observed with monoclonal antibodies to the early antigen complex or to gp350/220. The number of infected foci remained constant in a particular culture over time whereas the number of EBNA-positive cells per focus approximately doubled with each cell division of the L cells. The finding that expression of recombinant human CR2 renders murine L cells susceptible to stable, latent infection by EBV indicates that this single membrane protein can serve as one of the determinants of the narrow tissue tropism of this virus. However, the low efficiency of infection observed in the CR2-positive L cells as compared with human B lymphoblastoid cell lines indicates that additional factors are involved in determining EBV tropism.

References

Abrahamson DR, Fearon DT (1983) Endocytosis of the C3b receptor of complement within coated pits in human polymorphonuclear leukocytes and monocytes. Lab Invest 48: 162–168

Ahearn JM, Hayward SD, Hickey JC, Fearon DT (1988) Epstein-Barr virus (EBV) infection of murine L cells expressing recombinant human EBV/C3d receptor. Proc Natl Acad Sci USA 85: 9307–9311

Alsenz J, Schulz TF, Lambris JD, Sim RB, Dierich MP (1985) Structural and functional analysis of the complement component factor H with the use of different enzymes and monoclonal antibodies to factor H. Biochem J 232: 841–850

Appay MD, Mounier F, Gubler MC, Rouchon M, Beziau A, Kazatchkine MD (1985) Ontogenesis of the glomerular C3b receptor (CR1) in fetal human kidney. Clin Immunol Immunopathol 37: 103

Atkinson JP, Jones EA (1984) Biosynthesis of the human C3b/C4b receptor during differentiation of the HL-60 cell line. Identification and characterization of a precursor molecule. J Clin Invest 74: 1649

Barel M, Fiandino A, Delcayre AX, Lyamani F, Frade R (1988) Monoclonal and anti-idiotypic anti-EBV/C3d receptor antibodies detect two binding sites, one for EBV and one for C3d on glycoprotein 140, the EBV/C3dR, expressed on human B lymphocytes. J Immunol 141: 1590–1595

Barnum S, Kenney J, Kristensen T, Noack D, Seldon M, D'Eustachio P, Chaplin D, Tack B (1987) Chromosomal location and structure of the mouse C4BP gene. Complement 4: 131

Bartow T, Klickstein L, Wong W, Roux K, Fearon D (1989) Analysis of the tertiary structure of CR1 by electron microscopy. FASEB J 3: A501

Bohnsack JF, Cooper NR (1988) CR2 ligands modulate human B cell activation. J Immunol 141: 2569–2576

Bora NS, Lublin DM, Kumar BV, Hockett RD, Holers VM, Atkinson JP (1989) Structural gene for human membrane cofactor protein (MCP) of complement maps to within 100 kb of 3' end of the C3b/C4b receptor gene. J Exp Med 169: 597–602

Bottger EC, Hoffman T, Hadding U, Bitter-Suermann D (1985) Influence of genetically inherited complement deficiencies on humoral immune response in guinea pigs. J Immunol 135: 4100–4107

Carroll MC, Alicot EA, Katzman PJ, Klickstein LB, Smith JA, Fearon DT (1988) Organization of the gene encoding complement receptors type 1 and 2, decay accelerating factor and C4-binding protein in the RCA locus on human chromosome 1. J Exp Med 167: 1271

Carter RH, Spycher MO, Ng YC, Hoffman R, Fearon DT (1988) Synergistic interaction between complement receptor type 2 and membrane IgM on B lymphocytes. J Immunol 141: 457–463

Changelian PS, Fearon DT (1986) Tissue-specific phosphorylation of complement receptors CR1 and CR2. J Exp Med 163: 101–115

Chevalier J, Kazatchkine MD (1989) Distribution in clusters of complement receptor type one (CR1) on human erythrocytes. J Immunol 142: 2031–2036

Cohen JHM, Aubry JP, Kazatchkine MD, Banchereau J, Revillard JP (1987) CR1-bearing human T cells are exclusively found with the CD4+ subset (Abstract). Complement 4: 143

Daha MR, Bloem AC, Ballieux RE (1984) Immunoglobulin production by human peripheral lymphocytes induced by anti-C3 receptor antibodies. J Immunol 132: 1197–1201

Dykman TR, Cole JL, Iida K, Atkinson JP (1983) Polymorphism of human erythrocyte C3b/C4b receptor. Proc Natl Acad Sci USA 80: 1698–1702

Dykman TR, Hatch JA, Atkinson JP (1984) Polymorphism of the human C3b/C4b receptor. Identification of a third allele and analysis of receptor phenotypes in families and patients with systemic lupus erythematosus. J Exp Med 159: 691–703

Dykman TR, Hatch JA, Aqua MS, Atkinson JP (1985) Polymorphism of the C3b/C4b receptor (CR1): characterization of a fourth allele. J Immunol 134: 1787–1789

Eden A, Miller GW, Nussenzweig V (1973) Human lymphocytes bear membrane receptors for C3b and C3d. J Clin Invest 52: 3239–3242

Emancipator SN, Iida K, Nussenzweig V, Gallo GR (1983) Monoclonal antibodies to human complement receptor CR1) detect defects in glomerular diseases. Clin Immunol Immunopathol 27: 170–175

Erdei A, Melchers F, Schulz T, Dierich M (1985) The action of human C3 in soluble or cross-linked form with resting and activated murine B lymphocytes. Eur J Immunol 15: 184–188

Fearon DT (1979) Regulation of the amplification C3 convertase of human complement by an inhibitory protein isolated from the human erythrocyte membrane. Proc Natl Acad Sci USA 76: 5867–5871

Fearon DT (1980) Identification of the membrane glycoprotein that is the C3b receptor of the human erythrocyte, polymorphonuclear leukocyte, B lymphocyte, and monocyte. J Exp Med 152: 20–30

Fearon DT, Collins LA (1983) Increased expression of C3b receptors on polymorphonuclear leukocytes induced by chemotactic factors and by purification procedures. J Immunol 130: 370–375

Fearon DT, Kaneko I, Thomson GG (1981) Membrane distribution and adsorptive endocytosis by C3b receptors on human polymorphonuclear leukocytes. J Exp Med 153: 1615–1628

Fingeroth JD, Weis JJ, Tedder TF, Strominger JL, Biro PA, Fearon DT (1984) Epstein-Barr virus receptor of human B lymphocytes is the C3d receptor CR2. Proc Natl Acad Sci USA 81: 4510–4514

Fingeroth JD, Clabby ML, Strominger JD (1988) Characterization of a T-lymphocyte Epstein-Barr virus/C3d receptor (CD21). J Virol 62: 1442–1447

Fingeroth JD, Benedict MA, Levy DN, Strominger JL (1989) Identification of murine complement receptor type 2- Proc Natl Acad Sci USA 86: 242–246

Frade R (1986) Structure and functions of gp 140, the C3d/EBV receptor (CR2) of human B lymphocytes. Mol Immunol 23: 1249–1253

Frade R, Crevon MC, Barel M, Vazquez A, Krikorian L, Charriaut C, Galanaud P (1985) Enhancement of human B cell proliferation by an antibody to the C3d receptor, the gp 140 molecule. Eur J Immunol 15: 73–76

Fujisaku A, Harley JB, Frank MB, Gruner BA, Frazier B, Holers VM (1989) Genomic organization and polymorphisms of the human C3d/Epstein-Barr virus receptor. J Biol Chem 264: 2118–2125

Gelfand MC, Fran MM, Green I (1975) A receptor for the third component of complement in the human renal glomerulus. J Exp Med 142: 1029–1034

Hatzfeld A, Fischer E, Levesque J-P, Perrin R, Hatzfeld J, Kazatchkine MD (1988) Binding of C3 and C3dg to the CR2 complement receptor induces growth of an Epstein-Barr virus-positive human B cell line. J Immunol 140: 170–175

Holers VM, Chaplin DD, Leykam JF, Gruner BA, Kumar V, Atkinson JP (1987) Human complement C3b/C4b receptor (CR1) mRNA polymorphism that correlates with the CR1 allelic molecular weight polymorphism. Proc Natl Acad Sci USA 84: 2459

Hourcade D, Miesner DR, Atkinson JP, Holers VM (1988) Identification of an alternative polyadenylation site in the human C3b/C4b receptor (complement receptor type 1) transcriptional unit and prediction of a secreted form of complement receptor type 1. J Exp Med 168: 1255–1270

Hunter T, Cooper JA (1985) Protein-tyrosine kinases. Annu Rev Biochem 54: 897–930

Iida K, Nadler L, Nussenzweig V (1983) Identification of the membrane receptor for the complement fragment C3d by means of a monoclonal antibody. J Exp Med 158: 1021–1033

Jack RM, Fearon DT (1988) Selective synthesis of mRNA and protein by human peripheral blood neutrophils. J Immunol 140: 4286–4293

Jackson CG, Ochs HD, Wedgwood RJ (1979) Immune response of a patient with deficiency of the fourth component of complement and systemic lupus erythematosus. N Engl J Med 300: 1124–1129

Janatova J, Reid KBM, Willis AC (1988) Involvement of disulfide bonds in the structure of complement regulatory protein C4bp. FASEB J 2: A1832

Kazatchkine MD, Fearon DT, Appay MD, Mandet C, Bariety J (1982) Immunohistochemical study of the human glomerular C3b receptor in normal kidney and in seventy-five cases of renal diseases. J Clin Invest 69: 900–912

Kishimoto A, Nishiyama K, Nakanishi H, Uratsuji Y, Nomura H, Takeyama Y, Nishizuka Y (1985) Studies on the phosphorylation of myelin basic protein by protein kinase C and adenosine 3′:5′-monophosphate-dependent protein kinase. J Biol Chem 260: 12492–12499

Klaus GGB, Humphrey JH (1977) The generation of memory cells. I. The role of C3 in the generation of B memory cells. Immunology 33: 31–40

Klickstein LB, Wong WW, Smith JA, Weis JH, Wilson JG, Fearon DT (1987) Human C3b/C4b receptor (CR1): demonstration of long homologous repeating domains that are composed of the short consensus repeats characteristics of C3/C4 binding proteins. J Exp Med 165: 1095–1112

Klickstein LB, Bartow TJ, Miletic V, Rabson LD, Smith JA, Fearon DT (1988) Identification of distinct C3b and C4b recognition sites in the human C3b/C4b receptor (CR1, CD35) by deletion mutagenesis. J Exp Med 168: 1699–1717

Kristensen T, D'Eustachio P, Ogata RT, Chung LP, Reid KBM, Tack BF (1987) The superfamily of C3b/C4b binding proteins. Fed Proc 46: 2463–2469

Lambris JD, Ganu VS, Hirani S, Müller-Eberhard HJ (1985) Mapping of the C3d receptor (CR2)-binding site and a neoantigenic site in the C3d domain of the third component of complement. Proc Natl Acad Sci USA 82: 4235–4239

Lozier J, Takahaski N, Putnam FW (1984) Complete amino acid sequence of human plasma β_2-glycoprotein I. Proc Natl Acad Sci USA 81: 3640–3645

Lublin DM, Griffith RC, Atkinson JP (1986) Influence of glycosylation on allelic and cell-specific M_r variation, receptor processing, and ligand binding of the human complement C3b/C4b receptor. J Biol Chem 261: 5736

Lublin DM, Lemona RS, LeBeau MM, Holers VM, Tykocinski ML, Medof ME, Atkinson JP (1987) The gene encoding decay-accelerating factor (DAF) is located in the complement-regulatory locus on the long arm of chromosome 1. J Exp Med 165: 1731

Melchers F, Erdei A, Schulz T, Dierich MP (1985) Growth control of activated, synchronized murine B cells by the C3d fragment of human complement. Nature 317: 264–267

Menezes J, Seigneurin JM, Patel P, Bourkas A, Lenoir G (1977) Presence of Epstein-Barr virus receptors, but absence of virus penetration, in cells of an Epstein-Barr virus genome-negative human lymphoblastoid T line (Molt4). J Virol 22: 816–821

Mittler RS, Talle Ma, Carpenter K, Rao PE, Goldstein G (1983) Generation and characterization of monoclonal antibodies reactive with human B lymphocytes. J Immunol 131: 1754–1761

Mold C, Cooper NR, Nemerow GR (1986) Incorporation of the purified Epstein-Barr virus/C3d receptor (CR2) into liposomes and demonstration of its dual ligang binding functions. J Immunol 136: 4140–4145

Moore MD, Cooper NR, Tack BF, Nemerow GR (1987) Molecular cloning of the cDNA encoding the Epstein-Barr virus/C3d receptor (complement receptor type 2) of human B lymphocytes. Proc Natl Acad Sci USA 84: 9194–9198

Nadler LM, Stashenko, P, Hardy R, van Agthoven A, Terhorst C, Schlossman SF (1981) Characterization of a human B cell-specific antigen (B2) distinct from B1. J Immunol 126: 1941–1947

Nemerow GR, McNaughton ME, Cooper NR (1985a) Binding of monoclonal antibody to the Epstein-Barr virus (EBV)/CR2 receptor induces activation and differentiation of human B lymphocytes. J Immunol 135: 3068–3073

Nemerow GR, Wolfert R, McNaughton ME, Cooper NR (1985b) Identification and characterization of the Epstein-Barr virus receptor on human B lymphocytes and its relationship to the C3d complement receptor (CR2). J Virol 55: 347–351

Nemerow GR, Mold C, Schwend VK, Tollefson V, Cooper NR (1987) Identification of gp350 as the viral glycoprotein mediating attachment of Epstein-Barr virus (EBV) to the EBV/C3d receptor of B cells: sequence homology of gp350 and C3 complement fragment C3d. J Virol 61: 1416–1420

Nemerow GR, Houghten RA, Moore MD, Cooper NR (1989) Identification of an epitope in the major envelope protein of Epstein-Barr virus that mediates viral binding to the B lymphocyte EBV receptor (CR2). Cell 56: 369–377

Neuman E, Jack RM (1989) Transcription and translation of CR1 and class I by neutrophils (PMN) increases with GM-CSF. FASEB J 3: A368

Nickells MW, Seya T, Holers VM, Atkinson JP (1986) Analysis of C3b/C4b receptor (CR1) polymorphic variants by tryptic peptide mapping. Mol Immunol 23: 661–668

Nunez D, Charriaut-Marlangue C, Barel M, Benveniste J, Frade R (1987) Activation of human platelets through gp140, the C3d/EBV receptor (CR2). Eur J Immunol 17: 515–520

O'Neil KM, Ochs HD, Heller SR, Cork LC, Morris JM, Winkelstein JA (1988) Role of C3 in humoral immunity. Defective antibody production in C3-deficient dogs. J Immunol 140: 1939–1945

Perkins SJ, Chung LP, Reid KB (1986) Unusual ultrastructure of complement-component-C4b-binding protein of human complement by synchrotron X-ray scattering and hydrodynamic analysis. Biochem J 233(3): 799–807

Pernegger G, Schulz TF, Hosp M, Myones BL, Petzer AL, Eigentler A, Bock G, Wick G, Dierich MP (1988) Cell cycle control of a Burkitt lymphoma cell line: responsiveness to growth signals engaging the C3d/EBV receptor. Immunology 65: 237–241

Rao PE, Wright SD, Westberg EF, Goldstein G (1985) OKB7, a monoclonal antibody that reacts at or near the C3d binding site of human CR2. Cell Immunol 93: 549–555

Rey-Campos J, Rubinstein P, Rodriguez de Cordoba S (1987) Decay-accelerating factor genetic polymorphism and linkage to RCA (regulator of complement activation) gene cluster in humans. J Exp Med 166: 246

Rey-Campos J, Rubinstein P, Rodriguez de Cordoba S (1988) A physical map of the human regulator of complement activation gene cluster linking the complement genes CR1, CR2, DAF and C4bp. J Exp Med 167: 664

Reynes M, Aubert JP, Cohen JHM, Audouin J, Tricotet V, Diebold J, Kazatchkine MD (1985) Human follicular dendritic cells express CR1, CR2 and CR3 complement receptor antigens. J Immunol 135: 2687–2694

Rodriguez de Cordoba S, Rubinstein P (1986) Quantitative variations of the C3b/C4b receptor (CR1) in human erythrocytes are controlled by genes within the regulator of complement activation (RCA) gene cluster. J Exp Med 164: 1274–1283

Rodriguez de Corboda S, Rubinstein P (1987) New alleles of C4-binding protein and factor H and further linkage data in the regulator of complement activation (RCA) gene cluster in man. Immunogenetics 25: 267

Rodriguez de Cordoba S, Dykman TR, Ginsberg-Fellner F, Ercilla G, Aqua M, Atkinson JP, Rubinstein P (1984) Evidence for linkage between the loci coding for the binding protein for the fourth component of human complement (C4BP) and for the C3b/C4b receptor. Proc Natl Acad Sci USA 81: 7890–7892

Rodriguez de Cordoba S, Lublin DM, Rubinstein P, Atkinson JP (1985) Human genes for three complement components that regulate the activation of C3 are tightly linked. J Exp Med 161: 1189–1195

Ross GD, Polley MJ, Rabellino EM, Grey HM (1973) Two different complement receptors on human lymphocytes. One specific for C3b and one specific for C3b inactivator-cleaved C3b. J Exp Med 138: 798–811

Sixbey JW, Lemon SM, Pagano JS (1986) A second site for Epstein-Barr virus shedding: the uterine cervix. Lancet 2: 1122–1124

Sixbey JW, Davis DS, Young LS, Hutt-Fletcher L, Tedder TF, Rickinson AB (1987) Human epithelial cell expression of an Epstein-Barr virus receptor. J Gen Virol 68: 805–811

Tanner, J, Weis J, Fearon D, Whang Y, Kieff E (1987) Epstein-Barr virus gp350/220 binding to CR2 on the B lymphocyte mediates adsorption, capping and endocytosis. Cell 50: 203–213

Tanner J, Whang Y, Sample J, Sears A, Kieff E (1988) Soluble gp 350/220 and deletion mutant glycoproteins block Epstein-Barr virus adsorption to lymphocytes. J Virol 62: 4452–4464

Tatsumi E, Harada S, Kuszynski C, Volsky D, Minowada J, Purtilo DT (1985) Catalogue of Epstein-Barr virus (EBV) receptors on human malignant and nonmalignant hematopoietic cell lines. Leuk Res 9: 231–238

Tedder TF, Fearon DT, Gartland GL, Cooper MD (1983) Expression of C3b receptors on human B cells and myelomonocytic cells but not natural killer cells. J Immunol 130: 1668–1673

Tedder TF, Clement LT, Cooper MD (1984) Expression of C3d receptors during human B cell differentiation: immunofluorescence analysis with the HB-5 monoclonal antibody. J Immunol 133: 678–683

Tedder TF, Weis JJ, Clement LT, Fearon DT, Cooper MD (1986) The role of receptors for complement in the induction of polyclonal B-cell proliferation and differentiation. J Clin Immunol 6: 65–73

Tsokos GC, Thyphronitis G, Jack RM, Finkelman FD (1988) Ligand-loadded but not free complement receptors for C3b/C4b and C3d co-cap with cross-linked B cell surface IgM and IgD. J Immunol 141: 1261–1266

Tsoukas CD, Lambris JD (1988) Expression of CR2/EBV receptor on human thymocytes detected by monoclonal antibodies. Eur J Immunol 18: 1299–1302

van Dyne S, Holers VM, Lublin DM, Atkinson JP (1987) The polymorphism of the C3b/C4b receptor in the normal population and in patients with systemic lupus erythematosus. Clin Exp Immunol 68: 570

Vik DP and Fearon DT (1987) Cellular distribution of complement receptor type 4 (CR4): expression on human platelets. J Immunol 138: 254–258

Vik DP, Keeney JB, Bronson S, Westlund B, Kristensen T, Chaplin DD, Tack BF (1987) Analysis of the murine factor H gene and related DNA. Complement 4: 235

Vik DP, Keeney JB, Chaplin DD, Tack BF (1988) Structure of the murine factor H gene. FASEB J 2: A1643

Weis JH, Morton CC, Bruns GAP, Weis JJ, Klickstein LB, Wong WW, Fearon DT (1987) A complement receptor locus: genes encoding C3b/C4b receptor and C3d/Epstein-Barr virus receptor map to 1q32. J Immunol 138: 312–315

Weis JJ, Fearon DT (1985) The identification of N-linked oligosaccharides on the human CR2/Epstein-Barr virus receptor and their function in receptor metabolism, plasma membrane expression, and ligand binding. J Biol Chem 260: 13824–13830

Weis JJ, Tedder TF, Fearon DT (1984) Identification of a 145000M$_r$ membrane protein as the C3d receptor (CR2) of human B lymphocytes. Proc Natl Acad Sci USA 81: 881–885

Weis JJ, Fearon DT, Klickstein LB, Wong WW, Richards SA, deBruyn Kops A, Smith JA, Weis JH (1986) Identification of a partial cDNA clone for the C3d/Epstein-Barr virus receptor of human B lymphocytes: homology with the receptor for fragments C3b and C4b of the third and fourth components of complement. Proc Natl Sci USA 83: 5639–5643

Weis JJ, Toothaker LE, Smith JA, Weis JH, Fearon DT (1988) Structure of the human B lymphocyte receptor for C3d and the Epstein-Barr virus and relatedness to other members of the family of C3/C4 binding proteins. J Exp Med 167: 1047–1066

Weiss L, Del Fraissy J-F, Vazquez A, Wallon C, Galanaud P, Kazatchkine MD (1987) Monoclonal antibodies to the human C3b/C4b receptor (CR1) enhance specific B cell differentiation. J Immunol 138: 2988–2993

Wilson BS, Platt JL, Kay NE (1985) Monoclonal antibodies to the 140000 mol wt glycoprotein of B lymphocyte membranes (CR2 receptor) initiates proliferation of B cells in vitro. Blood 66: 824–829

Wilson JG, Wong WW, Schur PH, Fearon DT (1982) Mode of inheritance of decreased C3b receptors on erythrocytes of patients with systemic lupus erythematosus. N Engl J Med 307: 981–986

Wilson JG, Tedder TF, Fearon DT (1983) Characterization of human T lymphocytes that express the C3b receptor. J Immunol 131: 684–689

Wilson JG, Murphy EE, Wong WW, Klickstein LB, Weis JH, Fearon DT (1986) Identification of a restriction fragment length polymorphism by a CR1 cDNA that correlates with the number of CR1 on erythrocytes. J Exp Med 164: 50–59

Wong WW, Wilson JG, Fearon DT (1983) Genetic regulation of a structural polymorphism of human C3b receptor. J Clin Invest 72: 685–693

Wong WW, Klickstein LB, Smith JA, Weis JH, Fearon DT (1985) Identification of a partial cDNA clone for the human receptor for complement fragments C3b/C4b. Proc Natl Acad Sci USA 82: 7711–7715

Wong WW, Kennedy CA, Bonaccio ET, Wilson JG, Klickstein LB, Weis JH, Fearon DT (1986) Analysis of multiple restriction fragment length polymorphisms of the gene for the human complement receptor type 1: duplication of genomic sequences occurs in association with a high molecular weight receptor allotype. J Exp Med 164: 1531–1546

Wong WW, Cahill JM, Rosen MD, Kennedy CA, Bonaccio ET, Morris MJ, Wilson JG, Klickstein LB, Fearon DT (1989) Structure of the human CR1 gene: molecular basis of the structural and quantitative polymorphisms and identification of a new CR1-like allele. J Exp Med 169: 847–864

Yoon SH, Fearon DT (1985) Characterization of a soluble form of the C3b/C4b receptor (CR1) inhuman plasma. J Immunol 134: 3332–3333

Young LS, Clark D, Sixbey JW, Rickinson AB (1986) Epstein-Barr virus receptors on human pharyngeal epithelia. Lancet 1: 240–242

Yu GH, Holers M, Seya T, Ballard L, Atkinson JP (1986) Identification of a third component-binding glycoprotein of human platelets. J Clin Invest 78: 494–501

The Leukocyte Cell Surface Receptor(s) for the iC3b Product of Complement*

H. Rosen[1], and S. K. Alex Law[2]

1   Introduction   99
2   Structure   100
3   Ligands of CR3   103
3.1 Binding of iC3b to CR3 and p   103
3.1.1 Recognition Site on iC3b   105
3.2 Other iC3b-like Ligands for CR3   105
3.3 Non-protein Ligands for CR3   108
3.4 Non-protein Ligands Shared by CR3, p and LFA-1   108
3.5 Undefined Ligands for CR3 and p   109
4   Cellular Responses to CR3-Ligand Interactions   109
4.1 Phagocytosis   109
4.2 Surface Expression of CR3   110
4.3 Other Adhesive Activities of CR3   111
5   Genetics   112
5.1 Gene Locations and Restriction Fragment Length Polymorphism   112
5.2 Leukocyte Adhesion Deficiency   112
5.2.1 Defect at the Molecular Level   113
5.2.2 Cellular Defects in Leukocyte Adhesion Deficiency   113
5.2.3 Animal Models of Leukocyte Adhesion Deficiency   114
6   Summary   115
References   115

1 Introduction

The third component of complement (C3) may be activated on a cell surface by either the classical or the alternative complement pathways and is covalently bound to the cell surface as C3b. C3b is a cofactor in the activation of the terminal compoments of C5–C9. In addition, it serves as the ligand for the type 1 complement receptor (CR1). C3b loses both of these functions when it undergoes proteolytic cleavage to iC3b and C3f by factor I. The relatively stable iC3b opsonin may then undergo proteolysis to C3c and C3dg in the presence of serum. Both iC3b and C3dg remain surface bound and are recognized by the leukocyte type 3 complement receptor (CR3)

---

[1] Sir William Dunn School of Pathology, University of Oxford, South Parks Road, Oxford OX1 3RE, UK
[2] The MRC Immunochemistry Unit, Department of Biochemistry, University of Oxford, South Parks Road, Oxford OX1 3QU, UK

and the type 2 complement receptor (CR2), respectively. Of the surface-bound fragments of C3, iC3b is probably the most stable of the cell-associated C3 fragments and is thought to play a major role in the complement-mediated clearance of micro-organisms.

In structural terms, CR1 and CR3 belong to different protein families (LAW 1988). CR1 belongs to the family of regulators of complement activation which all have at least one copy of a 60 amino acid structural element with easily identifiable conserved residues including four cysteines (FEARON and AHEARN, this volume). Other members of the complement system in this family include the soluble control proteins factor H and C4BP, the serine proteases C1r, C1s, factor B and C2 as well as the membrane-bound CR2, decay-accelerating factor and membrane cofactor protein. CR3, together with p150,95 and the lymphocyte function-associated antigen-1 (LFA-1), are collectively known as the leukocyte cell adhesion molecules or leukocyte integrins. p150,95 can also bind iC3b under certain circumstances and is referred to as the type 4 complement receptor (CR4) by some workers. In this chapter we shall discuss the structure, functions, regulation and genetics of CR3 and p150,95 and attempt to correlate structure and function where the evidence permits.

2 Structure

CR3 belongs to the extended family of membrane glycoproteins now known as integrins (HYNES 1987). They are all heterodimeric cell surface receptors with divalent cation requirements for their adherence activities. When the primary structures of the integrins became available, it was clear that there were three distinct yet related β subunits. Each of the β subunits associates with one of a number of α subunits to form individual integrin complexes. The α subunits are also structurally related to each other, but they do not appear to be related to the β subunits. Hence, the integrin molecules can be subdivided into three groups according to their β subunits. The known integrin molecules in human, their $\alpha\beta$ composition, common names and known ligands are listed in Table 1.

CR3 shares its β subunit (95 kDa) with two other cell surface glycoproteins, p150,95 and LFA-1 (SANCHEZ-MADRID et al. 1983). They belong to the β_2 subgroup according to HYNES's (1987) classification and may be referred to as the leukocyte integrins. In the nomenclature on leukocyte differentiation antigens, the common β subunit is given the cluster differentiation number CD18, and the α subunits for LFA-1 (175 kDa), CR3 (165 kDa) and p150,95 (150 kDa) CD11a, CD11b and CD11c in descending order of their respective apparent molecular weight by sodium dodecyl sulphate polyacrylamide gel electrophoresis (SDS-PAGE; COBBOLD et al. 1987; HOGG and HORTON 1987).

The primary structures of the common subunit for the human leukocyte integrins (KISHIMOTO et al. 1987b; LAW et al. 1987) and the α subunits for CR3 (ARNAOUT et al. 1988a; CORBI et al. 1988a; HICKSTEIN et al. 1989), p150,95 (CORBI et al. 1987) and LFA-1 (LARSON et al. 1989) have been reported. All four polypeptides have a hydrophobic segment typical of membrane proteins near the C-terminal. Hence, the

Table 1. The human integrins

Common β subunits	α subunits	Nomenclature and names of αβ complexes	Ligands
β_1 (gpIIa)			
	α_1	VLA1	
	α_2 (gpIa)	VLA2, collagen receptor, gpIa/IIa	Collagen
	α_3	VLA3	Fibronectin, collagen, laminin
	α_4	VLA4, lymphocyte homing receptor	
	α_5 (gpIc)	VLA5, fibronectin receptor, gpIc/IIa	Fibronectin
	α_6 (gpIc)	VLA6, laminin receptor, gpIc/IIa	Laminin
β_2 (CD18)			
	α_L (CD11a)	LFA-1	ICAM-1, LPS
	α_M (CD11b)	Mac-1, Mo1, CR3	iC3b, LPS, β-glucan, fibrinogen, factor X
	α_X (CD11c)	p150,95, CR4	iC3b, LPS
β_3 (gpIIIa)			
	α_V	Vitronectin receptor	Vitronectin
	α_{IIb} (gpIIb)	gpIIb/IIIa	Fibronectin, vitronectin, von Willebrand factor, laminin

General properties of the β_1 and β_3 subgroup integrins can be found in HEMLER (1988), HEMLER et al. (1987, 1988), HOLZMANN et al. (1989), HYNES (1987), RUOSLAHTI (1988), RUOSLAHTI and PIERSCHBACHER (1987), SONNENBERG et al. (1988), TAKADA et al. 1987a, b, 1988) and WAYNER et al. (1988). See text for the integrins of the β_2 subgroup

two subunits for each of the three leukocyte integrins are independently anchored in the membrane. The primary structure of the α subunit of the mouse CR3 (PYTELA 1988) has also been reported, and it shares approximately 75% sequence identity with the human version.

The common β subunit is most unusual in the high level of cysteine residues, a total of 56 found in the putative mature form, which account for over 8% of the residues in the extracellular portion. In one region of 187 residues, 38 cysteine residues could be found, with three or four repeating structural elements. The most characteristic feature is the octet motif of Cys−X−Cys−X−X−Cys−X−Cys, which is found three times in the cysteine-rich region. Since they are extracellular, it is presumed that most of them are engaged in disulphide linkages. The primary structures of the β_1 (ARGRAVES et al. 1987) and β_3 (FITZGERALD et al. 1987b; ZIMRIN et al. 1988) subunits of human integrins are very similar to that of β_2 with all 56 cysteine residues conserved. Of particular interest is a region close to the N-terminal where homology is extensive and includes a stretch of 19 residues, 16 of which are identical in the three subunits. The three sequences of this highly conserved region and a possible unit arrangement of the repeating structures in the cysteine-rich region are shown in Fig. 1.

a
```
          1  FKRAE DYPI DLYYLMDLSYSMKDDLE NVKS LGT DLMNE MR R I TS DF RIGFGS FVE KTVMP
          2  FR RAKGYPI DLYYLMDLSYSML DDLR NVKKLGGDLL R AL NE I TES GRIGFGS FVDKTVL  P
          3  VR QVE DYPVDI YYLMDLSYSMKDDLWS I QNLGT KLA T QMR KLTS NL RIGFGAFVDKPVS  P

                  YP  D  YYLMDLSYSM  DDL        LG   L             T    RIGFG FV K V P
              FRRAED  I   L                K          NVK       MR   I S         S  D T
              313322    3 3                2          333       2 3   2 2 3 2   3   1 3
```

b   (−N−) C X C (−N−) C (−N−) C (−N−) C X C X X C X C (−N−)

Fig. 1a. The high-homology region of the β subunits of human integrins. The aligned sequences of the β subunits of the human integrins (residues 135–194 of $β_1$, 119–178 of $β_2$, and 131–190 of $β_3$ of the translated sequences) are shown. Residues identical in all three subunits are shown in the first line below the aligned sequences, and residues found in two out of three sequences and the subunit with outstanding residue in the lines below. **b** A possible repeating structure in the cysteine-rich region. C, Cysteine; X undefined amino acid residue; (−N−), a stretch of residues with variable lengths between 4 and 14 amino acids

The α subunit of CR3 (ARNAOUT et al. 1988a; CORBI et al. 1988a; HICKSTEIN et al. 1989) and p150,95 (CORBI et al. 1987) showed 62% overall post-alignment identity. The α subunit of LFA-1, however, showed only 35% and 36% post-alignment identity with those of CR3 and p150,95 respectively (LARSON et al. 1989). Seven repeating elements could be identified from the N terminal in all three sequences. Each of the last three elements contains a short segment with the sequence format of X–X–X–––X, where X could be any one of the oxygen-containing residues Asp, Glu, Asn, Gln, Ser or Thr. They resemble the Ca^{2+} binding motifs found in paravalbumin, calmodulin and troponin C (SZEBENYI et al. 1981), and they are therefore presumed to be the metal-binding motifs for the integrins. The consensus sequence of the metal-binding motifs of the leukocyte integrins is shown in Fig. 2.

```
Leukocyte       −Y F GA S L −−−−D V DGDG L T DL  −A I GA P  −G
Integrins              A           L N             V      V

All             S Y F GY S L −A L GD L DGDG Y −D L −A V GA P −G
Integrins             A A V   G V      V N    L       V I
                      S       A                       L

Ca²⁺ Binding                    D −DGDG −I D −−E
Consensus                       *  * *   *   *
```

Fig. 2. The metal-binding motifs in the α subunits of integrins. Consensus sequences for the leukocyte integrins is compiled from repeats 5, 6, and 7 of the α subunits of CR3, p150,95, and LFA-1. That of all integrins include the sequences of repeats 4, 5, 6, and 7 from human fibronectin receptor, vitronectin receptor, and platelet gpIIb/IIIa. Also shown is the consensus "loop" sequence of the EF-hand of Ca^{2+} binding proteins (SZEBENYI et al. 1981)

A spacer sequence of about 200 residues, referred to as the L domain or I domain, could be found between repeating units 2 and 3 and was shown to be homologous to the A domains of the von Willebrand factor (SHELTON-INLOES et al. 1986) as well as a region in factor B (MOLE et al. 1984) and C2 (BENTLEY 1986).

The primary structures of the α subunits of other integrin subgroups, namely that of the fibronectin receptor of the β_1 subgroup (ARGRAVES et al. 1987; FITZGERALD et al. 1987a) and those of the platelet gpIIb/IIIa (PONCZ et al. 1987) and the vitronectin receptor (SUZUKI et al. 1987) of the β_3 subgroup, are available. Upon comparison they were found to be more related to each other (sharing 40%–45% identity) than to the α subunits of CR3, p150,95 and LFA-1 (sharing 25%–30% identity). The major differences lie in the absence of the L domain, the additional Ca^{2+} binding motif in the fourth repeating unit, and the presence of an extra segment N-terminal to the hydrophobic region. The extra segment contains a site where the α subunits of the β_1 and β_3 integrin subgroups are cleaved into two polypeptides which are linked by disulphide bonds.

The α and β subunits for CR3, p150,95 and LFA-1 (SANCHEZ-MADRID et al. 1983) as well as those of other integrin molecules (see HYNES 1987) are non-covalently associated. The association of the α and β subunits of the leukocyte integrins does not require divalent cations, but it could be disrupted upon treatment at alkaline pH. Re-association of the subunits after such treatment has not been demonstrated. Electron microscopy has been used to study the structure of the fibronectin receptor (NERMUT et al. 1988). The resulting model appears like a two-pronged plug into the lipid bilayer. This model may serve to describe the ultrastructure for the leukocyte integrins with perhaps some modifications. A version for CR3 is presented in Fig. 3. (For a more detailed analysis of the structure of CR3 and related molecules, see LAW 1989.)

3 Ligands of CR3

3.1 Binding of iC3b to CR3 and p150,95

The cellular expression of CR3 and p150,95 has been studied using monoclonal antibodies. CR3 is found on cells of the myelomonocytic lineage including neutrophils (also referred to as polymorphonuclear leukocytes and granulocytes), monocytes, some macrophages and large granular lymphocytes in the human whilst p150,95 seems restricted to myelomonocytic cells (HOGG and HORTON 1987). CR3 was first identified as the Mac-1 myelomonocytic differentiation marker (SPRINGER et al. 1979) and was shown later (BELLER et al. 1982; WRIGHT et al. 1983b) to be the cell surface receptor for the cell-bound iC3b fragment of C3 when a monoclonal antibody specific for Mac-1 was shown to inhibit the binding of iC3b to murine macrophages and human neutrophils. This phagocyte-specific molecule functions as an opsonic receptor, promoting the binding and uptake of iC3b-derivatized particles and cells by myelomonocytic cells (WRIGHT and SILVERSTEIN 1982; WRIGHT and MEYER 1986). Binding of iC3b to its receptors requires the presence of extracellular divalent cations, particularly magnesium (LAY and NUSSENZWEIG 1968), in the mouse whilst human monocytes require both extracellular calcium and magnesium (WRIGHT and SILVERSTEIN 1982). CR3 is a receptor for surface-bound iC3b and does not bind sheep erythrocytes derivatized with C3b or with the further cleavage product C3d(g) (CARLO et al. 1979; ROSS and LAMBRIS 1982).

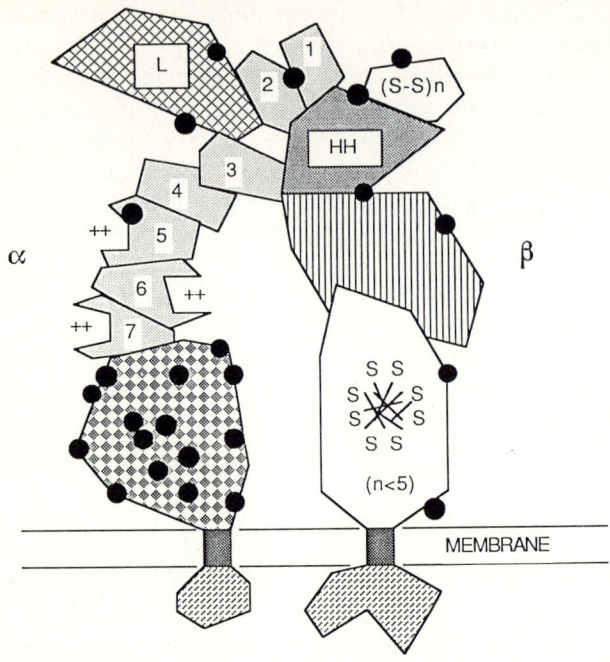

Fig. 3. Structural model of CR3. The domains are marked as discussed in the text. The seven repeating elements of the α subunit are numbered. *HH*, High-homology region found among the integrin β subunits; *L*, the L domain found only on the leukocyte integrins; ●, possible glycosylation sites; *S-S*, cysteine-rich regions

p150,95 has no unique function of its own, but like CR3 it could be isolated from extracts of either spleen or the monocyte-like U937 cell line by iC3b affinity chromatography (Micklem and Sim 1985; Malhotra et al. 1986). However, this interaction could be shown only under unphysiological low-salt conditions. p150,95-dependent rosetting of iC3b-coated erythrocytes (EC3bi) to U937 cells could not be demonstrated (Malhotra et al. 1986).

Myones et al. (1988) measured CR3 and p150,95 site number on neutrophils, monocytes and cultured monocytes. They found that neutrophils had about 50000 and fresh monocytes about 25000 molecules of CR3. In contrast, both cell types had only 5000–7000 molecules of p150,95. Cultured monocytes had about 130000 molecules of CR3 and 60000 molecules of p150,95. They measured the rosetting of EC3bi to neutrophils under low ionic strength conditions and found that antibodies to p150,95 inhibited rosette formation by less than 15% in fresh neutrophils from six donors and by over 23% in a further eight donors. Inhibition of rosetting of EC3bi to fresh monocytes by antibodies to p150,95 ranged between 6% and 14%. In contrast, when surface antigens were made unavailable selectively by adhering monocytes onto surfaces derivated with specific monoclonal antibodies, p150,95 could be shown to bind bacteria (Wright and Jong 1986) or yeast (Bullock and Wright 1987) but not EC3bi (Wright and Jong 1986). The role of p150,95 as a receptor for iC3b under normotonic conditions must therefore remain questionable.

3.1.1 Recognition Site on iC3b

A common feature of many but not all ligands of the various members of the extended integrin family is the presence of an Arg-Gly-Asp (RGD) sequence at the cell-attachment site of the ligand, and high concentrations of synthetic peptides containing the RGD motif may competitively block the interactions between the ligands and a variety of integrin molecules (PIERSCHBACHER and RUOSLAHTI 1984). Despite the presence of this common motif in many very different ligands, integrins can still maintain a fine specificity for ligand, presumably dependent upon the structural context within which the RGD sequence is found (RUOSLAHTI and PIERSCHBACHER 1987). For instance, the fibronectin receptor ($\alpha_5\beta_1$) does not recognize vitronectin whereas the vitronectin receptor ($\alpha_v\beta_3$) does not recognize fibronectin. The platelet IIb/IIIa glycoprotein ($\alpha_{IIb}\beta_3$) recognizes both fibronectin and vitronectin.

CR3 is able to recognize and be down-modulated (i.e., removal of receptor activity from the upper surface of an adherent cell) upon a 21-residue synthetic peptide derived from amino acid residues 1383–1403 of the human C3 (WRIGHT et al. 1987). This sequence contains an RGD motif although the murine C3 has LGD (Leu-Gly-Asp) in those positions (Table 2). A hexapeptide containing the RGD sequence is not recognized, suggesting that either the flanking sequences are of particular importance, or that the actual residues recognized are close to but are not the RGD motif. Binding of this 21-residue peptide to human monocytes was abolished by the OKM10 antibody known to inhibit iC3b binding (WRIGHT et al. 1983b) but not by the OKM1 antibody directed against another epitope on CR3.

Table 2. Sequences recognized by CR3

Protein	Sequence	Residues
Human C3	I –L –E –I –C –T –R –Y –R –G –D –Q	(1385–1397)
Murine C3	F –L –E –I –C –T –K –Y –L –G –D –V	(1361–1373)
Leishmania gp63	L –P –G –G –L –Q –Q –G –R –G –D –A	(366–378)
Fibrinogen	H –H –L –G –G – A –K –Q –A –G –D –V	(γ400–411)
Fibronectin	Y –A –V –T –G –R –G –D –S	(546–544)

3.2 Other iC3b-like Ligands for CR3

In addition to the well-characterized ability of CR3 to bind iC3b, CR3 binds a number of other soluble and integral membrane proteins that are iC3b related. These ligands are defined by the inhibition of their interactions with CR3 by soluble iC3b although such interactions are not necessarily blocked by a congruent set of monoclonal antibodies (Table 3). The relevant iC3b-like sequences are shown in Table 2. The three ligands so described are two components of the blood coagulation cascade, factor X and fibrinogen, and the gp63 glycoprotein of *Leishmania*.

Table 3. The ligands and binding domains of human CR3

Ligand	Monocyte						Neutrophil					References
	BIND	RGD	60.3	IB4	M1	M10	904	BIND	60.3	IB4	M1	M10	904	
iC3b	Y	+		+	−	+	−	Y					−	Wright et al. (1983b, 1989)
Factor X	Y	+	+		+	+		N						Altieri and Edgington (1988a)
Fibrinogen	Y	−			+	−		Y[b]		+	−	+		Altieri et al. (1988); Wright et al. (1988)
gp63	Y	+		+		+		N						Russell and Wright (1988)
β-Glucans	N							Y			+			Ross et al. (1987)
LPS	Y		+	+	−	−	+	N						Wright and Jong (1986); Wright et al. (1989)
Adhesion[a]	Y							Y	+		−	−	+	Dana et al. (1986)
Endothelium	Y		+		−	−		Y	+		−	−	+	Harlan et al. (1985); Arnaout et al. (1988); Vedder and Harlan (1988)

Antibodies	Epitope	Reference
60.3	β₂	Beatty et al. (1983)
IB4	β₂	Wright et al. (1983b)
OKM1	α_M	Breard et al. (1980)
OKM10	α_M	Wright et al. (1983b)
904	α_M	Dana et al. (1986)

Y, yes; N, not done; +/− refers to inhibition of binding.
[a] To protein-coated glass or plastic.
[b] Binding RGD inhibitable.

ALTIERI and EDGINGTON (1988a) showed that ADP-stimulated monocytes bind factor X of the coagulation cascade with a dissociation constant of about 40 nM. This binding was competitively inhibited by 0.1 µM soluble iC3b, suggesting that factor X and iC3b may share the same binding site. The binding of factor X, found only after stimulation, might be correlated with the appearance of CR3 molecules bearing the 7E3 antigenic epitope (COLLER 1985) which is more usually associated with the platelet IIb/IIIa integrin (ALTIERI and EDGINGTON 1988b).

Human monocytes (ALTIERI et al. 1988) bind fibrinogen following stimulation of cells with agonists such as ADP. This binding was inhibited by the monoclonal antibodies OKM1 and M1/70, both directed against the α-chain of CR3. OKM10, also directed against the CR3 α-chain, and RGD-containing peptides failed to inhibit binding. Human peripheral blood neutrophils also adhere to surfaces coated with fibrinogen (WRIGHT et al. 1988). This binding is inhibitable by millimolar concentrations of a range of RGD-containing synthetic peptides. In constrast to the human monocyte data of ALTIERI et al. (1988), neurophil adhesion to fibrinogen was inhibited by OKM10 and the anti-β-chain IB4 but not by OKM1. Both a synthetic decamer (0.4 mg/ml) derived from the fibrinogen γ-chain and the fibrinogen proteolytic fragment D (0.08 mg/ml) caused 50% inhibition of the binding of EC3bi to neutrophils (WRIGHT et al. 1988). The differences between receptor behaviour and antibody inhibition of fibrinogen binding in the two reports is not understood.

RUSSELL and WRIGHT (1988) showed that the major surface glycoprotein gp63 of *Leishmania* species, which contains an RGD sequence (BUTTON and MCMASTER 1988), binds to CR3. Two synthetic peptides based on human fibronogen and containing AGD (Ala-Gly-Asp) inhibited the binding of gp63-coated silica beads to macrophages. An RGD-containing synthetic peptide derived from gp63 could also inhibit the binding of EC3bi to macrophages. However, the binding and entry of *Leishmania* promastigotes may also be mediated by other receptors including the mannosyl-fucosyl receptor (KLEMPNER et al. 1983) and the macrophage receptor for advanced glycosylation end products (MOSSER et al. 1987). The relative contributions of these receptors in vivo remain uncertain.

The site of iC3b/fibrinogen binding to CR3 is not known. Cross-linking RGD-containing peptides to platelet IIb/IIIa glycoprotein labelled both α- and β-chains. This suggested that the binding site is formed by both chains (SANTORO and LAWING 1987). D'SOUZA et al. (1988) used bifunctional reagents to cross-link an RGD-containing heptapeptide, KYGRGDS, to the surface of platelets and found labelling predominantly in the region of residues 109–170, and especially Lys-125, of the integrin β_3 member gpIIIa. SMITH and CHERESH (1988) have identified the binding of another RGD-containing peptide to a similar region, namely residues 61–203, of the β_3-chain of the vitronectin receptor by photo-activatible cross-linking. It should be noted that this region of the β-chain is highly conserved between all three β subunits of the extended integrin family (see Fig. 1). These results therefore suggest that this region may form the interface with the α subunits.

There is no evidence yet that p150,95 shares this ability to bind fibrinogen, factor X and gp63. The complete inhibition of fibrinogen binding by antibodies specific for the CR3 suggests that p150,95 does not play a role (WRIGHT et al. 1988).

Table 4. Inhibitory effects of monoclonal antibodies to CR3 on leukocyte functions in vivo

In vivo model	Antibody	Reference
Inhibition of PMN recruitment to inflammatory stimuli in rabbit skin	60.3 (β)	Arfors et al. (1987)
Inhibition of murine peritoneal myelomonocytic exudates in response to thioglycolate broth	5C6 (α)	Rosen and Gordon (1987)
Inhibition of murine delayed hypersensitivity responses	5C6	Rosen et al. (1989a)
Potentiation of murine listeriosis	5C6	Rosen et al. (1989b)
Decreased ischemia-reperfusion injury	Mo1 (α)	Simpson et al. (1988)
	904 (α)	Todd et al. (1989)
Promotion of survival in rabbits	60.3	Vedder et al. (1988)
Inhibition of Wallerian degeneration of sectioned murine sciatic nerve	5C6	Lunn et al. (1989)

PMN, Polymorphonuclear leukocytes and granulocytes

3.3 Non-protein Ligands for CR3

There is a growing body of evidence that CR3 has a well-defined lectin-like capacity. Monoclonal antibody inhibition studies (Ezekowitz et al. 1984) showed that antibodies to CR3 partially inhibit the binding of particles such as zymosan. Ross et al. (1985a) showed that CR3 is able to bind unopsonized zymosan in a divalent cation-dependent manner. They also showed that OKM1 blocked neutrophil zymosan binding only and not iC3b binding whilst the anti-Leu-15 antibody, also restricted to CR3, blocked iC3b binding alone. This suggested that the lectin-like binding site of CR3 is structurally distinct from the iC3b/RGD-binding domain. The lectin-like binding site has a specificity for β-glucans, a major carbohydrate component of zymosan, and was inhibited by soluble iC3b (Ross et al. 1987). Antibodies to LFA-1 and p150,95 do not inhibit the binding of yeast. The uptake of zymosan is a complex event associated with secretion of superoxide anions by monocytes (Ross et al. 1985a) and its possible ligation to other receptors including the mannosyl-fucosyl receptor. On the other hand, phagocytosis via CR3 alone is not coupled to other secretory events (Wright and Silverstein 1982; Aderem et al. 1985).

3.4 Non-protein Ligands Shared by CR3, p150,95 and LFA-1

A range of receptor activities are shared by all three leukocyte integrins having the common $β_2$-chain. Specifically, antibodies directed against either CR3 alone or against all three family members have inhibited, at least in part, the uptake or binding of unopsonised *Leishmania* (Blackwell et al. 1985; Mosser and Edelson (1985), *Staphylococcus epidermidus* (Ross et al. 1985b) and *Histoplasma capsulatum* (Bullock and Wright 1987). A more detailed study by Wright and Jong (1986) showed that the binding of certain rough strains of *Escherichia coli* and *Salmonella typhimurium* to monocytes requires both calcium and magnesium for maximum efficiency and was inhibitible by surface-bound lipopolysaccharide (LPS). This binding was lost in smooth bacterial strains which express the additional O antigen. Further, sheep erythrocytes derivatized with LPS mimicked the binding of the bacteria. This binding of bacteria to the cell surface was lost on down-modulation of CR3, p150,95 and LFA-1 by culturing cells on a surface derivatized with the anti-β-chain monoclonal antibody IB4. Down-modulation of each member singly or any two together had no effect whilst down-modulation of all three using a surface derivatized with a mixture of anti-α-chain reagents led to effective inhibition of LPS binding. WRIGHT et al. (1989) have subsequently suggested that CR3 expresses two distinct binding sites, the first for RGD-containing ligands and inhibitable by the OKM10 but not the 904 antibody and the second, for LPS, inhibitible by the monoclonal antibody 904 and not by OKM10.

3.5 Undefined Ligands for CR3 and p150,95

The role of CR3 and p150,95 in cell-cell interactions and specifically in leukocyte-endothelial interactions has been of considerable interest and was reviewed by HARLAN (1985). Antibody inhibition studies using monoclonal antibodies to CR3 and p150,95 (ARNAOUT et al. 1988b; KEIZER et al. 1987; TONNESEN et al. 1989) to quantitate the inhibition of neutrophil adhesion to human umbilical vein endothelium showed that both CR3 and p150,95 play a role in cell adhesion. The nature of the ligand(s) for CR3 and p150,95 mediating attachment to endothelium or other cell surfaces remain(s) obscure. There is a possibility that the induceable endothelial leukocyte adhesion molecule (ELAM-1; BEVILACQUA et al. 1987) could be a ligand for CR3 and p150,95. POHLMAN et al. (1986) described the induction of endothelial surface factors in response to LPS, IL-1 and tumor-necrosis factor that enhanced CR3-dependent attachment of neutrophils. Further experimentation to clarify this is required.

4 Cellular Responses to CR3-Ligand Interactions

4.1 Phagocytosis

The best characterized function of CR3 is its ability to bind iC3b-opsonized particles and mediate their phagocytic uptake (BIANCO et al. 1975; GRIFFIN et al. 1975). Initial adherence is followed by the encirclement of the particle by pseudopodia with formation of an impermeant space (black hole) between the plasma membrane and the particle (WEIGHT and SILVERSTEIN 1984). The pseudopodia fuse to engulf the particle in a membrane-bound intracellular vesicle called the phagosome which then fuses with a primary lysozome to form the phagolysozome (SILVERSTEIN et al. 1977). Unlike the phagocytic receptors for the Fc portion of IgG (DIVIRGILLIO et al. 1988) which do not require extracellular calcium for phagocytosis, phagocytosis of iC3b-opsonized particles requires extracellular divalent cations (WRIGHT and SILVERSTEIN 1982). Cultured monocytes avidly bind erythrocytes coated with 5000 molecules of iC3b/cell but do not ingest them (WRIGHT and SILVERSTEIN 1982; WRIGHT et al.

1983a). Additional signals are required to activate phagocytosis such as binding of monocytes to fibronectin or serum amyloid P (WRIGHT et al. 1983a) or stimulation of cells with phorbol esters (WRIGHT and SILVERSTEIN 1982). This coupling of receptors to promote phagocytosis required the continuous presence of a fibronectin substratum or phorbol esters (WRIGHT et al. 1984). In contrast, stimulation of neutrophils by phorbol esters causes an increase in CR3 expression on the cell surface accompanied by the activation of phagocytosis. This activation of phagocytosis peaked at 20 min and then declined to below initial levels when the cells were continually exposed to phorbol esters for 80 min. However, the level of surface CR3 remained elevated WRIGHT and MEYER 1986).

Culture of monocytes in the presence of interferon-γ depressed the binding activity of CR3 almost completely whilst receptor number remained unchanged (WRIGHT et al. 1986). This down-regulation of binding could be reversed by culturing those cells on a fibronectin-coated surface. Stimulation of phagocytic capacity required an additional signal provided by phorbol esters. It should be noted that exposure of monocytes cultured in the absence of interferon-γ to a substratum of fibronectin is sufficient to activate both binding and phagocytic functions of CR3 (WRIGHT et al. 1984). This work illustrates that, at least in this one case, the binding and phagocytic activities of CR3 can be independently regulated.

Phagocytosis of immunoglobulin-opsonized particles is accompanied by the secretion of microbicidal reactive oxygen intermediates (superoxide anion and hydrogen peroxide) as well as an array of inflammatory mediators derived from the metabolism of arachidonate, thus setting in train the acute inflammatory response. CR3-mediated phagocytosis, however, does not appear to trigger a secretory response, and neither reactive oxygen intermediates (WRIGHT and SILVERSTEIN 1983; YAMAMOTO and JOHNSTON 1984) nor arachidonate metabolites (ADEREM et al. 1985) are released. This suggests that iC3b-mediated phagocytosis is a clearance mechanism rather than a means of inducing or perpetuating the inflammatory response. It is perhaps also advantageous for an adhesion-promoting receptor used in migration to avoid secretion of inflammatory and cytotoxic mediators en route.

4.2 Surface Expression of CR3

Resting neutrophils have a low capacity to bind and phagocytose iC3b-opsonized particles. Many agents including chemotaxins such as C5a or fMLP (formyl-Met-Leu-Phe) and secretory agonists such as phorbol esters are capable of enhancing CR3 function and surface expression. Exposure of neutrophils to such agents causes the sudden degranulation of cytoplasmic granules which fuse with plasma membrane and induce a five- to tenfold increase in the site number of CR3 on the cell surface (BERGER et al. 1984). The exact source of this CR3 may vary depending upon the stimulus and the degree of degranulation taking place. Intracellular pools of CR3 in the neutrophil have been located in gelatinase (PETREQUIN et al. 1987) and in peroxidase-negative (BAINTON et al.) or specific granules (TODD et al. 1984; O'SHEA et al. 1985). Degranulation of the former, alkaline-phosphatase-positive, population occurs at nanomolar concentrations of fMLP (BORREGAARD et al. 1987) whilst degranulation of the specific granules requires 0.1–1 μM fMLP (PAINTER et al. 1984).

The up-regulation of surface CR3 from the gelatinase/alkaline-phosphatase granules might occur in the vascular compartment prior to the egress of neutrophils from the circulation. The monocyte also has a latent intracellular pool of CR3 and p150,95 in peroxidase-negative cytoplasmic granules that moves to the surface on stimulation of cells with C5a, leukotriene B4, tumour necrosis factor and platelet-derived growth factor (MILLER et al. 1987).

Although increased surface expression of CR3 correlates temporally with enhancement of capacity to bind and internalize iC3b-opsonized material, there is no evidence supporting a causal relationship. WRIGHT and MEYER (1986) could transiently down-regulate CR3 function with phorbol esters at a time when surface expression of CR3 remained constant. Furthermore, BUYON et al. (1988) showed that increased surface expression of CR3 could be dissociated from homotypic neutrophil aggregation. Antibody blockade of pre-existing surface CR3 alone was sufficient to inhibit neutrophil aggregation, even after exposure to degranulating agonists. VEDDER and HARLAN (1988) showed that agonists like fMLP increased surface expression of CR3 which could be prevented by pretreatment of neutrophils with the anion-channel blocker 4,4'diisocyanostilbene-2,2'-disulphonic acid. Neutrophil adhesion to cultured endothelium was enhanced by agonist stimulation but was not dependent on an increase in surface CR3 from intracellular sites.

One potential mechanism by which the higher order organization of CR3 could be altered is by clustering receptors within the plane of the plasma membrane. DETMERS et al. (1987) showed alterations in the two-dimensional distribution of CR3 in response to phorbol myristic acetate but not fMLP using immunoelectron microscopy. Receptors moved from a random distribution into a series of small, tight clusters. This corresponded temporally with changes in phagocytic activity although there are no insights into the molecular mechanisms underlying this redistribution. It should be noted that fMLP does not induce the enhancement of EC3bi binding and phagocytic activity of neutrophils (DETMERS et al. 1987). Additional evidence in favour of molecular changes associated with receptor activation is the finding by ALTIERI and EDGINGTON (1988b) that ADP induces an activation epitope on CR3 that correlates with functional receptor changes.

4.3 Other Adhesive Activities of CR3

The elucidation of the clinical syndrome of leukocyte adhesion deficiency (LAD; ARNAOUT et al. 1982) as a heritable deficiency of the leukocyte-specific integrins has led to the characterization of CR3 and p150,95 as divalent cation-dependent adhesion-promoting receptors. ANDERSON et al. (1986) studied the contributions of CR3 and p150,95 to a range of adherence-dependent neutrophil functions including adhesion to protein-coated glass, aggregation, chemotaxis, zymosan and oil red-O phagocytosis as well as intracellular killing. The study was performed using subunit-specific monoclonal antibodies. Granulocyte adhesion was significantly inhibited by anti-β_2 > anti-CR3-α > anti-p150-α. ROSEN and GORDON (1987) found that a monoclonal antibody directed to the murine CR3 was able to inhibit attachment and detach adherent neutrophils from protein-coated tissue culture plastic. The spreading of murine elicited macrophages on protein-coated glass was also inhibited. The murine receptor has an absolute requirement for 0.1 mM magnesium and not calcium for adhesion. In fact, adhesion can be demonstrated in the presence of magnesium and ethyleneglycoltetraacetate (ROSEN and GORDON 1987). Similar results were obtained in the adhesion studies to endothelial monolayers cited above.

5 Genetics

5.1 Gene Locations and Restriction Fragment Length Polymorphism

The gene for the β subunit of the leukocyte integrins has been located to chromosome 21 (MARLIN et al. 1986) in band q22 (CORBI et al. 1988b; SOLOMON et al. 1988). By analysing an extended panel of cell hybrids containing defined regions of chromosome 21 with pulsed field gel electrophoresis, GARDINER et al. (1988) refined the location of the $β_2$ gene to band q22.3. The genes for the three α subunits are found in a cluster on chromosome 16 between bands p11 and p13.1 (CORBI et al. 1988b). Their relative order and orientation have not been determined.

The genes for $β_1$ and $β_3$ are not linked to that of $β_2$. They have been located to chromosomes 10 (PETERS et al. 1984) and 17 (SOSNOSKI et al. 1988), respectively. The genes for other α subunits of the integrins are located in various parts of the genome (see LAW 1989).

The gene structure of the β subunit of the leukocyte integrins has been studied and an RFLP was initially found with the *Bgl*II restriction enzyme with the near full-length cDNA probe J9 (LAW et al. 1987) among normal individuals (WELLS and LAW 1987). This RFLP is also observed when the enzyme *Bam*HI was used in addition to *Bgl*II in the digestion. Three different patterns were seen: those with three major fragments of sizes 14, 8.5 and 5 kb, and those with the 5 kb fragment, and those with either the 14 kb or the 8.5 kb fragment. Subsequently, another pattern was observed from a healthy individual who has three J9 positive fragments with sizes of 14, 11 and 5 kb (S. K. A. LAW, unpublished result). It appears that whereas the 5-kb fragment is found in all individuals observed, the 14-, 11- and 8.5-kb fragments is present either singly or in pairs. Possibly, they represent allelic polymorphism of the gene encoding the β subunit. Similar studies have been carried out with three unrelated LAD patients. The deficiency cannot be associated with the presence of any of the three variable fragments (S. K. A. LAW, unpublished results).

5.2 Leukocyte Adhesion Defiency

A set of patients were found to be defective in the leukocyte integrins (see TODD and FREYER 1988 for a historical account). The absence of all three antigens on their leukocytes suggested that the deficiency lies in the common β subunit (SPRINGER et al. 1984). Clinically, these patients suffer from recurrent infection and usually die at an early age. The deficiency is inherited as an autosomal recessive trait, which is in agreement with the structural gene being located on chrmosome 21. More detailed accounts of the physiological and clinical aspects of this deficiency can be found elsewhere (SPRINGER and ANDERSON 1986; ANDERSON and SPRINGER 1987; TODD and FREYER 1988; FISCHER et al. 1988).

5.2.1 Defect at the Molecular Level

The LAD deficiency has been studied both at the mRNA and at the protein level. EBV-transformed B-lymphocytes or phytohaemagglutinin (PHA) activated T-lymphocytes of the patients were usually employed. Since these cells derived from normal individuals express only LFA-1 and not CR3 or p150,95, the studies at the protein level are done for the LFA-1 antigen. Five distinct phenotypic subgroups have been described. They include: (a) the inability to transcribe a detectable mRNA; (b) the expression of a low but detectable level of mRNA; (c) the synthesis of normal level and size of mRNA but an aberrantly small protein precursor; (d) the synthesis of normal level and size of mRNA but an aberrantly large protein precursor; and (e) the synthesis of normal level and size of mRNA and a normal size protein precursor. In all cases, the failure to express a mature β subunit on the cell surface seems to be the common defective feature (DANA et al. 1987; KISHIMOTO et al. 1987a; DIMANCHE-BOITREL et al. 1988). Expression of the α subunit on leukocyte surfaces is also not observed in these patients although synthesis of the α subunit precursors appears to be normal (SPRINGER et al. 1984; ANDERSON et al. 1985; DIMANCHE et al. 1987). To exclude the possibility that there are defects in the α subunits, MARLIN et al. (1986) fused the PHA lymphoblasts from an LAD patient with a mouse thymoma line and detected the expression of a $\alpha_{human}/\beta_{mouse}$ hybrid complex, thus emphasizing that a defective β subunit is sufficient to account for the failure to express the LFA-1 antigen. These results strongly suggest that the biosynthesis and maturation of CR3 and p150,95 is by a similar process, and that the expression of their α subunits is dependent on the presence of a normal β subunit.

5.2.2 Cellular Defects in Leukocyte Adhesion Deficiency

A small group of receptors play a critical role in the recruitment of phagocytes to sites of inflammation. This evidence has emerged from the study of patients with LAD who invariably have recurrent, life-threatening bacterial and fungal sepsis with leukocytosis but do not form pus at sites of infection. Their leukocytes are defective in mediating a range of activities that require adherence, including iC3b binding, leukocyte chemotaxis, leukocyte infiltration of extravascular sites, T-lymphocyte cytotoxicity and natural killer activity (ANDERSON et al. 1984; DANA et al. 1984; KRENSKY et al. 1985). Neutrophils from these patients are defective in adhesion assays in vitro, failing to adhere to protein-coated substrata (ANDERSON et al. 1985) or to endothelial cells (HARLAN et al. 1985) or LPS (WRIGHT et al. 1989) and also showing poor migration in subagarose but not Boyden chamber chemotaxis assays.

Because patients are deficient in a group of surface receptors, the patient data cannot give insights into the contribution of individual receptors to leukocyte adhesion and recruitment in vivo.

5.2.3 Animal Models of Leukocyte Adhesion Deficiency

Monoclonal antibody studies in vivo have allowed some insights into the role of CR3 alone in phagocyte recruitment in vivo. The various animal models studied thus far with monoclonal antibodies are shown in Table 3.

ARFORS et al. (1987) showed that the 60.3 antibody directed against the human β_2 chain could inhibit neutrophil accumulation and plasma membrane leakage in the rabbit in response to a range of inflammatory stimuli injected intradermally. PRICE et al. (1987) showed that 60.3 inhibited neutrophil accumulation in a dose-dependent fashion in subcutaneous polyvinyl sponges placed in rabbits. VEDDER et al. (1988) showed that injection of 60.3 also protected rabbits against haemorrhagic shock. Administration of 60.3 causes a leukocytosis in peripheral blood.

Using human neutrophils to induce injury in perfused rat lungs, ISMAIL et al. (1987) prevented pulmonary injury by pretreating the neutrophils with the Mo1 antibody directed to the CR3 α-chain. The importance of CR3-dependent adhesion in endothelial injury and tissue damage was shown by SIMPSON et al. (1988) who, injecting the Mo1 antibody in the dog, reduce the reperfusion injury following myocardial infarction.

In the mouse, intravenous injection of the 5C6 antibody to CR3 is able to inhibit recruitment of macrophages and neutrophils to thioglycolate broth in the peritoneal cavity (ROSEN and GORDON 1987). Unlike the reagents to the common β_2-chain, anti-CR3 antibodies do not seem to induce a leukocytosis. In addition, injection of 5C6 is able to inhibit the T-lymphocyte-dependent delayed-type hypersensitivity response (ROSEN et al. 1989a), Wallerian degeneration of peripheral nerve (LUNN et al. 1989) and potentiate infection with *Listeria* monocytogenes a million-fold (ROSEN et al. 1989b). The 5C6 antibody does not potentiate fungal, viral or malarial infections in the mouse (H. ROSEN, unpublished results).

Antibodies to CR3 function in vivo by impairing myelomonocytic adhesion and migration and not by depleting leukocytes from the circulation. Direct proof of this is the study by TODD et al. (1989) which shows that pepsin fragments derived from the 904 antibody to CR3 inhibit ischaemic reperfusion injury in the canine myocardium. As mentioned above, 904 blocks CR3 adhesion functions alone and not iC3b binding (DANA et al. 1986).

The differences between some of the animal models and the human patients could be explained by the combined deficiency of LFA-1, CR3 and p150,95 in the human patients. The antibody studies with CR3-restricted antibodies highlight the critical role of CR3-dependent recruitment for some, but not all, facets of host defense to pathogens. ZIMMERMAN and MCINTYRE (1988) demonstrated the ability of certain inflammatory mediators such as thrombin and leukotriene C4 to facilitate CR3/LFA-1-independent adhesion of neutrophils to endothelium. These alternative adherence mechanisms could play a role in vivo. Cells undergoing specific antibody-mediated blockade of CR3 are certainly less profoundly disturbed than cells lacking all three leukocyte integrins.

It is clear that blocking CR3 in vivo might produce some beneficial anti-inflammatory effects through the inhibition of myelomonocytic cell recruitment to extravascular sites, and that CR3 might be an useful target for anti-inflammatory drug therapy.

6 Summary

CR3 is probably the major adhesion molecule on monocytes and neutrophils. Its function as a phagocytic receptor for iC3b-coated particles has been well characterized. CR3 also has binding affinity for other ligands, including those that compete with iC3b such as fibrinogen, factor X, and β-glucan, and those that do not such as bacterial LPS. CR3 binding to endothelial cells probably plays an important role in the extravascular migration of monocytes and neutrophils, but the ligand that it recognizes on endothelial cells has not been identified. Structurally CR3 belongs to the integrin family, and it shares a common subunit with p150,95 and LFA-1. The expression of these three membrane antigens appear to be limited to leukocytes, and they are sometimes referred to collectively as the leukocyte integrins. All three antigens have a common binding affinity for bacterial LPS. p150,95 also has affinity for iC3b, but p150,95/iC3b-dependent cellular responses has not been demonstrated. Its status as a complement receptor therefore awaits further experimental support.

Acknowledgements. Hugh Rosen is a Junior Research Fellow of Jesus College, Oxford, and is supported by the Medical and the Arthritis and Rheumatism Research Councils.

References

Aderem AA, Wright SD, Silverstein SC, Cohn ZA (1985) Ligated complement receptors do not activate the arachidonate acid cascade in resident peritoneal macrophages. J Exp Med 161: 617–622

Altieri DC, Edgington TS (1988a) The saturable high affinity association of factor X to ADP-stimulated monocytes defines a novel function of the Mac-1 receptor. J Biol Chem 263: 7007–7015

Altieri DC, Edgington TS (1988b) A monoclonal antibody reacting with distinct adhesion molecules defines a transition in the functional state of the receptor CD11b/CD18 (Mac-1) J Immunol 141: 2656–2660

Altieri DC, Bader R, Mannucci PM, Edgington TS (1988) Oligospecificity of the cellular adhesion receptor Mac-1 encompasses an inducible recognition specificity for fibrinogen. J Cell Biol 107: 1893–1900

Anderson DC, Springer TA (1987) Leukocyte adhesion deficiency: an inherited defect in the Mac-1, LFA-1, and p150,95 glycoproteins. Annu Rev Med 38: 175–194

Anderson DC, Schmalstieg FC, Arnaout MA, Kohl S, Tosi MF, Dana N, Buffone GJ, Hughes BJ, Brinkley BR, Dickey WD, Abramson JS, Springer TA, Boxer LA, Hollers JM, Smith CW (1984) Abnormalities of polymorphonuclear leukocytes associated with a heritable deficiency of high molecular weight surface glycoproteins (GP138): common relationship to diminished cell adherence. J Clin Invest 74: 536–551

Anderson DC, Schmalstieg FC, Finegold MJ, Hughes BJ, Rothlein R, Miller LJ, Kohl S, Tosi MF, Jacobs RL, Waldrop TC, Goldman AS, Shearer WT, Springer TA (1985) The severe and moderate phenotypes of heritable Mac-1, LFA-1 deficiency: their quantitative definition and relation to leukocyte dysfunction and clinical features. J Infect Dis 152: 668–689

Anderson DC, Miller LJ, Schmalstieg FC, Rothlein R, Springer TA (1986) Contribution of the Mac-1 glycoprotein family to adherence-dependent granulocyte functions: structure-function assessments employing subunit-specific monoclonal antibodies. J Immunol 137: 15–27

Arfors K-E, Lindberg C, Lindblom L, Lundnerg K, Beatty PG, Harlan JM (1987) A monoclonal antibody to the membrane glycoprotein complex CD18 inhibits polymorphonuclear leukocyte accumulation and plasma leakage in vivo. Blood 69: 338–340

Argraves WS, Suzuki S, Arai H, Thompson K, Pierschbacher MD, Ruoslahti E (1987) Amino acid sequence of the human fibronectin receptor. J Cell Biol 105: 1183–1190

Arnaout MA, Pitt J, Cohen HJ, Melamed J, Rosen FS, Schlossman SF, Colten HR (1982) Deficiency of a granulocyte membrane glycoprotein in a boy with recurrent bacterial infections. N Engl J Med 306: 693–699

Arnaout MA, Gupta SK, Pierce MW, Tenen DG (1988a) Amino acid sequence of the alpha subunit of human leukocyte adhesion receptor Mo1 (complement receptor type 3). J Cell Biol 106: 2153–2158

Arnaout MA, Lanier LL, Faller DV (1988b) Relative contribution of the leukocyte molecule Mo1, LFA-1, and p150,95 (LeuM5) in adhesion of granulocytes and monocytes to vascular endothelium is tissue- and stimulus-specific. J Cell Physiol 137: 305–309

Bainton DF, Miller LJ, Kishimoto TK, Springer TA (1987) Leukocyte adhesion receptors are stored in peroxidase-negative granules of human neutrophils. J Exp Med 166: 1641–1653

Beatty PG, Ledbetter JA, Martin PJ, Price TH, Hansen JA (1983) Definition of a common leukocyte-surface antigen associated with diverse cell-mediated immune function. J Immunol 131: 2913–2916

Beller DI, Springer TA, Schreiber RD (1982) Anti-Mac-1 selectively inhibits the mouse and human type three complement receptor. J Exp Med 156: 1000–1009

Bennett JS, Vilaire G, Cines DB (1982) Identification of the fibrinogen receptor on human platelets by photoaffinity labeling. J Biol Chem 257: 8049–8054

Bentley DR (1986) Primary structure of human complement component C2: homology to two unrelated protein families. Biochem J 239: 339–345

Berger M, O'Shea J, Cross AS, Folks TM, Chused TM, Brown EJ, Frank MM (1984) Human neutrophils increase expression of C3bi as well as C3b receptors upon activation. J Clin Invest 74: 1566–1571

Bevilacqua MP, Pober JS, Mendrick DL, Cotran RS, Gimbrone MA (1987) Identification of an inducible endothelial-leukocyte adhesion molecule. Proc Natl Acad Sci USA 84: 9238–9242

Bianco C, Griffin FM, Silverstein SC (1975) Studies of the macrophage complement receptor: alteration of receptor function upon macrophage activation. J Exp Med 141: 1279–1290

Blackwell JM, Ezekowitz RAB, Roberts MB, Channon JY, Sim RB, Gordon S (1985) Macrophage complement and lectin-like receptors bind *Leishmania* in the absence of serum. J Exp Med 162: 324–331

Borregaard N, Miller LJ, Springer TA (1987) Chemoattractant-regulated mobilization of a novel intracellular compartment in human neutrophils. Science 237: 1204–1206

Breard J, Reinhertz EL, Kung PC, Goldstein G, Schlossman SF (1980) A monoclonal antibody reactive with human peripheral blood monocytes. J Immunol 124: 1943–1946

Bullock WE, Wright SD (1987) Role of the adherence-promoting receptors, CR3, LFA-1 and p150,95 in binding of *Histoplasma capsulatum* by human macrophages. J Exp Med 165: 195–210

Button LL, McMaster WR (1988) Molecular cloning of the major surface antigen of *Leishmania*. J Exp Med 167: 724–729

Buyon JP, Abramson SB, Philips MR, Slade SG, Ross GD, Weissmann G, Winchester RJ (1988) Dissociation between increased surface expression of gp165/95 and homotypic neutrophil aggregation. J Immunol 140: 3156–3160

Carlo JR, Ruddy S, Studer E, Conrad DH (1979) Complement receptor binding of C3b-coated cells treated with C3b-inactivator, β1H-globulin and trypsin. J Immunol 123: 523–528

Cobbold S, Hale G, Waldmann H (1987) Non-lineage, LFA-1 family, and leucocyte common antigens: new and previously defined clusters- In: McMichael AJ (ed) Leucocyte typing III: white cell differentiation antigens. Oxford University Press, Oxford, pp 788–801

Coller BS (1985) A new murine monoclonal antibody reports an activation-dependent change in the conformation and/or microenvironment of the platelet glycoprotein IIb/IIIa complex. J Clin Invest 76: 101–108

Corbi AL, Miller LJ, O'Connor K, Larson RS, Springer TA (1987) cDNA cloning and complete primary structure of the α subunit of a leukocyte adhesion glycoprotein, p150,95. EMBO J 6: 4023–4028

Corbi AL, Kishimoto TK, Miller LJ, Springer TA (1988a) The human leukocyte adhesion glycoprotein Mac-1 (complement receptor type 3, CD11b) α subunit: cloning, primary structure, and relation to the integrins, von Willebrand factor and factor B. J Biol Chem 263: 12403–12411

Corbi AL, Larson RS, Kishimoto TK, Springer TA, Morgan CC (1988b) Chromosomal location of the genes encoding the leukocyte adhesion receptors LFA-1, Mac-1 and p195,95: identification of a gene cluster involved in cell adhesion. J Exp Med 167: 1597–1607

Dana N, Todd RF, Pitt J, Springer TA, Arnaout MA (1984) Deficiency of a surface membrane glycoprotein (Mo1) in man. J Clin Invest 73: 153–159

Dana N, Styrt B, Griffin JD, Todd RF, Klempner MS, Arnaout MA (1986) Two functional domains in the phagocyte membrane glycoprotein Mo1 identified with monoclonal antibodies. J Immunol 137: 3259–3263

Dana N, Clayton LK, Tennen DG, Pierce MW, Lachmann PJ, Law SA, Arnaout MA (1987) Leukocytes from four patients with complete or partial Leu-CAM deficiency contain the common β-subunit precursor and β-subunit messenger RNA. J Clin Invest 79: 1010–1015

Detmers PA, Wright SD, Olsen E, Kimball B, Cohn ZA (1987) Aggregation of complement receptors on human neutrophils in the absence of ligand. J Cell Biol 105: 1137–1145

Dimanche MT, Deist FL, Fischer A, Arnaout MA, Griscelli C, Lisowska-Grospierre B (1987) LFA β-chain synthesis and degradation in patients with leukocyte-adhesive proteins deficiency. Eur J Immunol 17: 417–419

Dimanche-Boitrel MT, Guyot A, de Saint-Basile G, Fischer A, Groscelli C, Lisowska-Grispierre B (1988) Heterogeneity in the molecular defect leading to the leukocyte adhesion deficiency. Eur J Immunol 18: 1575–1579

DiVirgillio F, Meyer BC, Greenburg S, Silverstein SC (1988) Fc-receptor mediated phagocytosis occurs in macrophages at vanishingly low cytosolic calcium levels. J Cell Biol 106: 675–686

D'Souza SE, Ginsberg MH, Burke TA, Lam SCT, Flow EF (1988) Localization of an Arg-Gly-Asp recognition site with an integrin adhesion receptor. Science 242: 91–93

Ezekowitz RAB, Sim RB, Hill M, Gordon S (1984) Local opsonisation by secreted macrophage complement components: role of receptors for complement in the uptake of zymosan. J Exp Med 159: 244–260

Fischer A, Lisowska-Grospierre B, Anderson DC, Springer TA (1988) Leukocyte adhesion defiency: molecular basis and functional consequences. Immunodeficiency Rev 1: 39–54

Fitzgerald LA, Poncz M, Steiner B, Rall SC, Bennett JS, Phillips DR (1987a) Comparison of cDNA-derived protein sequences of the human fibronectin and vitronectin receptor α-subunits and platelet glycoprotein IIb. Biochemistry 26: 8158–8165

Fitzgerald LA, Steiner B, Rall SC, Lo S-S, Phillips DR (1987b) Protein sequence of endothelial glycoprotein IIIa derived from a cDNA clone: identity with platelet glycoprotein IIIa and similarity to "integrin". J Biol Chem 262: 3936–3939

Gardiner K, Watkins P, Munke M, Drabkin H, Jones C, Patterson D (1988) Partial physical map of human chromosome 21. Somatic Cell Mol Genet 14: 623–638

Griffin FM, Bianco C, Silverstein SC (1975) Characterization of the macrophage receptor for complement and demonstration of its functional independence from the receptor for the Fc portion of immunoglobulin G. J Exp Med 141: 1269–1277

Harlan JM (1985) Leukocyte-endothelial interactions. Blood 65: 513–525

Harlan JM, Killen PD, Senecal FM, Schwartz BR, Yee EK, Taylor RF, Beatty PG, Price TH, Ochs HD (1985) The role of neutrophil membrane glycoprotein GP-150 in neutrophil adherence to endothelium in vitro. Blood 66: 167–178

Hemler ME (1988) Adhesive protein receptors on hematopoietic cells. Immunol Today 9: 109–113

Hemler ME, Huang C, Takada Y, Schwarz L, Strominger JL, Clabby ML (1987) Characterization of five distinct cell surface heterodimers each with a common 130000 molecular weight β subunit. J Biol Chem 262: 3300–3309

Hemler ME, Crouse C, Takada Y, Sonnenberg A (1988) Multiple very late antigen (VLA) heterodimers on platelets: evidence for distinct VLA-2, VLA-5 (fibronectin receptor), and VLA-6 structures. J Biol Chem 263: 7660–7665

Hickstein DD, Hickey MJ, Ozols J, Baker DM, Back AL, Roth GJ (1989) cDNA sequence for the aM subunit of the human neutrophil adherence receptor indicates homology to integrin α subunits. Proc Natl Acad Sci USA 86: 257–261

Hogg N, Horton MA (1987) Myeloid antigens: new and previously defined clusters. In: McMichael AT (ed) Leucocyte typing III: white cell differentiation antigens. Oxford University Press, Oxford, pp 576–602

Holzman B, McIntyre BW, Weissman IL (1989) Identification of a murine Peyer's patch-specific lymphocyte homing receptor as an integrin molecule with an α chain homologous to human VLA-4α. Cell 56: 37–46

Hynes RO (1987) Integrins: a family of cell surface receptors. Cell 48: 549–554

Ismail G, Morgenroth ML, Todd RF, Boxer LA (1987) Prevention of pulmonary injury in isolated perfused rat lungs by activated human neutrophils preincubated with anti-Mo-1 monoclonal antibody. Blood 69: 1167–1174

Keizer GD, Te Velde AA, Schwarting R, Figdor CG, de Vries JE (1987) Role of p150,95 in adhesion, migration, chemotaxis and phagocytosis of human monocytes. Eur J Immunol 17: 1317–1322

Kishimoto TK, Hollander N, Roberts TM, Anderson DC, Springer TA (1987a) Heterogeneous mutations in the β subunit common to the LFA-1, Mac-1, and p150,95 glycoproteins cause leukocyte adhesion deficiency. Cell 50: 193–202

Kishimoto TK, O'Connor K, Lee A, Roberts TM, Springer TA (1987b) Cloning of the β subunit of the leukocyte adhesion proteins: homology to an extracellular matrix receptor defines a novel supergene family. Cell 48: 681–690

Klempner MS, Cendron M, Wyler DJ (1983) Attachment of plasma membrane vesicles of human macrophages to *Leishmania tropica* promastigotes. J Infect Dis 148: 377–384

Krensky AM, Mentzer SJ, Clayberger C, Anderson DC, Schmalstieg FC, Burakoff SJ, Springer TA (1985) Heritable lymphocyte function-associated antigen-1 defiency: abnormalities of cytotoxicity and proliferation associated with abnormal expression of LFA-1. J Immunol 135: 3102–3108

Larson RS, Corbi AL, Berman L, Springer T (1989) Primary structure of the leukocyte function-associated molecule-1 α subunit: an integrin with an embedded domain deifining a protein superfamily. J Cell Biol 108: 703–712

Law SKA (1988) C3 receptors on macrophages. In: Gordon S (ed) Macrophage plasma membrane receptors: structure and function. J Cell Sci (Suppl 9): 67–97

Law SKA (1989) Complement receptor type III (CR3) and related proteins. In: Sim RB (ed) Biochemistry and molecular biology of complement. MTP Press, Lancester (in press)

Law SKA, Gagnon J, Hildreth JEK, Wells CE, Willis AC, Wong AJ (1987) The primary structure of the β-subunit of the cell surface adhesion glycoproteins LFA-1, CR3 and p150,95 and its relationship to the fibronectin receptor. EMBO J 6: 915–919

Lay WH, Nussenzweig V (1968) Receptors for complement on leukocytes. J Exp Med 128: 991–1010

Lisowska-Grospierre B, Bohler MC, Fischer A, Mawas C, Springer TA, Giscelli C (1986) Defective membrane expression of the LFA-1 complex may be secondary to the absence of the β chain in a child with recurrent bacterial infection. Eur J Immunol 16: 205–208

Lunn ER, Perry VH, Brown MC, Rosen H, Gordon S (1989) Absence of Wallerian degeneration does not hinder regeneration in peripheral nerve. Eur J Neurosci 1: 27–33

Malhotra V, Hogg N, Sim RB (1986) Ligand binding by the p150,95 antigen of U937 monocytic cells: properties in common with complement receptor type 3 (CR3). Eur J Immunol 16: 1117–1123

Marlin SD, Morton CC, Anderson DC, Springer TA (1986) LFA-1 immunodefixiency disease: definition of the gentic defect and chromosomal mapping of the α and β subunits of the lymphocyte function associated antigen 1 (LFA-1) by complementation in hybrid cells. J Exp Med 164: 855–867

Micklem KJ, Sim RB (1985) Isolation of complement-fragment-iC3b-binding proteins by affinity chromatography: the identification of p150,95 as an iC3b-binding protein. Biochem J 231: 233–236

Miller LJ, Bainton DF, Borregaard N, Springer TA (1987) Stimulated mobilization of monocyte Mac-1 and p150,95 adhesion proteins from an intracellular vesicular compartment to the cell surface. J Clin Invest 80: 535–554

Mole JE, Anderson JK, Davison EA, Woods DE (1984) Complete primary structure for the zymogen of human complement factor B. J Biol Chem 259: 3407–3412

Mosser DM, Edelson PJ (1985) The mouse macrophage receptor for iC3b is a major mechanism in the phagocytosis of *Leishmania* promastigotes. J Immunol 135: 2785–2788

Mosser DM, Vlassara H, Edelson PJ, Cerami A (1987) *Leishmania* promastigotes are recognized by the macrophage receptor for advanced glycosylation end products. J Exp Med 165: 140–145

Myones BL, Daizell JG, Hogg N, Ross GD (1988) Neutrophil and monocyte cell surface p150,95 has iC3b-receptor (CR4) activity resembling CR3. J Clin Invest 82: 640–651

Nermut MX, Green NM, Eason P, Yamada KM (1988) Electron microscopy and structural model of human fibronectin receptor. EMBO J 7: 4095–4099

O'Shea JJ, Brown EJ, Seligmann BE, Metcalf JA, Frank MM, Gallin JI (1985) Evidence for distinct intracellular pools of receptors for C3b and C3bi in human neutrophils. J Immunol 134: 2580–2587

Painter RG, Sklar LA, Jesaitis AJ, Schmitt M, Cochrane CG (1984) Activation of neutrophils by N-formyl chemotactic peptides. Fed Proc 43: 2737–2743

Peters MS, Kamarck ME, Hemler ME, Strominger JL, Ruddle FH (1984) Genetic and biological characterization of human lymphocyte cell surface antigens: the A-1A5 and A-3A4 determinants. J Exp Med 159: 1441–1445

Petrequin PR, Todd RF, Devaili LJ, Boxer LA, Curnutte JT (1987) Association between gelatinase release and increased plasma membrane expression of the Mo1 glycoprotein. Blood 69: 605–610

Piershbacher MD, Ruoslahti E (1984) The cell attachment activity of fibronectin can be duplicated by small synthetic fragments of the molecule. Nature 309: 30–33

Pohlman TH, Stanness KA, Beatty PG, Ochs HD, Harlan JM (1986) An endothelial cell surface factor(s) induced in vitro by lipopolysacchardie, interleukin 1, and tumor necrosis factor-α increases neutrophil adherence by a CDw18-dependent mechanism. J Immunol 136: 4548–4553

Poncz M, Eisman R, Heidenreich R, Silver SM, Vilaire G, Surrey S, Schwartz E, Bennet JS (1987) Structure of the platelet membrane glycoprotein IIb: homology to the α subunits of the vitronectin and fibronectin membrane receptors. J Biol Chem 262: 8476–8482

Price TH, Beatty PG, Corpuz SR (1987) In vivo inhibition of neutrophil function in the rabbit using monoclonal antibody to CD18. J Immunol 139: 4174–4177

Pytela R (1988) Amino acid sequence of the murine Mac-1 α chain reveals homology with the integrin family and an additional domain related to von Willebrand factor. EMBO J 7: 1371–1378

Rosen H, Gordon S (1987) Monoclonal antibody to the murine type 3 complement receptor inhibits adhesion of myelomonocytic cells in vitro and inflammatory cell recruitment in vivo. J Exp Med 166: 1685–1701

Rosen H, Milon G, Gordon S (1989a) Antibody to the murine type 3 complement receptor inhibits T lymphocyte-dependent recruitment of myelomonocytic cells in vivo. J Exp Med 169: 535–549

Rosen H, Gordon S, North RJ (1989b) Exacerbation of murine listeriosis by a monoclonal antibody specific for the type 3 complement receptor of myelomonocytic cells. J Exp Med 170: 27–39

Ross GD, Lambris JD (1982) Identification of the C3bi-specific membrane complement receptor that is expressed on lymphocytes, monocytes, neutrophils and erythrocytes. J Exp Med 155: 96–110

Ross GD, Cain JA, Lachmann PJ (1985a) Membrane complement receptor type three (CR3) has lectin-like properties analogous to bovine conglutinin and functions as a receptor for zymosan and rabbit erythrocytes as well as a receptor for iC3b. J Immunol 134: 3307–3315

Ross GD, Thompson RA, Walport MJ, Springer TA, Watson JV, Ward RHR, Lida J, Newman SL, Harrison RA, Lachmann PJ (1985b) Characterization of patients with an increased susceptibility to bacterial infections and a genetic deficiency of leukocyte membrane complement receptor type 3 and the related membrane antigen LFA-1. Blood 66: 882–890

Ross GD, Cain JA, Myones BL, Newman SL, Lachmann PJ (1987) Specificity of membrane complement receptor type three (CR3) for β-glucans. Complement 4: 61–74

Ruoslahti E (1988) Fibronectin and its receptors. Annu Rev Biochem 57: 375–413
Ruoslahti E, Pierschbacher MD (1987) New perspectives in cell adhesion: RGD and integrins. Science 238: 491–497
Russell DG, Wright SD (1988) Complement receptor type 3 (CR3) binds to an Arg-Gly-Asp-containing region of the major surface glycoprotein, gp63, of *Leishmania* promastigotes. J Exp Med 168: 279–292
Sanchez-Madrid F, Nagy JA, Robbins E, Simon P, Springer TA (1983) A human leukocyte differentiation antigen family with distinct α-subunits and a common β-subunit: the lymphocyte function-associated antigen (LFA-1), the C3bi complement receptor (OKM1/Mac-1), and the p150,95 molecule. J Exp Med 158: 1785–1803
Santoro SA, Lawing WL (1987) Competition for related but nonidentical binding sites on the glycoprotein IIb-IIIa complex by peptides derived from platelet adhesive proteins. Cell 48: 867–873
Shelton-Inloes BB, Titani K, Sadler JE (1986) cDNA sequences for human von Willebrand factor reveal five types of repeated domains and five possible sequence polymorphisms. Biochemistry 25: 3164–3171
Silverstein SC, Steinman RM, Cohn ZA (1977) Endocytosis. Annu Rev Biochem 46: 699–722
Simpson PJ, Todd RF, Fantone JC, Mickelson JK, Griffin FM, Lucchesi BR (1988) Reduction of experimental canine myocardial reperfusion injury by a monoclonal antibody that inhibits leukocyte adhesion. J Clin Invest 81: 624–629
Smith JW and Cheresh DA (1988) The Arg-Gly-Asp binding domain of the vitronectin receptor: photoaffinity cross-linking implicates amino acid residues 61-203 of the β subunit. J Biol Chem 263: 18726–18731
Solomon E, Palmer R, Hing S, Law SKA (1988) Regional localization of CD18, the β-subunit of the cell surface adhesion molecule LFA-1, on human chromosome 21 by in situ hybridization. Ann Human Genet 52: 123–128
Sonnenberg A, Modderman PW, Hogervorst F (1988) Laminin receptor on platelets is the integrin VLA-6. Nature 336: 487–489
Sosnoski DM, Emanuel BS, Hawkins AL, van Tuinen P, Ledbetter DH, Nussbaum RL, Kaos FT, Schwartz E, Phillips D, Bennett JS, Fitzgerald LA, Poncz M (1988) Chromosomal localization of the genes for the vitronectin and fibronectin receptors α subunits and for platelet glycoproteins IIb and IIIa. J Clin Invest 81: 1993–1998
Springer TA, Anderson DC (1986) Leukocyte complement receptors and adhesion proteins in the inflammatory response: insights from an experiment of nature. Biochem Soc Symp 51: 47–57
Springer T, Galfre G, Secher DS, Milstein C (1979) Mac-1: a macrophage differentiation antigen identified by a monoclonal antibody. Eur J Immunol 9: 301–309
Springer TA, Thompson WS, Miller LJ, Schmalstieg FC, Anderson DC (1984) Inherited deficiency of the Mac-1, LFA-1 p150,95 glycoprotein family and its molecular basis. J Exp Med 160: 1901–1918
Suzuki S, Argraves WS, Arai H, Languino LR, Pierschbacher MD, Ruoslahti E (1987) Amino acid sequence of the vitronectin receptor α subunit and comparative expression of adhesion receptor mRNAs. J Biol Chem 262: 14080–14085
Szebenyi DME, Obendorf SK, Moffat K (1981) Structure of vitamin D-dependent calcium-binding protein from bovine intestine. Nature 294: 327–332
Takada Y, Huang C, Hemler ME (1987a) Fibronectin receptor structures in the VLA family of heterodimers. Nature 326: 607–609
Takada Y, Strominger JL, Hemler ME (1987b) The very late antigen family of heterodimers is part of a superfamily of molecules involved in adhesion and embryogenesis. Proc Natl Acad Sci USA 84: 3229–3243
Takada Y, Wayner EA, Carter WG, Hemler ME (1988) Extracellular matrix receptors, ECMRII and ECMRI, for collagen and fibronectin correspond to VLA-2 and VLA-3 in the VLA family of heterodimers. J Cell Biochem 37: 385–393
Todd RF, Freyer DR (1988) The CD11/CD18 leukocyte glycoprotein deficiency. In: Curnutte JT (ed) Phagocytic defects. Saunders, New York, pp 13–31

Todd RF, Arnaout MA, Rosin RE, Crowley CA, Peters WA, Babior BM (1984) Subcellular localization of the large subunit of Mo1 (Mo1α; formerly gp 110), a surface glycoprotein associated with neutrophil adhesion. J Clin Invest 74: 1280–1290

Todd RF, Simpson PJ, Lucchesi BR (1989) The anti-inflammatory properties of monoclonal anti-Mo1 antibodies in vitro and in vivo. In: Rosenthal AS, Springer TA, Anderson DC, Rothlein R (eds) Structure and function of molecules involved in leukocyte adhesion. Springer, New York Berlin Heidelberg

Tonnesen MG, Anderson DC, Springer TA, Knedler A, Avdi N, Henson PM (1989) Adherence of neutrophils to cultured human microvascular endothelial cells: stimulation by chemotactic peptides and lipid mediators and dependence upon the Mac-1, LFA-1, p150,95 glycoprotein family. J Clin Invest 83: 637–646

Vedder NB, Harlan JM (1988) Increased surface expression of CD11b/CD18 (Mac-1) is not required for stimulated neutrophil adherence to cultured endothelium. J Clin Invest 81: 676–682

Vedder NM, Winn RK, Rice CL, Chi EY, Arfors K-E, Harlan JM (1988) A monoclonal antibody to the adherence-promoting leukocyte glycoprotein CD18 reduces organ injury and improves survival from hemorrahgic shock in rabbits. J Clin Invest 81: 939–944

Wayner EA, Carter WG, Piotrowicz RS, Kunicki TJ (1988) The function of multiple extracellular matrix receptors in mediating cell adhesion to extracellular matrix: preparation of monoclonal antibodies to fibronectin receptor that specifically inhibit cell adhesion to fibronectin and react with platelet glycoproteins Ic–IIa. J Cell Biol 107: 1881–1891

Wells CE, Law SKA (1987) RFLP of the β-subunit of the cell surface adhesion glycoproteins (Abstract). Complement 4: 238

Wright SD, Jong MTC (1986) Adhesion-promoting receptors on human macrophages recognize *Escherichia coli* by binding to lipopolysaccharide. J Exp Med 164: 1876–1888

Wright SD, Meyer BC (1986) Phorbol esters cause sequential activation and deactivation of complement receptors on polymorphonuclear leukocytes. J Immunol 136: 1759 to 1764

Wright SD, Silverstein SC (1982) Tumor-promoting phorbol esters stimulate C3b and C3b' receptor-mediated phagocytosis in cultured human monocytes. J Exp Med 156: 1149–1164

Wright SD, Silverstein SC (1983) Receptors for C3b and C3bi promote phagocytosis but not the release of toxic oxygen from human phagocytes. J Exp Med 158: 2016–2023

Wright SD, Silverstein SC (1984) Phagocytosing macrophages exclude proteins from the zone of contact with opsonized targets. Nature 309: 359–361

Wright SD, Craigmyle LS, Silverstein SC (1983a) Fibronectin and serum amyloid P component stimulate C3b- and C3bi-mediated phagocytosis in cultured human monocytes. J Exp Med 158: 1338–1343

Wright SD, Rao PE, Van Voorhis WC, Craigmyle LS, Iida K, Talle MA, Westberg EF, Gold-stein G, Silverstein SC (1983b) Identification of the C3bi receptor of human monocytes and macrophages by using monoclonal antibodies. Proc Natl Acad Sci USA 80: 5699–5703

Wright SD, Licht MR, Craigmyle LS, Silverstein SC (1984) Communication between receptors for different ligands on a single cell: ligation of fibronectin receptors induces a reversible alteration in the function of complement receptors on cultured human monocytes. J Cell Biol 99: 336–339

Wright SD, Detmers PA, Jong MTC, Eyer BC (1986) Interferon-gamma depresses the binding of ligand by C3b and C3bi receptors on cultured human monocytes, an effect reversed by fibronectin. J Exp Med 163: 1245–1259

Wright SD, Reddy PA, Jong MTC, Erickson BW (1987) C3bi receptor (complement receptor type 3) recognizes a region of complement protein C3 containing the sequence Arg-Gly-Asp. Proc Natl Acad Sci USA 84: 1965–1968

Wright SD, Weitz JI, Huang AJ, Levin SM, Silverstein SC, Loike JD (1988) Complement receptor type three (CD11b/CD18) of human polymorphonuclear leukocytes recognizes fibrinogen. Proc Natl Acad Sci USA 85: 7734–7738

Wright SD, Levin SM, Jong MTC, Chad Z, Kabbash LG (1989) CR3 (CD11b/CD18) expresses on binding site for Arg-Gly-Asp-containing peptides and a second site for bacterial lipopolysaccharide. J Exp Med 169: 175–183

Yamamoto K, Johnston RB (1984) Dissociation of phagocytosis from stimulation of the oxiative metabolic burst in macrophages. J Exp Med 159: 405–416

Zimmerman GA, McIntyre TM (1988) Neutrophil adherence to human endothelium in vitro occurs by CDw18 (Mo1, Mac-1/LFA-1/gp 150,95) glycoprotein-dependent and -independent mechanisms. J Clin Invest 81: 531–537

Zimrin AB, Eisman R, Vilaire G, Schwartz E, Bennett JS, Poncz M (1988) Structure of platelet glycoprotein IIIa: a common subunit for two different membrane receptors. J Clin Invest 81: 1470–1475

Decay-Accelerating Factor and Membrane Cofactor Protein*

D. M. Lublin[1], and J. P. Atkinson[2]

1   Decay-Accelerating Factor   124
1.1 Identification and Purification   124
1.2 Biosynthesis and Glycosylation of DAF   124
1.3 Glycophospholipid Anchor of Membrane DAF   126
1.4 Sites of Expression and Alternate Forms of DAF   127
1.5 Cloning of DAF cDNA   128
1.6 DAF Gene   131
1.7 Biochemical Activities and Physiological Roles   131
1.8 Paroxysmal Nocturnal Hemoglobinuria   132
1.9 Blood Group Antigens on DAF   133
1.10 DAF in Other Species   133

2   Membrane Cofactor Protein   133
2.1 Identification and Purification   133
2.2 Structural Variations of MCP on Human Cells   134
2.3 Biosynthesis and Glycosylation of MCP   136
2.4 Cloning of MCP cDNA   136
2.5 MCP Gene   138
2.6 Sites of Expression   139
2.7 Homologous Proteins in Other Species   139
2.8 Physiological Role   139

3   DAF and MCP   140

References   142

Cells exposed to plasma proteins are frequently under attack from the complement system. This can arise either as a bystander process to the classical or alternative pathways of activation initiated during the immune response to foreign particles and organisms or from the constant tick-over of the alternative pathway. Thus, it is critical for the cell to regulate the complement pathway on its own surface. The plasma proteins, H and C4 binding protein (C4bp), in conjugation with the serine protease I, function to this end. Additionally, cells possess a number of membrane proteins

---

[1] Division of Laboratory Medicine, Departments of Pathology and Medicine, Washington University School of Medicine, St. Louis, MO 63110, USA
[2] Division of Rheumatology, Department of Medicine, and Howard Hughes Medical Institute Laboratories, Washington University School of Medicine, St. Louis, MO 63110, USA
* Portions of this material appear in *Annual Review of Immunology,* Volume 7, 1989. This represents an update on a chapter in Sim R (ed), *Biochemistry and Molecular Biology of Complement* MTP Press, 1989.

to regulate complement deposited on their surfaces; the largest group, focused on C3 and the C3 convertases, consists of the C3b/C4b receptor (CR1), decay-accelerating factor (DAF), and membrane cofactor protein (MCP). CR1, although it has both decay-accelerating activity and serves as a cofactor for the I-mediated cleavage of C3b and C4b, acts mainly extrinsically as a receptor for C3b-bearing immune complexes. DAF exerts its decay-accelerating activity intrinsically on the cell itself (see below). Indeed, the lack of DAF in the membrane of blood cells in the disease paroxysmal nocturnal hemoglobinuria (PNH) leads to an increased complement sensitivity of these cells. Purified MCP can also regulate C3 and the C3 convertases through cofactor I activity. MCP has the same approximate size and overall structure as DAF, and hence it might also function intrinsically to control C3 convertases formed on the same cell. This chapter reviews in detail the structure, both at the protein and DNA levels, of these two complement regulatory membrane glycoproteins, DAF and MCP, and discusses their physiological roles in protecting cells from damage by autologous complement.

1 Decay-Accelerating Factor

1.1 Identification and Purification

In 1969 HOFFMANN reported that a substance remaining in the aqueous phase from an extraction of human erythrocyte stroma with n-butanol could inhibit the complement-mediated hemolysis of antibody-coated sheep erythrocytes (HOFFMANN 1969a). Furthermore, this inhibition involved an acceleration in the decay of EAC14b2a to EAC14b (HOFFMANN 1969b). It was over a decade later when NICHOLSON-WELLER and colleagues purified an intrinsic membrane glycoprotein from guinea pig and human erythrocyte (E) stroma by butanol extraction followed by sequential chromatography on DEAE-Sephacel, hydroxylapatite, phenyl-Sepharose, and trypan blue Sepharose (NICHOLSON-WELLER et al. 1981, 1982). This protein was purified during this scheme by monitoring its ability to accelerate the decay of the classical pathway C3 convertase, and hence named decay-accelerating factor. This resulted in a membrane protein, DAF, with an M_r of 60000 (guinea pig) or 70000 (human) on sodium dodecyl sulfate polyacrylamide gel electrophoresis (SDS-PAGE) under reducing conditions. Staining of human DAF with periodic acid Schiff's reagent demonstrated that it was a glycoprotein.

1.2 Biosynthesis and Glycosylation of DAF

DAF undergoes several important posttranslational modifications to attain its final overall structure in the cell membrane. These have been elucidated for DAF (all references are to human DAF unless otherwise noted) by studying the biosynthesis of DAF in tissue culture and by chemical and enzymatic analysis of purified DAF.

Analysis of the oligosaccharide structure of DAF by endo- and exoglycosidase digestions showed that DAF from E has one N-linked complex-type oligosaccharide (accounting for approximately 3000M_r on SDS-PAGE) and multiple, highly sialylated

O-linked oligosaccharides (accounting for approximately $26000 M_r$; LUBLIN et al. 1986b), as shown in Fig. 1. Similar results were obtained for DAF from peripheral blood granulocytes and cell lines such as HL-60, except for partial resistance to enzymatic removal of O-linked oligosaccharides. This suggests that the higher M_r of DAF on white blood cells versus E (see below) might arise from differences in O-linked glycosylation, but this point requires further investigation. Studies of the biosynthesis of DAF in the HL-60 and K-562 cell lines showed an intracellular pro-DAF form of M_r 46000 that possesses one N-linked high-mannose unit, added cotranslationally, but no O-linked oligosaccharides (Fig. 2; LUBLIN et al. 1986b). A similar pro-DAF species (assigned M_r 48000) was found in HeLa cells (MEDOF et al. 1986). Brief pulse labelings of less than 15 min demonstrate an additional species of M_r 43000 (Fig. 2) with the same N- and O-linked oligosaccharides as pro-DAF. Kinetic studies suggest that this is an earlier species of DAF, but the structural difference between these two DAF intracellular forms is unknown (LUBLIN et al. 1986b). Pro-DAF proceeds through the Golgi complex, where the one N-linked oligosaccharide is modified to a complex type and multiple O-linked oligosaccharides are added to produce the mature form of the protein seen on the cell surface. All forms of DAF have a slower migration on SDS-PAGE under reducing conditions compared to nonreducing, indicating the presence of intrachain disulfide bonds.

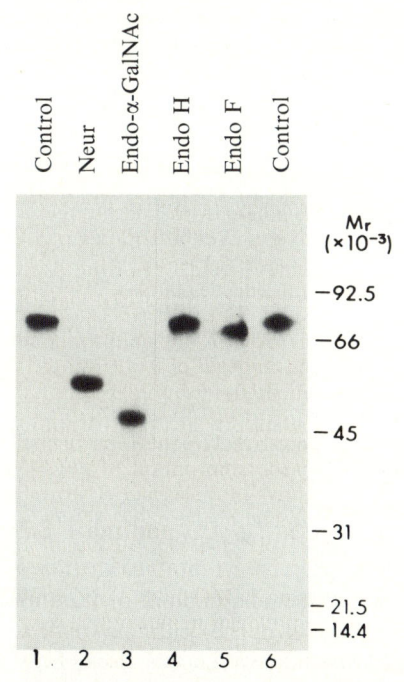

Fig. 1. N- and O-linked oligosaccharide structure of erythrocyte membrane DAF. Erythrocytes were prepared from peripheral blood of a healthy human donor and were then surface-labeled with ^{125}I. DAF was immunoprecipitated from detergent lysates and then divided into equal aliquots for treatment with enzymes. The samples were analyzed by SDS-PAGE (on a 9% gel under reducing conditions) and autoradiography. Enzyme treatments are neuraminidase (*lane 2*), neuraminidase plus endo-α-N-acetylgalactosaminidase (*lane 3*), endoglycosidase H (*lane 4*), endoglycosidase F (*lane 5*), or buffer alone (*lanes 1, 6*). (Reprinted from LUBLIN et al. 1986b)

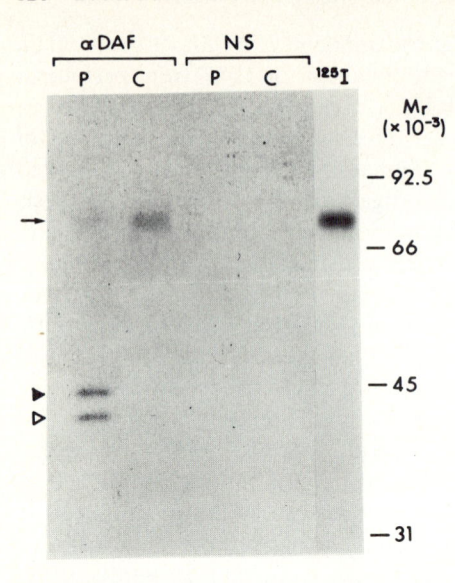

Fig. 2. Biosynthetic labeling of DAF in HL-60 cells. HL-60 cells (differentiated for 48 h with vitamin D to increase DAF expression) were biosynthetically labeled with [^{35}S]methionine during a 10-min pulse (*P*) followed by a 60-min chase (*C*) with unlabeled methionine. The detergent lysate from each condition was divided in half and was immunoprecipitated with either anti-DAF antibody (α*DAF*) or nonspecific (*NS*) control nonimmune rabbit Ig and was then analyzed by SDS-PAGE (under reducing conditions) and fluorography. Another aliquot of HL-60 cells was surface labeled with ^{125}I, immunoprecipitated with anti-DAF antibody, and analyzed by SDS-PAGE and autoradiography (*lane 5*). *Arrow*, the position of mature DAF (M_r 80000); *open* and *solid arrowheads,* positions of the DAF species of 43000 and 46000 M_r, respectively. (Reprinted from LUBLIN et al. 1986b)

1.3 Glycophospholipid Anchor of Membrane DAF

A striking observation concerning DAF was that when purified DAF from the E membrane was added back to a cell suspension, it reincorporated in the membrane, apparently as an integral membrane protein, and it displayed functional activity (MEDOF et al. 1984). This physicochemical property was one reason prompting an examination of the membrane-anchoring domain of DAF by two groups. DAF was found to belong to a recently described class of membrane proteins that are anchored by covalent attachment of the carboxy-terminus to a glycophospholipid containing phosphatidylinositol (PI) that is inserted in the outer leaflet of the lipid bilayer (this class of membrane compounds is reviewed in FERGUSON and WILLIAMS 1988; LOW and SALTIEL 1988). This anchoring was first shown by DAVITZ and colleagues, who demonstrated the release of DAF from peripheral blood cells following treatment with phosphatidylinositol-specific phospholipase C (PI-PLC; DAVITZ et al. 1986). Specifically, 60%–80% of membrane DAF was released from leukocytes by PI-PLC, although only 10% of E membrane DAF was removed. This partial resistance to PI-PLC has been found in other glycophospholipid-anchored proteins, and it may represent selective acylation of the inositol ring (ROBERTS et al. 1988). DAF released by PI-PLC has lost its hydrophobic character and its ability to reincorporate into cell membranes, and it thus could not intrinsically inhibit assembly of the C3 convertase on the cell surface (MEDOF et al. 1986; DAVITZ et al. 1986). However, this hydrophilic form of DAF could still accelerate the decay of preformed C4b2a, albeit at a much reduced efficiency, and thus the functional site on DAF can be separated from the glycophospholipid anchor (MEDOF et al. 1986).

The complete structure of the glycophospholipid anchor has been determined for the trypanosome variant surface glycoprotein (FERGUSON et al. 1988) and for Thy-1 antigen from rat brain (HOMANS et al. 1988), and these anchors show an identical backbone with variation in the side-chain groups. Chemical analysis of the anchor from E DAF, although less detailed, is consistent with these structures. These studies by MEDOF and coworkers (MEDOF et al. 1986; WALTER et al. 1987) demonstrated the presence of ethanolamine and glucosamine (1.8 and 0.8 moles per mole of DAF protein, respectively) in the carboxy terminus of the protein, as well as inositol (0.7 mole) and a mixture of saturated and unsaturated fatty acids (0.7 and 1.2 moles, respectively). In addition, analysis by thin layer chromatography of labeled anchor fragments released by nitrous acid deamination (which cleaves at the nonacetylated glucosamine) revealed the presence of inositol phospholipids other than PI. This could explain the partial resistance to PI-PLC described previously.

1.4 Sites of Expression and Alternate Forms of DAF

DAF is present on virtually all peripheral blood cells: E, granulocytes, T- and B-lymphocytes, monocytes, and platelets (KINOSHITA et al. 1985; NICHOLSON-WELLER et al. 1985a). The DAF molecule from leukocytes has a 3000–9000 higher M_r than E DAF. Interestingly, DAF is absent on natural killer cells (NICHOLSON-WELLER et al. 1986). DAF has also been found on bone marrow mononuclear cells and erythroid progenitors (MOORE et al. 1985). It is present on the epithelial surface of cornea, conjuctiva, oral and gastrointestinal mucosa, exocrine glands, renal tubules, ureter and bladder, cervical and uterine mucosa, and pleural, pericardial and synovial serosa (MEDOF et al. 1987b), as well as on cultured umbilical vein endothelial cells (ASCH et al. 1986). Overall, what is clear is that DAF has a very wide tissue distribution, supporting the important role of this protein in controlling the complement system.

Soluble forms of DAF have been found in extracellular fluids and tissue culture supernatants. DAF was detected in body fluids using a two-site radioimmunometric assay (MEDOF et al. 1987b). DAF antigen was present in plasma, urine, tears, saliva, synovial fluid, and cerebrospinal fluid, with levels ranging from 40 to 400 ng/ml. Analysis by immunoprecipitation and Western blot showed that the DAF from plasma, tears, and saliva had an apparent M_r of approximately 100000, whereas the urine DAF had a M_r of 67000, slightly lower than E membrane DAF. The urine DAF species was less hydrophobic than membrane DAF, did not inhibit the intrinsic assembly of C3 convertases on the cell surface, but could accelerate the decay of preformed C4b2a with an efficiency comparable to that of C4bp. Urine DAF is thus similar to PI-PLC released membrane DAF (MEDOF et al. 1986). A DAF species of the same size as urine DAF was also detected in the culture supernatants of the HeLa epithelial cell line (MEDOF et al. 1987b), prompting the suggestion that urine DAF might be synthesized by the adjacent urethelium.

Alternate forms of the membrane DAF molecule have also been described. A larger variant, designated DAF-2, was detected on E membranes by Western blot (KINOSHITA et al. 1987). DAF-2 possesses a M_r of 140000 and represents less than 10% of membrane DAF. This variant has functional activity for accelerating the decay of the C3 convertase and also shares with DAF the ability to reincorporate into E membranes, suggesting the presence of the glycophospholipid anchor. The apparent M_r of DAF-2 is twice that of DAF, raising the possibility that it is a dimer of DAF, although neither reduction with 2-mercaptoethanol nor denaturation in SDS could separate DAF-2 into two components. The structure of DAF-2 thus remains unexplained.

Lower M_r degradation fragments of membrane DAF have been produced in vitro by treatment of DAF with PI-PLC (MEDOF et al. 1986; DAVITZ et al. 1986; SEYA et al. 1987), a PI-specific phospholipase D from serum (DAVITZ et al. 1987), or papain (MEDOF et al. 1986; SEYA et al. 1987). Interestingly, incubation of surface-labeled E with leukocytes led to release of a fragment of equal size to a papain-derived fragment (SEYA et al. 1987). It is unknown whether any of these degradative processes are relevant in vivo.

1.5 Cloning of DAF cDNA

Two groups have independently cloned DAF cDNAs (MEDOF et al. 1987a; CARAS et al. 1987a). Both groups utilized oligonucleotide probes based on the amino-terminal sequence of immunoaffinity-purified E DAF to isolate cDNA clones. These clones were derived from libraries constructed with mRNA from either the HeLa epithelial cell line (MEDOF et al. 1987a; CARAS et al. 1987a) or the HL-60 promyelocytic leukemia cell line (MEDOF et al. 1987a). The nucleotide and derived amino acid sequences for DAF are shown in Fig. 3. There is a single long open reading frame beginning with an initiation methionine codon and extending 1143 bp; this is surrounded by 5′ and 3′ untranslated regions, the latter ending in a poly(A) track. The deduced amino acid sequence yields a protein of 381 amino acids including a 34 amino acid signal peptide. Starting at the amino terminus of the mature protein, there are four contiguous short consensus repeat (SCR) units of approximately 60 amino acids. These SCRs, which contain conserved residues of cysteine, proline, tryptophan, and several other amino acids, are homologous to domains found in other complement regulatory proteins, including CR1, CR2, C4bp, and H, as well as in several noncomplement proteins. The SCRs are followed by a 70 amino acid region that is rich in serine and threonine residues (45%). A similar serine- and threonine-rich region, located just extracellular to the plasma membrane, is the site of clustered O-glycosylation in the low-density lipoprotein receptor (DAVIS et al. 1986). This is consistent with the large amount of O-linked oligosaccharide previously identified in DAF (LUBLIN et al. 1986b); the deduced protein sequence also shows one site for N-linked glycosylation, again as expected (LUBLIN et al. 1986b).

```
GCTGCGACTCGGCGGAGTCCCGGCGGCGCGTCCTTGTTCTAACCCGGCGCGCCATGACCGTCGCGCGGCCGAGCGTGCCCGCGGCGCTG     89
                                                     M  T  V  A  R  P  S  V  P  A  A  L     -23
                                                    -34

CCCCTCCTCGGGGAGCTGCCCCGGCTGCTGCTGCTGGTGCTGTTGTGCCTGCCGGCCGTGTGGGGTGACTGTGGCCTTCCCCCAGATGTA    179
 P  L  L  G  E  L  P  R  L  L  L  L  V  L  L  C  L  P  A  V  W  G  D  C  G  L  P  P  D  V      8
                                                               -1 +1

CCTAATGCCCAGCCAGCTTTGGAAGGCCGTACAAGTTTTCCCGAGGATACTGTAATAACGTACAAATGTGAAGAAAGCTTTGTGAAAATT    269
 P  N  A  Q  P  A  L  E  G  R  T  S  F  P  E  D  T  V  I  T  Y  K  C  E  E  S  F  V  K  I     38

CCTGGCGAGAAGGACTCAGTGATCTGCCTTAAGGGCAGTCAATGGTCAGATATTGAAGAGTTCTGCAATCGTAGCTGCGAGGTGCCAACA    359
 P  G  E  K  D  S  V  I  C  L  K  G  S  Q  W  S  D  I  E  E  F  C  N  R  S  C  E  V  P  T     68
                                                                    ▲

AGGCTAAATTCTGCATCCCTCAAACAGCCTTATATCACTCAGAATTATTTTCCAGTCGGTACTGTTGTGGAATATGAGTGCCGTCCAGGT    449
 R  L  N  S  A  S  L  K  Q  P  Y  I  T  Q  N  Y  F  P  V  G  T  V  V  E  Y  E  C  R  P  G     98

TACAGAAGAGAACCTTCTCTATCACCAAAACTAACTTGCCTTCAGAATTTAAAATGGTCCACAGCAGTCGAATTTTGTAAAAAGAAATCA    539
 Y  R  R  E  P  S  L  S  P  K  L  T  C  L  Q  N  L  K  W  S  T  A  V  E  F  C  K  K  K  S    128

TGCCCTAATCCGGGAGAAATACGAAATGGTCAGATTGATGTACCAGGTGGCATATTATTTGGTGCAACCATCTCCTTCTCATGTAACACA    629
 C  P  N  P  G  E  I  R  N  G  Q  I  D  V  P  G  G  I  L  F  G  A  T  I  S  F  S  C  N  T    158

GGGTACAAATTATTTGGCTCGACTTCTAGTTTTTGTCTTATTTCAGGCAGCTCTGTCCAGTGGAGTGACCCGTTGCCAGAGTGCAGAGAA    719
 G  Y  K  L  F  G  S  T  S  S  F  C  L  I  S  G  S  S  V  Q  W  S  D  P  L  P  E  C  R  E    188

ATTTATTGTCCAGCACCACCACAAATTGACAATGGAATAATTCAAGGGGAACGTGACCATTATGGATATAGACAGTCTGTAACGTATGCA    809
 I  Y  C  P  A  P  P  Q  I  D  N  G  I  I  Q  G  E  R  D  H  Y  G  Y  R  Q  S  V  T  Y  A    218

TGTAATAAAGGATTCACCATGATTGGAGAGCACTCTATTTATTGTACTGTGAATAATGATGAAGGAGAGTGGAGTGGCCCACCACCTGAA    899
 C  N  K  G  F  T  M  I  G  E  H  S  I  Y  C  T  V  N  N  D  E  G  E  W  S  G  P  P  P  E    248

TGCAGAGGAAAATCTCTAACTTCCAAGGTCCCACCAACAGTTCAGAAACCTACCACAGTAAATGTTCCAACTACAGAAGTCTCACCAACT    989
 C  R  G  K  S  L  T  S  K  V  P  P  T  V  Q  K  P  T  T  V  N  V  P  T  T  E  V  S  P  T    278

TCTCAGAAAACCACCACAAAAACCACCACACCAAATGCTCAAGCAACACGGAGTACACCTGTTTCCAGGACAACCAAGCATTTTCATGAA   1079
 S  Q  K  T  T  T  K  T  T  T  P  N  A  Q  A  T  R  S  T  P  V  S  R  T  T  K  H  F  H  E    308

ACAACCCCAAATAAAGGAAGTGGAACCACTTCAGGTACTACCCGTCTTCTATCTGGGCACACGTGTTTCACGTTGACAGGTTTGCTTGGG   1169
 T  T  P  N  K  G  S  G  T  T  S  G  T  T  R  L  L  S  G  H  T  C  F  T  L  T  G  L  L  G    338

ACGCTAGTAACCATGGGCTTGCTGACTTAGCCAAAGAAGAGTTAAGAAGAAAATACACACAAGTATACAGACTGTTCCTAGTTTCTTAGA   1259
 T  L  V  T  M  G  L  L  T  *                                                                 347

CTTATCTGCATATTGGATAAAATAAATGCAATTGTGCTCTTCATTTAGGATGCTTTCATTGTCTTTAAGATGTGTTAGGAATGTCAACAG   1349

AGCAAGGAGAAAAAAGGCAGTCCTGGAATCACATTCTTAGCACACCTACACCTCTTGAAAATAGAACAACTTGCAGAATTGAGAGTGATT   1439

CCTTTCCTAAAAGTGTAAGAAAGCATAGAGATTTGTTCGTATTTAGAATGGGATCACGAGGAAAAGAGAAGGAAAGTGATTTTTTTCCAC   1529

AAGATCTGTAATGTTATTTCCACTTATAAAGGAAATAAAAAATGAAAAACATTATTTGGATATCAAAAGCAAATAAAAACCCAATTCAGT   1619

CTCTTCTAAGCAAAATTGCTAAAGAGAGATGAACCACATTATAAAGTAATCTTTGGCTGTAAGGCATTTTCATCTTTCCTTCGGGTTGGC   1709

AAAATATTTTAAAGGTAAAACATGCTGGTGAACCAGGGGTGTTGATGGTGATAAGGGAGGAATATAGAATGAAAGACTGAATCTTCCTTT   1799

GTTGCACAAATAGAGTTTGGAAAAAGCCTGTGAAAGGTGTCTTCTTTGACTTAATGTCTTTAAAAGTATCCAGAGATACTACAATATTAA   1889

CATAAGAAAAGATTATATATTATTTCTGAATCGAGATGTCCATAGTCAAATTTGTAAATCTTATTCTTTTGTAATATTTATTTATATTTA   1979

TTTATGACAGTGAACATTCTGATTTTACATGTAAAACAAGAAAAGTTGAAGAAGATATGTGAAGAAAAATGTATTTTTCCTAAATAGAAA   2069

TAAATGATCCCATTTTTTGGTAAAAAAAAAAA                                                             2101
```

Fig. 3. Nucleotide and derived amino acid sequences of DAF cDNA. The nucleotide sequence is numbered from the most 5′ nucleotide, and the derived amino acid sequence, numbered from the first amino acid of the mature protein, is shown below using single-letter codes. *Asterisk*, stop codon; *arrow*, the single N-glycosylation site; *underline*, a serine/threonine-rich region (probable site of O-linked glycosylation); *box*, a carboxy-terminal hydrophobic region (replaced posttranslationally with a glycophospholipid anchor); *dashed lines*, potential polyadenylation signals

The deduced protein structure ends in a 24 amino acid segment of markedly hydrophobic character; the series of basic residues (that act as a stop anchor sequence) and the cytoplasmic tail that are present in polypeptide-anchored membrane proteins are not seen in DAF. However, this carboxy-terminal hydrophobic peptide is similar to extension peptides encoded by the cDNAs for other glycophospholipid-anchored membrane proteins such as the trypanosome variant surface glycoproteins (BOOTHROYD et al. 1981) and Thy-1 (TSE et al. 1985). These extension peptides of 17–31 amino acids are removed posttranslationally, and the carboxy-terminal glycophospholipid anchor is attached (reviewed in FERGUSON and WILLIAMS 1988; LOW and SALTIEL 1988). A similar processing presumably leads to the attachment of the DAF glycophospholipid anchor, perhaps with the hydrophobic extension peptide acting as a transient membrane anchor in the endoplasmic reticulum. Studies in HeLa cells demonstrated that the major intracellular pro-DAF species incorporated ethanolamine, a component of the glycophospholipid anchor (MEDOF et al. 1986). In addition, treatment of this precursor species with PI-PLC removed its hydrophobic domain, as judged by partitioning in Triton X-114 detergent (D. M. LUBLIN, unpublished results). These results demonstrate that the glycophospholipid anchor is already attached to this DAF precursor.

The signal for attachment of the glycophospholipid anchor to a given protein is still unclear. However, two groups have made and expressed mutant cDNAs containing the carboxy-terminal segment of DAF attached to the amino-terminal segment of another protein (CARAS et al. 1987b; TYKOCINSKI et al. 1988). These mutant cDNAs, when expressed in transfected cell lines, led to the production of a membrane protein that was glycophospholipid anchored, thus establishing that the carboxy-terminal amino acids contain the signal for attachment of a glycophospholipid anchor. The actual nature of that signal is still unknown.

The DAF cDNA detects several bands on Northern blot analysis of mRNA from various cell lines. The major species are reported as 2.0 and 2.7 kb (MEDOF et al. 1987a) or 1.5 and 2.2 kb (CARAS et al. 1987a), apparently simply reflecting M_r standardization differences. These two species of mRNA are products of alternative polyadenylation (CARAS et al. 1987a). Relative levels of DAF mRNA in the cell lines HeLa, HL-60, and HSB-2 correlated with the levels of DAF protein detected by immunoradiometric assay (MEDOF et al. 1987a), suggesting tissue-specific transcriptional control of DAF expression.

One group also found a second class of DAF cDNA clones (CARAS et al. 1987a). These contained an extra 118-bp insertion near the end of the coding region; the ensuing frame shift results in a longer encoded protein (440 amino acids including signal peptide) that now has a hydrophilic carboxy-terminus. They speculated that the 118-bp insertion represents an unspliced intron. A probe based on this sequence detected a minor species of DAF on Northern analysis of HeLa cell RNA. Transfection of these two types of cDNA into Chinese hamster ovary cells resulted in the production of DAF immunoreactive material, but only the cDNA encoding in the hydrophobic extension peptide produced surface DAF. Their suggestion was that the spliced and unspliced species of cDNA encode membrane and secreted DAF species, respectively. However, subsequent work has cast doubt on this hypothesis (D. M. LUBLIN, unpublished results; V. NUSSENZWEIG, personal communication). Chinese hamster ovary cells transfected with the spliced (regular or hydrophobic) DAF cDNA produce both membrane DAF attached by a glycophospholipid anchor and a secreted form of DAF approximately 5000 lower in M_r. In addition, antibodies raised against the carboxy-terminal hydrophilic peptide (encoded only by the alternate, unspliced cDNA) did not recognize the soluble DAF species in HeLa cell culture supernatants. The physiological relevance of this alternate DAF cDNA species, along with the origin of the secreted form of DAF, remains unclear.

1.6 DAF Gene

Southern blot analysis of human DNA shows a relatively simple pattern generated from restriction digests (LUBLIN et al. 1987; REY-CAMPOS et al. 1987; STAFFORD et al. 1988), suggesting that DAF is a single-copy gene. This was supported by hybridizations with DAF-specific oligonucleotide probes (STAFFORD et al. 1988). Three restriction fragment length polymorphisms (RFLPs) have been identified in the DAF gene: two for the enzyme *Hind*III and one for *Bam*HI (REY-CAMPOS et al. 1987; STAFFORD et al. 1988). All are located in the noncoding region of the gene. Cloning of the human DAF gene demonstrates it to be composed of ten exons spread over 50 kb of DNA (POST et al. 1989). These exons encode (sequentially in a 5′ to 3′ direction) the 5′ untranslated region/signal peptide, SCR I, SCR II, SCR III split between two equally sized exons, SCR IV, three exons encoding the serine/threonine-rich region, and the carboxy-terminal hydrophobic extension peptide/3′-untranslated region together on the final exon.

The chromosomal location of the DAF gene is on the long arm of human chromosome 1, band q3.2. This was derived from analysis of a panel of hamster × human somatic cell hybrids and by in situ hybridization of the DAF cDNA to human metaphase cells (LUBLIN et al. 1987). The same result was obtained by segregation analysis of the DAF RFLPs in families that are informative for segregation of alleles at the CR1, C4bp, and H loci (REY-CAMPOS et al. 1987). The latter three complement proteins were already known to be located at the regulator of complement activation (RCA) gene cluster at 1q3.2, so the DAF gene is added to this group. Furthermore, recombinations within the RCA locus demonstrated that DAF maps closer to the CR1/C4bp loci than to the H locus. Subsequent detailed mapping of the RCA gene cluster by pulsed-field gel electrophoresis has shown the order of the genes to be MCP-CR1-CR2-DAF-C4bp, located within an 800-kb segment of DNA on the long arm of chromosome 1 (REY-CAMPOS et al. 1988; CARROLL et al. 1988; BORA et al. 1989).

1.7 Biochemical Activities and Physiological Roles

Several lines of evidence indicate that DAF serves to protect cells from damage by autologous complement proteins deposited on their surfaces; specifically, this protection involves prevention of the assembly of the C3 and C5 convertases of the classical or alternative pathways, which act as amplification steps in the complement cascade. DAF was initially purified based on its ability to accelerate the spontaneous decay of the preformed classical C3 convertase, C4b2a (NICHOLSON-WELLER et al. 1982).

Blocking with antibodies to DAF showed that it was responsible for this function on intact E, for both the classical and alternative pathways, but that it lacked any cofactor activity for I-mediated cleavage of C4b or C3b (Pangburn et al. 1983). Pivotal insights into the role of DAF came from studies using DAF reincorporated into sheep E (Medof et al. 1984). These studies demonstrated that DAF inhibits the formation of the C3 and C5 convertases; this effect was reversible, as DAF did not affect the structure of C4b or C3b. In addition, DAF only exerted this effect intrinsically, i.e., on C3 convertases being assembled on the same cell as the DAF. Another group of investigators narrowed the site of action further by showing that DAF does not prevent the initial binding of C2 or B to the cell (containing C4b or C3b, respectively), but that it rapidly dissociates C2a or Bb from their binding sites, thus preventing the assembly of the C3 convertase (Fujita et al. 1987). The precise mechanism underlying this interference with the C3 convertase and the specific binding sites for DAF on the C3 convertase are still unclear. One group utilized a homobifunctional cross-linking reagent to show an endogenous association of DAF with C4b and C3b on the cell surface (Kinoshita et al. 1986). Another investigation using fluid-phase competitive inhibition suggested that the primary interaction of DAF with C3 convertases is with the C2a or Bb components (Pangburn 1986). The experimental systems of these two groups are quite different, but the discrepancy in their results has not been resolved.

1.8 Paroxysmal Nocturnal Hemoglobinuria

The physiological role of DAF has been elucidated by studies of PNH, an acquired hemolytic anemia in which the affected blood cells show increased complement sensitivity, partly due to an increased uptake of C3b (reviewed in Rosse and Parker 1985). Three groups of investigators found that these cells lack DAF (Pangburn et al. 1983; Nicholson-Weller et al. 1983, 1985b; Kinoshita et al. 1985). Furthermore, this defect was causally related to increased complement sensitivity, since reincorporation of purified DAF into these cells normalized their C3b uptake and partially corrected their complement sensitivity (Medof et al. 1985). Thus, DAF is clearly seen to be critical in vivo for protection of host cells from damage by autologous complement.

Investigations of DAF have also shed light on the underlying lesion in PNH. Southern and Northern analyses utilizing leukocytes from PNH patients revealed a normal DAF gene and mRNA transcripts (Stafford et al. 1988). Indeed, affected cells of PNH patients lack not only DAF, but also acetylcholinesterase (Auditore and Hartmann 1959), alkaline phosphatase (Lewis and Dacie 1965), lymphocyte function-associated antigen 3 (Selvaraj et al. 1987), F_c receptor type III (Selvaraj et al. 1988), and homologous restriction factor (Zalman et al. 1987). All of these proteins have been shown to be anchored to the cell membrane by a glycophospholipid anchor (reviewed in Ferguson and Williams 1988; Low and Saltiel 1988). The fact that these otherwise unrelated proteins are all absent in a clonal disorder strongly suggests that the lesion in PNH must involve their only common element, the glycophospholipid anchor (Davitz et al. 1986). The nature of this defect in the pathway for biosynthesis or attachment of the anchor structure is unknown.

1.9 Blood Group Antigens on DAF

An interesting recent report demonstrated that the Cromer-related human blood group antigens Cr[a] and Tc[a] reside on the DAF molecule (TELEN et al. 1988). Antibodies to Cr[a] and Tc[a] recognized purified DAF on Western blots, and these antisera had reduced or absent reactivity with PNH E that lack DAF. Moreover, cells of the rare Cromer-related null phenotype Inab did not react with antiserum to DAF by direct binding or Western blotting. The reason underlying lack of expression of DAF in this null phenotype is unknown.

1.10 DAF in Other Species

The critical role of DAF in protecting host tissues from damage by autologous complement leads to the expectation that a DAF or DAF-like molecule would exist in any species with a complement system. To date, only the DAF from guinea pig (NICHOLSON-WELLER et al. 1981) and rabbit (HORSTMANN and MÜLLER-EBERHARD 1986; SUGITA et al. 1987) E have been isolated. The guinea pig DAF was actually purified before human DAF (NICHOLSON-WELLER et al. 1981); the same investigators then utilized this scheme to isolate human DAF (NICHOLSON-WELLER et al. 1982). The guinea pig DAF has a M_r of 60000 on SDS-PAGE under reducing conditions. Decay-accelerating activity was also found on rabbit E (HORSTMANN and MÜLLER-EBERHARD 1986; SUGITA et al. 1987). The rabbit DAF was purified, yielding a protein with an M_r of 66000 on SDS-PAGE under nonreducing conditions (SUGITA et al. 1987). Rabbit DAF has an amino acid composition resembling human DAF, and it can spontaneously and selectively reincorporate into sheep E, suggesting that it possesses a glycophospholipid anchor similar to human DAF.

2 Membrane Cofactor Protein

2.1 Identification and Purification

MCP, a regulatory protein of the complement system, was discovered during an analysis by affinity chromatography of iC3/C3b binding proteins of human peripheral blood mononuclear cells (COLE et al. 1985). Originally named gp45-70 to indicate its M_r on SDS-PAGE, MCP was identified along with CR1 and CR2 by this approach (Fig. 4; COLE et al. 1985). Preliminary functional analysis indicated that it had cofactor activity for C3b/iC3 (TURNER 1984), and this activity was employed to develop a purification scheme (SEYA et al. 1986). The purification involved solubilization of the cells with NP-40 followed by several sequential column chromatographic steps which included iC3 affinity chromatography. The purified protein was utilized to obtain amino-terminal sequence (LUBLIN et al. 1988) and to raise a polyclonal antibody (BALLARD et al. 1987). The unique amino-terminal sequence (LUBLIN et al. 1988) and the lack of cross-reactivity of the antisera to other known complement receptors and inhibitors (BALLARD et al. 1987; SEYA et al. 1988) confirmed that it was an additional regulatory protein of the complement system.

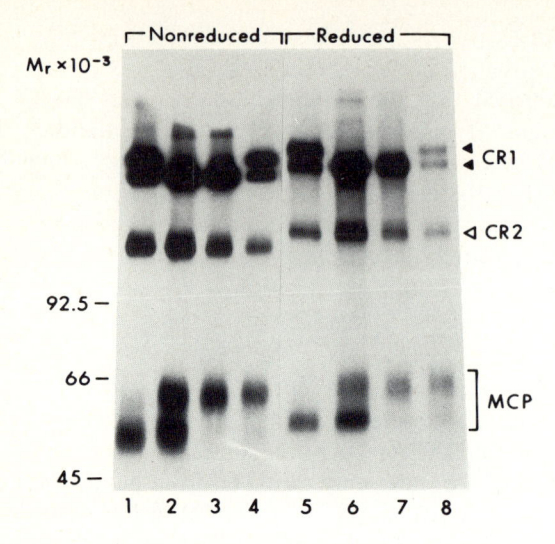

Fig. 4. Autoradiograph of eluates of iC3-Sepharose to which ^{125}I-surface labeled solubilized preparations of human mononuclear cells were applied. The same four donors (*lanes 1–4* and *lanes 5–8*) were analyzed under reducing and nonreducing conditions on a 7.5% gel. The three major iC3 binding proteins of mononuclear cells are visualized. In *lanes 1 and 4*, heterozygotes for the two most common CR1 alleles with M_r of 190000 and 220000 are demonstrated. All donors also display CR2 with an M_r of 140000. The donor in *lane 1* has the lower-band predominant form of MCP, the donor in *lane 2* the form in which there is approximately equal distribution in each band, and the donors in *lanes 3 and 4* have the upper-band predominant form of MCP. All three classes of C3 binding proteins (CR1, CR2, and MCP) increase in M_r with reduction (*lanes 5 to 8*). (Reprinted from BALLARD et al. 1987)

2.2 Structural Variations of MCP on Human Cells

An unusual feature of MCP of human peripheral blood mononuclear cells and platelets is that on SDS-PAGE it consists of two relatively broad protein species (Fig. 4; COLE et al. 1985; SEYA et al. 1986; YU et al. 1986; BALLARD et al. 1987; SEYA et al. 1988; BALLARD et al. 1988). More than 100 individuals have been evaluated for MCP, and all expressed both species (BALLARD et al. 1987). The quantity of each expressed is under autosomal codominant control. In a population survey of 74 unrelated individuals, three stable patterns of expression were noted: an upper-form predominant pattern in 65%; approximately equal expression of the upper and lower species in 29%; and lower-form predominance in 6%. Family studies further support a two-allelic codominantly inherited system regulating expression as none of the 29 offspring from ten matings had a phenotype that deviated from expected. MCP of human peripheral blood mononuclear cells (B and T cells and monocytes) and platelets aligns on analysis by SDS-PAGE, and a given individual expresses the same phenotype in these cell populations (YU et al. 1986; BALLARD et al. 1987).

The heterogeneity within each band is in part secondary to posttranslational modifications, especially glycosylation (see below), since the two precursors and the neuraminidase-treated mature forms of MCP focus more sharply (YU et al. 1986; BALLARD et al. 1988). The unfolding of the molecule upon reduction, consistent with the presence of intrachain disulfide bonds, probably accounts for the increase in M_r on SDS-PAGE upon reduction (see Fig. 4). The two species of MCP are functionally, genetically, and structurally similar proteins (Table 1). The only difference so far detected, other than in M_r, is the increased amount of sialic acid on the upper species (see below).

Table 1. A comparison of the two mature species of MCP of human peripheral blood mononuclear cells and platelets

	Upper	Lower
M_r (mean)		
Reducing	66000	58000[a]
Nonreducing	61000	53000[a]
C3/C4 binding		
iC3/C3b	+++	+++
iC4/C4b	+	+
C3bi	−	−
C3dg	−	−
Cofactor activity		
iC3/C3b	+++	+++
iC4/Cb4	+	+
Oligosaccharides		
N-linked	2, 3	2, 3
O-linked	Multiple[b]	Multiple

Data are derived from COLE et al. (1985), SEYA et al. (1986), YU et al. (1986), BALLARD et al. (1987), SEYA et al. (1988) and BALLARD et al. (1988)
[a] The lower species focuses more sharply, especially under reducing conditions.
[b] Neuraminidase produces a greater shift in M_r of the upper than the lower species, suggesting that there are more O-linked units on the upper form

Two other types of structural variations of MCP have been described. The first relates to slightly different M_r for MCP of T, B, and mononuclear cell lines (COLE et al. 1985). For example, both species on MCP of U937 (a human monocyte cell line) are several thousand larger in M_r than the corresponding molecule on peripheral blood cells (COLE et al. 1985; BALLARD et al. 1988). While this possibly reflects variation in the structure of complex N-linked sugars, as it does for CR1 on this cell line (LUBLIN et al. 1986a), this point has not been definitively addressed. The second variation concerns MCP on granulocytes. By iC3 affinity chromatography, only a small amount of MCP was isolated from granulocytes (COLE et al. 1985). However, by immunoprecipitation and fluorescence-activated cell sorter analysis using the polyclonal antibody, at least as much MCP was expressed by granulocytes as by mononuclear cells (SEYA et al. 1988). Moreover, the autoradiographs showed a broad single species that was not resolvable into two distinct species (SEYA et al. 1988). The nature of this structural variation of granulocyte MCP and the reason for the reduced affinity for iC3 are unknown.

2.3 Biosynthesis and Glycosylation of MCP

Because sugar precursors are incorporated into the molecule, and it is altered in M_r by treatment with endoglycosidases (YU et al. 1986; BALLARD et al. 1988), MCP is a glycoprotein. The precursors of MCP decrease in M_r by approximately 6000 following an incubation with endoglycosidase H (BALLARD et al. 1988). This enzyme cleaves asparagine-linked high-mannose (simple) sugars, and the decrease is consistent with the presence of two or three such units on MCP. Additionally, the mature molecule is insensitive to endoglycosidase H, demonstrating that the high-mannose units are processed to complex oligosaccharides. MCP precursors bind iC3/C3b, indicating that this processing to a complex unit is not required for ligand binding (BALLARD et al. 1988). These complex sugars are only partially removed by endoglycosidase F and N-glycanase, enzymes which digest complex oligosaccharides. N-linked units on CR1 are also partially resistant to such treatments (LUBLIN et al. 1986a).

The mature forms of MCP decrease in M_r upon treatment with neuraminidase (YU et al. 1986; BALLARD et al. 1988), an enzyme which removes terminal sialic acid residues of oligosaccharides. Treatment of MCP with O-glycanase, which cleaves core disaccharides linked to a serine or threonine residue, establishes the presence of O-linked sugars on MCP.

Biosynthetic analyses identify two high-mannose containing precursors that chase into the mature forms with different half-lives ($t_{1/2}$; BALLARD et al. 1988). Thus, for the U937 cell line the upper precursor species chases with a $t_{1/2}$ of approximately 30 min while the lower species chases with a $t_{1/2}$ of approximately 90 min. This differential rate of processing may account for the upper-form predominant phenotype of U937. It is unclear, however, from the pulse-chase experiments whether the upper and lower precursors chase only into the larger and smaller mature forms, respectively. Since there appears to be only one structural gene (see below) but two precursors and two mature forms, two separate mRNAs, possibly derived by alternative splicing, may explain these results.

MCP is not released from cells by PI-PLC, as is DAF (see above). This observation is consistent with a polypeptide membrane anchor and cytoplasmic tail in MCP (see below) as compared to the glycophospholipid anchor of DAF.

2.4 Cloning of MCP cDNA

One cDNA for MCP (MCP-9) contains a long open reading frame encoding 384 amino acids beginning with an initiation methionine codon (Fig. 5; LUBLIN et al. 1988). The first 34 amino acids represent a signal peptide. This is followed by 350 amino acids that encode a polypeptide of 39 kDa. The amino-terminal 252 amino acids of the mature protein consist of four contiguous repeats of 59–65 residues each.

```
TCTGCTTTCCTCCGGAGAAATAACAGCGTCTTCCGCGCCGCGCATGGAGCCTCCCGGCCGCCGCGAGTGTCCCTTTCCTTCCTGGCGC   88
                                                M  E  P  P  G  R  R  E  C  P  F  P  S  W  R    -20
                                               -34
TTTCCTGGGTTGCTTCTGGCGGCCATGGTGTTGCTGCTGTACTCCTTCTCCGATGCCTGTGAGGAGCCACCAACATTTGAAGCTATGGAG   178
 F  P  G  L  L  L  A  A  M  V  L  L  L  Y  S  F  S  D  A  C  E  E  P  P  T  F  E  A  M  E    11
                                              -1  +1
CTCATTGGTAAACCAAAACCCTACTATGAGATTGGTGAACGAGTAGATTATAAGTGTAAAAAAGGATACTTCTATATACCTCCTCTTGCC   268
 L  I  G  K  P  K  P  Y  Y  E  I  G  E  R  V  D  Y  K  C  K  K  G  Y  F  Y  I  P  P  L  A    41
ACCCATACTATTTGTGATCGGAATCATACATGGCTACCTGTCTCAGATGACGCCTGTTATAGAGAAACATGTCCATATATACGGGATCCT   358
 T  H  T  I  C  D  R  N  H  T  W  L  P  V  S  D  D  A  C  Y  R  E  T  C  P  Y  I  R  D  P    71
                    ▲
TTAAATGGCCAAGCAGTCCCTGCAAATGGGACTTACGAGTTTGGTTATCAGATGCACTTTATTTGTAATGAGGGTTATTACTTAATTGGT   448
 L  N  G  Q  A  V  P  A  N  G  T  Y  E  F  G  Y  Q  M  H  F  I  C  N  E  G  Y  Y  L  I  G    101
                   ▲
GAAGAAATTCTATATTGTGAACTTAAAGGATCAGTAGCAATTTGGAGCGGTAAGCCCCCAATATGTGAAAAGGTTTTGTGTACACCACCT   538
 E  E  I  L  Y  C  E  L  K  G  S  V  A  I  W  S  G  K  P  P  I  C  E  K  V  L  C  T  P  P    131
CCAAAAATAAAAAATGGAAAACACACCTTTAGTGAAGTAGAAGTATTTGAGTATCTTGATGCAGTAACTTATAGTTGTGATCCTGCACCT   628
 P  K  I  K  N  G  K  H  T  F  S  E  V  E  V  F  E  Y  L  D  A  V  T  Y  S  C  D  P  A  P    161
GGACCAGATCCATTTTCACTTATTGGAGAGAGCACGATTTATTGTGGTGACAATTCAGTGTGGAGTCGTGCTGCTCCAGAGTGTAAAGTG   718
 G  P  D  P  F  S  L  I  G  E  S  T  I  Y  C  G  D  N  S  V  W  S  R  A  A  P  E  C  K  V    191
GTCAAATGTCGATTTCCAGTAGTCGAAAATGGAAAACAGATATCAGGATTTGGAAAAAAATTTTACTACAAAGCAACAGTTATGTTTGAA   808
 V  K  C  R  F  P  V  V  E  N  G  K  Q  I  S  G  F  G  K  K  F  Y  Y  K  A  T  V  M  F  E    221
TGCGATAAGGGTTTTTACCTCGATGGCAGCGACACAATTGTCTGTGACAGTAACAGTACTTGGGATCCCCCAGTTCCAAAGTGTCTTAAA   898
 C  D  K  G  F  Y  L  D  G  S  D  T  I  V  C  D  S  N  S  T  W  D  P  P  V  P  K  C  L  K    251
                                              ▲
GTGTCGACTTCTTCCACTACAAAATCTCCAGCGTCCAGTGCCTCAGGTCCTAGGCCTACTTACAAGCCTCCAGTCTCAAATTATCCAGGA   988
 V  S  T  S  S  T  T  K  S  P  A  S  S  A  S  G  P  R  P  T  Y  K  P  P  V  S  N  Y  P  G    281
TATCCTAAACCTGAGGAAGGAATACTTGACAGTTTGGATGTTTGGGTCATTGCTGTGATTGTTATTGCCATAGTTGTTGGAGTTGCAGTA   1078
 Y  P  K  P  E  E  G  I  L  D  S  L  D │V  W  V  I  A  V  I  V  I  A  I  V  V  G  V  A  V    311
ATTTGTGTTGTCCCGTACAGATATCTTCAAAGGAGGAAGAAGAAAGGGAAAGCAGATGGTGGAGCTGAATATGCCACTTACCAGACTAAA   1168
│I  C  V  V  P  Y│ R  Y  L  Q  R  R  K  K  K  G  K  A  D  G  G  A  E  Y  A  T  Y  Q  T  K    341
TCAACCACTCCAGCAGAGCAGAGAGGCTGAATAGATTCCACAACCTGGTTTGCCAGTTCATCTTTTGACTCTATTAAAATCTTCAATAGT   1258
 S  T  T  P  A  E  Q  R  G  *                                                                350
TGTTATTCTGTAGTTTCACTCTCATGAGTGCAACTGTGGCTTAGCTAATATTGCAATGTGGCTTGAATGTAGGTAGCATCCTTTGATGCT   1348
TCTTTGAAACTTGTATGAATTTGGGTATGAACAGATTGCCTGCTTTCCCTTAAATAACACTTAGATTTATTGGACCAGTCAGCACAGCAT   1438
GCCTGGTTGTATTAAAGCAGGGATATGCTGTATTTTATAAAATTGGCAAAATTAGAGAAATATAGTTCACAATGAAATTATATTTTCTTT   1528
GTAAAAAAAAAAAAAAAA                                                                           1546
```

Fig. 5. Nucleotide and derived amino acid sequences of MCP cDNA clone. The complete nucleotide sequence of the MCP cDNA is displayed, numbered from the most 5′ nucleotide. The derived amino acid sequence, numbered from the first amino acid of the mature protein, is shown below using single-letter codes. *Asterisk*, stop codon; *arrows*, three potential N-glycosylation sites; *underline*, a serine/threonine-rich region; *box*, the hydrophobic transmembrane domain; *dashed lines*, potential polyadenylation signals

The position and number of cysteines as well as other highly conserved amino acids within each repeat match those of the SCR sequence that has been found in other members of the multigene family of C3b/C4b binding proteins (HOURCADE et al. 1989). These four repeating units show 18%–35% amino acid homology, similar to the degree of homology for these SCRs in other members of this family. MCP binds C3b and, with a lower affinity, C4b (COLE et al. 1985; SEYA et al. 1986). Other receptor/regulatory proteins that interact with C3b and C4b also possess this structural motif.

Further, for CR1, CR2, C4bp, H, and DAF, as few as four to as many as 30 or more of these repeats constitute at least two-thirds of the structure of the protein and are contiguously arranged beginning at the amino terminus. MCP therefore belongs to this family of C3b/C4b-binding receptor and regulatory proteins, and presumably the binding domain(s) lies within this part of the protein.

There are three sites for N-linked glycosylation: one site is located near the end of the first repeat, another in the middle of the second unit, and the third near the end of the fourth repeat. Based on the results with endoglycosidase digestions (YU et al. 1986; BALLARD et al. 1988), it is likely that two or all three of these sites are glycosylated.

Following the four tandemly arranged SCRs, lie a grouping of amino acids (253–277) in which 12 of 25 are serines or threonines, a likely site of extensive O-linked glycosylation. In this case, as with several other membrane proteins such as the interleukin 2 and low-density lipoprotein receptors, this region is close to the membrane spanning domain. Following this, between amino acids 278 and 294, is a region of unknown significance. The next segment, between amino acids 295 and 317, is rich in hydrophobic residues, typical of a transmembrane spanning domain. The last 33 residues, between amino acids 318 and 350, constitute a cytoplasmic tail.

The derived amino acid sequence is based on a clone that was obtained from a cDNA library prepared from U937 cells. U937 is a human mononuclear cell line that expresses a relatively large amount of MCP. Northern blots of the mRNA of U937 and HeLa (both of which express MCP) demonstrate a dominant MCP mRNA at 4.2 kb. Raji, a human B cell precursor cell line, is negative for surface expression of MCP and for a mRNA band on a Northern blot.

This MCP cDNA contains a full-length coding region (Fig. 5) matching the expected size of the MCP polypeptide. Recent characterization of additional MCP cDNAs indicates that the discrepancy in size between the 1.5-kb cDNA and the 4.2-kb mRNA reflects differential polyadenylation at the 3' end.

2.5 MCP Gene

Southern blots employing several restriction enzymes suggested that the MCP gene is contained within 40–50 kb of DNA (unpublished data). The structure of the MCP gene has been partially characterized. The exons have been located (POST and ATKINSON 1989). The first exon encodes the 5' untranslated region and signal peptide. SCRs I, III, and IV are encoded by separate exons. SCR II is split, being formed by two exons. The serine/threonine region is encoded by two approximately equally sized small exons, and the adjacent region of unknown function is encoded by a separate exon. The exon/intron structure of the hydrophobic region and basic amino acid anchor is presently being investigated. The cytoplasmic tail and the 3' untranslated region are encoded by a single large exon.

Southern blots of hamster-human somatic cell hybrids place the structural gene for MCP on human chromosome 1 (LUBLIN et al. 1988). In situ hybridizations with both the full-length MCP cDNA probe and probes not containing any SCR sequences positioned the gene at 1q3.2. To further localize the structural gene, pulse-field gel electrophoresis was performed. The order of the structural genes on an approximately

800-kb fragment was MCP-CR1-CR2-DAF-C4bp (REY-CAMPOS et al. 1988; CARROLL et al. 1988; BORA et al. 1989). The MCP gene lies within 100 kb of the 3' end of the CR1 gene (BORA et al. 1989).

2.6 Sites of Expression

MCP was initially identified on mononuclear cells including T-lymphocytes, B-lymphocytes, and monocytes, and on T, B, and mononuclear cell lines (COLE et al. 1985; TURNER 1984; SEYA et al. 1986). It was later found on platelets (YU et al. 1986). In the initial report (COLE et al. 1985) a small amount of MCP was identified on granulocytes. The production of a rabbit polyclonal antibody to MCP has permitted a more critical analysis of its distribution. MCP is uniformly present and in roughly equivalent amounts on T cells including both the T helper and T suppressor/cytotoxic subpopulations, natural killer cells, B-lymphocytes, monocytes, platelets, and granulocytes (SEYA et al. 1988). It was not expressed by E. These results, except for the presence of as much MCP on granulocytes as on mononuclear cells, are consistent with those obtained by iC3 affinity chromatography. As noted above, a comparison of MCP isolated by affinity chromatography and immunoprecipitation demonstrated that granulocyte MCP is a single broad species wich has reduced affinity for iC3 (SEYA et al. 1988).

Recent experiments further addressed the question of tissue distribution of MCP (ATKINSON et al. 1989). MCP was found to be expressed on cells of epithelial, endothelial, and mesenchymal origin including endothelial cells derived from human umbilical veins, a variety of epithelial cells and cell lines including HeLa, Hep-2, and low-passage human keratinocytes, and several types of fetal and adult fibroblast cell lines. For several of these, MCP was shown to possess cofactor activity and to be synthesized by the cell in question.

2.7 Homologous Proteins in Other Species

To date, no proteins homologous to MCP have been definitively identified in other species. C3b-binding membrane proteins, however, of similar M_r have been reported for rabbit alveolar and peritoneal macrophages (M_r of 64000) (SCHNEIDER et al. 1981; DIXIT et al. 1981) and on mouse cells (M_r of 65000) (WONG and FEARON 1985). However, the rabbit protein was not present on peripheral blood mononuclear cells or platelets (MANTHEI et al. 1988), and thus its tissue distribution does not resemble that of human MCP. The mouse protein was identified through a cross-reaction with a rabbit polyclonal antibody to human CR1 and has a broad tissue distribution.

2.8 Physiological Role

The biological function of MCP can be suggested from its affinity for C3b, cofactor activity, and tissue distribution and through a comparison to other related complement receptor and regulatory proteins. First, MCP and DAF are the only two members of this group with a broad tissue distribution. DAF is an intrinsic regulatory protein.

Second, cells that possess MCP or DAF but no CR1, CR2, or CR3 do not rosette with C3b-coated ligands. These data suggest that, like DAF, MCP is primarily an intrinsically rather than extrinsically acting protein. Its major function may be to down-regulate complement activation on autologous tissue rather than complement components bound to extrinsic material such as immune complexes. MCP may bind efficiently to C3b or C4b attached to the cell upon which the MCP itself is inserted. In fact, recent evidence indicates that purified MCP, like DAF, is a relatively weak extrinsic cofactor (SEYA and ATKINSON 1989). A direct demonstration, however, of the intrinsic cofactor activity of MCP has not been obtained, and there is no disease condition, such as PNH, to facilitate the understanding of its role.

3 DAF and MCP

The protein and DNA data presented here show that DAF and MCP have very similar structures (Fig. 6). Our working hypothesis is that DAF and MCP function jointly to down-regulate complement activation on host tissue. Their broad tissue distributions and complementary functional profiles (shown in Fig. 7) make them well suited for this task. MCP and DAF are present on the membranes of invading inflammatory cells and on tissues at sites of inflammation. DAF can prevent C3 convertase formation and dissociate preformed convertases but cannot inactivate cell-bound C3b or C4b. In contrast, C3b or C4b that is cleaved by MCP and I can no longer associate with the activating proteins. This combination of DAF, as a very mobile membrane protein to prevent amplification, and MCP, as perhaps a more slowly moving but more permanently destructive cofactor protein, could minimize the damage to host tissue at sites of complement activation.

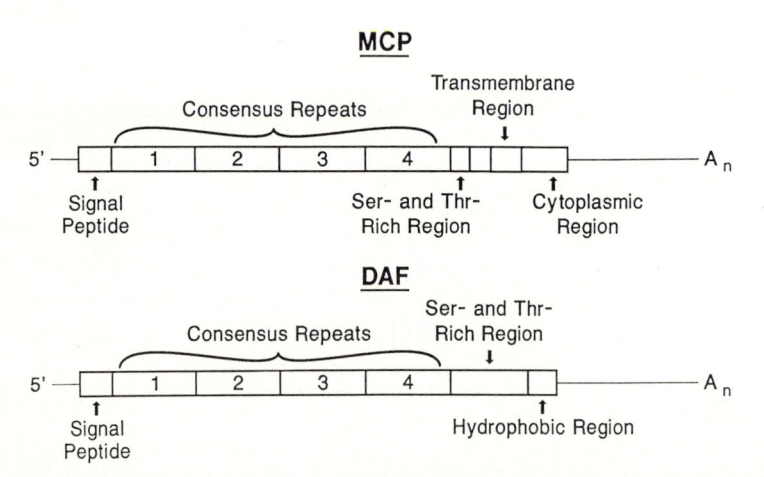

Fig. 6. Comparison of MCP and DAF cDNAs. The MCP cDNA and the DAF cDNA are displayed, highlighting their very similar structures. *Boxes,* coding regions; *lines,* 5' and 3' untranslated regions. The carboxy-terminal hydrophobic region of DAF is replaced post-translationally with a glycophospholipid anchor

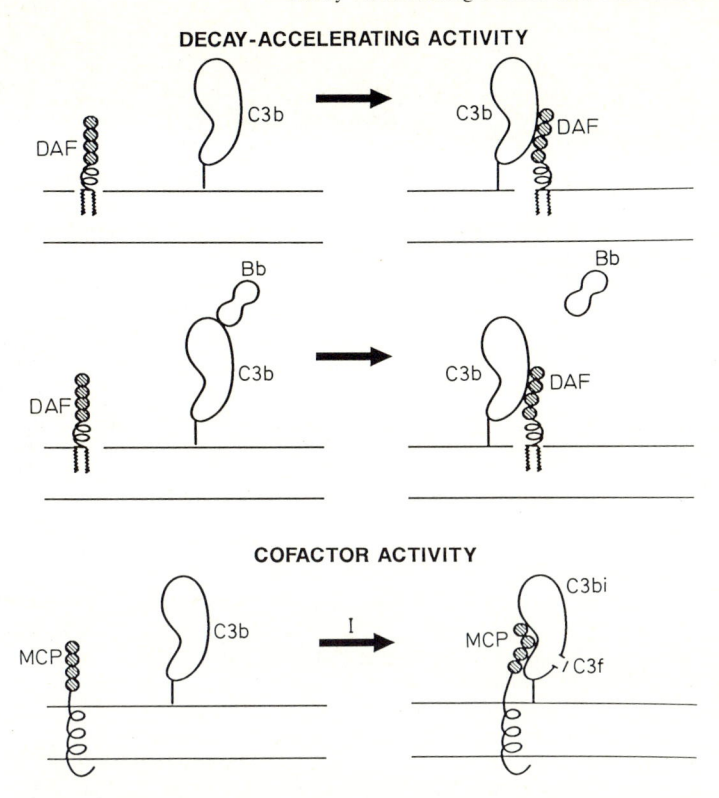

Fig. 7. Functional profile of DAF and MCP. Schematic representation of decay-accelerating activity (*top two panels*) and cofactor activity is shown for alternative pathway C3 convertase, C3bBb. DAF can prevent the formation of the C3 convertase (*top panel*) or dissociate a preformed C3 convertase (*middle panel*). MCP can act as a cofactor for the factor I mediated cleavage of C3b (*bottom panel*).

The discovery of DAF and MCP as two additional regulatory proteins of the complement system was unanticipated. Many investigators felt that C4bp and H could account for the inhibition of complement activation in plasma and on host tissue. Their discovery has led us to speculate that the major function of the inhibitory proteins of plasma, C4bp and H, is to down-regulate fluid-phase C4b and C3b, while MCP and DAF are more important for intrinsic regulation on cells (Holers et al. 1985; Atkinson and Farries 1987).

A second speculation about the function of these two proteins relates to the "activation" of the alternative pathway. In this pathway, a small percentage of C3 is constantly being activated in plasma and indiscriminately attacks both host and foreign tissue. The discovery of DAF and MCP provides an explanation for why human cells are not injured by this continuously turning-over alternative pathway (Atkinson and Farries 1987). Amplification on self-tissue is *not* permitted by DAF and MCP but is allowed on foreign surfaces which lack such regulatory proteins.

References

Asch AS, Kinoshita T, Jaffe EA, Nussenzweig V (1986) Decay-accelerating factor is present on cultured human umbilical vein endothelial cells. J Exp Med 163: 221–226

Atkinson JP, Farries T (1987) Separation of self from non-self in the complement system. Immunol Today 8: 212–215

Auditore JV, Hartmann RC (1959) Paroxysmal nocturnal hemoglobinuria II. Erythrocyte acetylcholinesterase defect. Am J Med 27: 401–410

Ballard L, Seya T, Teckman J, Lublin DM, Atkinson JP (1987) A polymorphism of the complement regulatory protein MCP (membrane cofactor protein or gp45-70). J Immunol 138: 3850–3855

Ballard LL, Bora NS, Yu GH, Atkinson JP (1988) Biochemical characterization of membrane cofactor protein of the complement system. J Immunol 141: 3923–3929

Boothroyd JC, Paynter CA, Cross GAM, Bernards A, Borst P (1981) Variant surface glycoproteins of *Trypanosoma brucei* are synthesized with cleavable hydrophobic sequences at the carboxy and amino termini. Nucl Acids Res 9: 4735–4743

Bora NS, Lublin DM, Kumar BV, Hockett RD, Holers VM, Atkinson JP (1989) Structural gene for human membrane cofactor protein (MCP) of complement maps to within 100 kb of the 3' and of the C3b/C4b receptor gene. J Exp Med 169: 597–602

Caras IW, Davitz MA, Rhee L, Weddell G, Martin DW, Nussenzweig V (1987a) Cloning of decay-accelerating factor suggests novel use of splicing to generate two proteins. Nature 325: 545–549

Caras IW, Weddell GN, Davitz MA, Nussenzweig V, Martin DW (1987b) Signal for attachment of a phospholipid membrane anchor in decay-accelerating factor. Science 238: 1280–1283

Carroll MC, Alicot EM, Katzman PJ, Klickstein LB, Smith JA, Fearon DT (1988) Organization of the genes encoding complement receptors type 1 and 2, decay accelerating factor, and C4-binding protein in the RCA locus on human chromosome 1. J Exp Med 167: 1271–1280

Cole J, Housley GA, Dykman TR, MacDermott RP, Atkinson JP (1985) Identification of an additional class of C3-binding membrane proteins of human peripheral blood leukocytes and cell lines. Proc Natl Acad Sci USA 83: 859–863

Davis CG, Elhammer A, Russell DW, Schneider WJ, Kornfeld S, Brown MS, Goldstein JL (1986) Deletion of clustered O-linked carbohydrates does not impair function of low density lipoprotein receptor in transfected fibrolasts. J Biol Chem 262: 2828–2838

Davitz MA, Low MG, Nussenzweig V (1986) Release of decay-accelerating factor (DAF) from the cell membrane by phosphatidylinositol-specific phospholipase C (PIPLC). Selective modification of a complement regulatory protein. J Exp Med 163: 1150–1161

Davitz MA, Hereld D, Shak S, Krakow J, Englund PT, Nussenzweig V (1987) A glycan-phosphatidylinositol-specific phospholipase D in human serum. Science 238: 81–84

Dixit R, Schneider R, Law SK, Kulczycki A, Atkinson JP (1981) Ligand binding specificity of a rabbit alveolar macrophage receptor for C3b. J Biol Chem 257: 1595–1597

Ferguson MAJ, Homans SW, Dwek RA, Rademacher TW (1988) Glycosyl-phosphatidylinositol moiety that anchors *Trypanosoma brucei* variant surface glycoprotein to the membrane. Science 239: 753–759

Ferguson MAJ, Williams AF (1988) Cell-surface anchoring of proteins via glycosyl-phosphatidyl-inositol structure. Annu Rev Biochem 57: 285–320

Fujita T, Inoue T, Ogawa K, Iida K, Tamura N (1987) The mechanism of action of decay-accelerating factor (DAF). DAF inhibits the assembly of C3 convertases by dissociating C2a and Bb. J Exp Med 167: 1221–1228

Hoffmann EM (1969a) Inhibition of complement by a substance isolated from human erythrocytes. I. Extraction from human erythrocyte stromata. Immunochemistry 6: 391–403

Hoffmann EM (1969b) Inhibition of complement by a substance isolated from human erythrocytes. II. Studies on the site and mechanism of action. Immunochemistry 6: 405–419

Holers VM, Cole JL, Lublin DM, Seya T, Atkinson JP (1985) Human C3b- and C4b-regulatory proteins: a new multi-gene family. Immunol Today 6: 188–192

Homans SW, Ferguson MAJ, Dwek RA, Rademacher TW, Anand R, Williams AF (1988) Complete structure of the glycosyl phosphatidylinositol membrane anchor of rat brain Thy-1 glycoprotein. Nature 333: 269–272

Horstmann RD, Müller-Eberhard HJ (1986) Demonstration of C3b receptor-like activity and of decay-accelerating factor-like activity on rabbit erythrocytes. Eur J Immunol 16: 1069–1073

Hourcade D, Holers VM, Atkinson JP (1989) The regulators of complement activation (RCA) gene cluster. Adv Immunol 45: 381–416

Kinoshita T, Medof ME, Silber R, Nussenzweig V (1985) Distribution of decay-accelerating factor in the peripheral blood of normal individuals and patients with paroxysmal nocturnal hemoglobinuria. J Exp Med 162: 75–92

Kinoshita T, Medof ME, Nussenzweig V (1986) Endogenous association of decay-accelerating factor (DAF) with C4b and C3b on cell membranes. J Immunol 136: 3390–3395

Kinoshita T, Rosenfeld SI, Nussenzweig V (1987) A high m.w. form of decay-accelerating factor (DAF-2) exhibits size abnormalities in paroxysmal nocturnal hemoglobinuria erythrocytes. J Immunol 138: 2994–2998

Lewis SM, Dacie JV (1965) Neutrophil (leucocyte) alkaline phosphatase in paroxysmal nocturnal haemoglobinuria. Br J Haematol 11: 549–556

Low MG, Saltiel AR (1988) Structural and functional roles of glycosyl-phosphatidylinositol in membranes. Science 239: 268–275

Lublin DM, Griffith R, Atkinson JP (1986a) Influence of glycosylation on allelic and cell-specific M_r variation, receptor processing, and ligand binding of the human complement C3b/C4b receptor (CR1). J Biol Chem 261: 5736–5744

Lublin DM, Krsek-Staples J, Pangburn MK, Atkinson JP (1986b) Biosynthesis and glycosylation of the human complement regulatory protein decay-accelerating factor. J Immunol 137: 1629–1635

Lublin DM, Lemons RS, Le Beau MM, Holers VM, Tykocinski M, Medof ME, Atkinson JP (1987) The gene encoding decay-accelerating factor (DAF) is located in the complement-regulatory locus on the long arm of chromosome 1. J Exp Med 165: 1731–1736

Lublin DM, Liszewski MK, Post TW, Arce MA, LeBeau MM, Rebentisch MB, Lemons RS, Seya T, Atkinson JP (1988) Molecular cloning and chromosomal localization of human membrane cofactor protein (MCP): evidence for inclusion in the multi-gene family of complement-regulatory proteins. J Exp Med 168: 181–194

Manthei U, Nickells MW, Barnes SH, Ballard LL, Cui W, Atkinson JP (1988) Identification of a C3b/iC3 binding protein of rabbit platelets and leukocytes: a CR1-like candidate for the immune adherence receptor. J Immunol 140: 1228–1235

McNearney T, Ballard L, Seya T, and Atkinson JP (1989) Membrane cofactor protein of complement is present on human fibroblast, epithelial and endothelial cells. J Clin Invest 84: 538–545

Medof ME, Kinoshita T, Nussenzweig V (1984) Inhibition of complement activation on the surface of cells after incorporation of decay-accelerating factor (DAF) into their membranes. J Exp Med 160: 1558–1578

Medof ME, Kinoshita T, Silber R, Nussenzweig V (1985) Amelioration of lytic abnormalities of paroxysmal nocturnal hemoglobinuria with decay-accelerating factor. Proc Natl Acad Sci USA 82: 2980–2984

Medof ME, Walter EI, Roberts WL, Haas R, Rosenberry TL (2986) Decay accelerating factor of complement is anchored to cells by a C-terminal glycolipid. Biochemistry 25: 6740–6747

Medof ME, Lublin DM, Holers VM, Ayers DJ, Getty RR, Leykam JF, Atkinson JP, Tykocinski ML (1987a) Cloning and characterization of cDNAs encoding the complete sequence of decay-accelerating factor of human complement. Proc Natl Acad Sci USA 84: 2007–2011

Medof ME, Walter EI, Rutgers JL, Knowles DM, Nussenzweig V (1987b) Identification of the complement decay-accelerating factor (DAF) on epithelium and glandular cells and in body fluids. J Exp Med 165: 848–864

Moore JG, Frank MM, Müller-Eberhard HJ, Young NS (1985) Decay-accelerating factor is present on paroxysmal nocturnal hemoglobinuria erythroid progenitors and lost during erythropoiesis in vitro. J Exp Med 162: 1182–1192

Nicholson-Weller A, Burge J, Austen KF (1981) Purification from guinea pig erythrocyte stroma of a decay-accelerating factor for the classical C3 convertase, C4b,2a. J Immunol 127: 2035–2039

Nicholson-Weller A, Burge J, Fearon DT, Weller PF, Austen KF (1982) Isolation of a human erythrocyte membrane glycoprotein with decay-accelerating activity for C3 convertases of the complement system. J Immunol 129: 184–189

Nicholson-Weller A, March JP, Rosenfeld SI, Austen KF (1983) Affected erythrocytes of patients with paroxysmal nocturnal hemoglobinuria are deficient in the complement regulatory protein decay-accelerating factor. Proc Natl Acad Sci USA 80: 5066–5070

Nicholson-Weller A, March JP, Rosen CE, Spicer DB, Austen KF (1985a) Surface membrane expression by human blood leukocytes and platelets of decay-accelerating factor, a regulatory protein of the complement system. Blood 65: 1237–1244

Nicholson-Weller A, Spicer DB, Austen KF (1985b) Deficiency of the complement regulatory protein, "decay-accelerating factor", on membranes of granulocytes, monocytes, and platelets in paroxysmal nocturnal hemoglobinuria. N Engl J Med 312: 1091–1097

Nicholson-Weller A, Russian DA, Austen KF (1986) Natural killer cells are deficient in the surface expression of the complement regulatory protein, decay-accelerating factor (DAF). J Immunol 137: 1275–1279

Pangburn MK (1986) Differences between the binding sites of the complement regulatory proteins DAF, CR1, and factor H on C3 convertases. J Immunol 136: 2216–2221

Pangburn MK, Schreiber RD, Müller-Eberhard HJ (1983) Deficiency of an erythrocyte membrane protein with complement regulatory activity in paroxysmal nocturnal hemoglobinuria. Proc Natl Acad Sci USA 80: 5430–5434

Post TW, Atkinson JP (1989) The structure and organization of the MCP gene. FASEB J 3: A368

Post TW, Arce MA, Liszewski MK, Atkinson JP, Lublin DM (1989) Structure of human DAF gene. FASEB J 3: A798

Rey-Campos J, Rubinstein P, Rodriguez-de-Cordoba S (1987) Decay-accelerating factor. Genetic polymorphism and linkage to the RCA (regulator of complement activation) gene cluster in humans. J Exp Med 166: 246–252

Rey-Campos J, Rubinstein P, De Cordoba SR (1988) A physical map of the human regulator of complement activation gene cluster linking the complement genes CR1, CR2, DAF, and C4BP. J Exp Med 167: 664–669

Roberts WL, Santikarn S, Reinhold VN, Myher JJ, Kuksis A, Rosenberry TL (1988) Variations in the structure of glycoinositol phospholipids covalently attached to protein C-termini. FASEB J 2: A988

Rosse WF, Parker CJ (1985) Paroxysmal nocturnal haemoglobinuria. Clin Haematol 14: 105–125

Selvaraj P, Dustin ML, Silber R, Low MG, Springer TA (1987) Deficiency of lymphocyte function-associated antigen 3 (LFA-3) in paroxysmal nocturnal hemoglobinuria. J Exp Med 166: 1011–1025

Selvaraj P, Rosse WF, Silber R, Springer TA (1988) The major F_c receptor in blood has a phosphatidyl-inositol anchor and is deficient in paroxysmal nocturnal haemoglobinuria. Nature 333: 565–567

Seya T, Atkinson JP (1989) Characterization of the functional activity of membrane cofactor protein (MCP) of complement. Biochem J (in press)

Seya T, Turner J, Atkinson JP (1986) Purification and characterization of a membrane protein (gp45-70) which is a cofactor for cleavage of C3b and C4b. J Exp Med 163: 837–855

Seya T, Farries T, Nickells M, Atkinson JP (1987) Additional forms of human decay-accelerating factor (DAF). J Immunol 139: 1260–1267

Seya T, Ballard L, Bora N, McNearney T, Atkinson JP (1988) Distribution of membrane cofactor protein (MCP) of complement on human peripheral blood cells. Eur J Immunol 18: 1289–1294

Schneider RJ, Kulczycki A, Law SK, Atkinson JP (1981) Isolation of a biologically active macrophage receptor for the third component of complement. Nature 290: 789–792

Stafford HA, Tykocinski ML, Lublin DM, Holers VM, Rosse WF, Atkinson JP, Medof ME (1988) Normal polymorphic variations and transcription of the decay-accelerating factor gene in paroxysmal nocturnal hemoglobinuria cells. Proc Natl Acad Sci USA 85: 880–884

Sugita Y, Uzawa M, Tomita M (1987) Isolation of decay-accelerating factor (DAF) from rabbit erythrocyte membranes. J Immunol Methods 104: 123–130

Telen MJ, Hall SE, Green AM, Moulds JJ, Rosse WF (1988) Identification of human erythrocyte blood group antigens on decay-accelerating factor (DAF) and an erythrocyte phenotype negative for DAF. J Exp Med 167: 1993–1998

Tse AGD, Barclay AN, Watts A, Williams AF (1985) A glycophospholipid tail at the carboxyl terminus of the Thy-1 glycoprotein of neurons and thymocytes. Science 230: 1003–1008

Turner JR (1984) Structural and functional studies of the C3b and C4b binding proteins of a human monocyte-like cell line (U937). Masters thesis, Washington University, St Louis

Tykocinski ML, Shu HK, Ayers DJ, Walter EI, Getty RR, Groger RK, Hauer CA, Medof ME (1988) Glycolipid reanchoring of T-lymphocyte surface antigen CD8 using the 3' end sequence of decay-accelerating factor's mRNA. Proc Natl Acad Sci USA 85: 3555–3559

Walter EI, Roberts WF, Rosenberry TL, Medof ME (1987) Analysis of fatty acids and inositol in the membrane anchor of human erythrocyte decay accelerating factor (DAF) (Abstract). Fed Proc 46: 772

Wong WM, Fearon DT (1985) A C3b-binding protein on murine cells that shares antigenic determinants with the human C3b receptor (CR1) and is distinct from immune C3b receptors. J Immunol 134: 4048–4056

Yu G, Holers VM, Seya T, Ballard L, Atkinson JP (1986) Identification of a third component of complement-binding glycoprotein of human platelets. J Clin Invest 78: 494–501

Zalman LS, Wood LM, Frank MM, Müller-Eberhard HJ (1987) Deficiency of the homologous restriction factor in paroxysmal nocturnal hemoglobinuria. J Exp Med 165: 572–577

Factor H*

D. P. VIK[1], P. MUÑOZ-CÁNOVES[1], D. D. CHAPLIN[2], and B. F. TACK[1]

1   Functions of Factor H   147
1.1 C3 Convertase Regulatory Activity   147
1.2 Binding of H to Components of C3 and C5 Convertases   149
1.3 Binding Domains of H and C3   150
1.4 H Deficiencies   151

2   Structure and Variant forms of Factor H   151

3   Sequence Analysis and Genomic Organization   152
3.1 Characterization of the cDNA   152
3.2 Genomic Organization and Structure   153
3.3 Linkage Studies   154

4   H-Related Molecules   155
4.1 A Truncated Form of H in Humans   155
4.2 mRNA, cDNA, and Genomic DNA Homologous, to H in Mice   156

5   Summary   157

References   158

1 Functions of Factor H

1.1 C3 Convertase Regulatory Activity

Complement component factor H (H) is a 155-kDa plasma protein. It was first observed by NILSSON and MÜLLER-EBERHARD (1965) as a minor component in C3 preparations, and its function was not known. They termed this protein β_1H globulin because of its β-electrophoretic mobility in agarose gels. The first function for this protein was recognized by WHALEY and RUDDY (1976a). They isolated a protein from human serum based on its ability to accelerate the activity of C3b inactivator (now termed factor I) and found that this protein was biochemically and immunologically identical to β_1H globulin. At the time, it was thought that I alone cleaved C3b into two

---

* This work was supported by the United States Public Health Service awards AI19222 and AI17354 (to B.F.T.), training grant AI07706 (to D.P.V.), and the Howard Hughes Medical Institute
[1] Department of Immunology, Scripps Clinic and Research Foundation, La Jolla, CA 92037, USA
[2] Department of Internal Medicine and the Howard Hughes Medical Institute, Washington University School of Medicine, St. Louis, MO 63110, USA

fragments, C3c and C3d. Addition of H synergistically enhanced the activity of I-mediated cleavage of C3b. H not only enhanced the activity of I (termed its cofactor activity) but was also able to accelerate the decay of the C3 convertases of the alternative pathway of complement activation C3b,Bb or C3b,Bb,P (Fig. 1) and to inhibit the formation of these convertases (WHALEY and RUDDY 1976a).

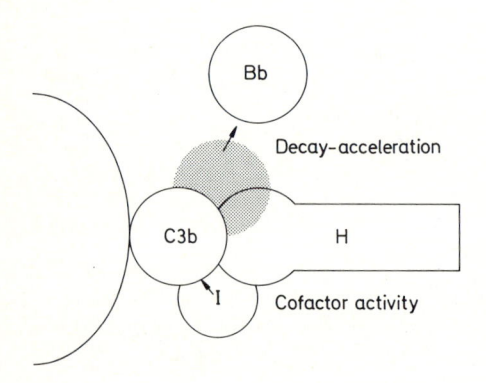

Fig. 1. Schematic representation of the two functions of H. In its role of decay-acceleration, H binds to C3b and displaces Bb. In its cofactor activity, H promotes I-mediated cleavage of C3b (cleavage is represented by *arrowhead*)

These results were confirmed and extended by WEILER et al. (1976) who showed that H not only enhanced the effects of I-mediated cleavage of C3b and accelerated the functional decay of the properdin-stabilized C3 convertase but also caused the physical decay of the complexes, accelerating release of ^{125}I-labeled Bb from these convertases. This indicated that H may be in competition with Bb for binding to C3b. H was much less effective in displacing Bb from C3 nephritic factor stabilized convertases, indicating that this factor, which is isolated from the sera of patients with pathologic conditions, may stabilize the C3 convertase more effectively than alternative pathway component P.

Initial efforts to define the exact role of H in I-mediated cleavage of C3b were hampered by incomplete purification of the isolated components, with small amounts of H contaminating the C3b preparations. Subsequent studies by PANGBURN et al. (1977) have established that H is a necessary cofactor for I. When fluid-phase ^{125}I-labeled C3b was incubated with either purified H or I alone, no cleavage of C3b was observed. However, when H and I were incubated simultaneously with the ^{125}I-labeled C3b, the α'-chain of C3b was cleaved into two fragments, of 67 and 40 kDa, which remained covalently bound to the β-chain.

Early experiments indicated that H was binding to C3b (WHALEY and RUDDY 1976b; WEILER et al. 1976). The first line of evidence was that the erythrocyte intermediate cells EAC43 but not EAC4 were able to absorb out H activity. In addition, EAC43 cells incubated with H could be easily agglutinated with anti-H antibody, whereas EAC4 incubated with H required 50- to 100-fold more antibody to attain comparable agglutination levels (WHALEY and RUDDY 1976b). Furthermore, H apparently could displace Bb bound to C3b (WEILER et al. 1976). Later experiments demonstrated that ^{125}I-labeled H bound to EAC1423 cells, but not EAC142 cells, in a dose-dependent, saturable manner (CONRAD et al. 1978), confirming the interaction between H and C3b.

1.2 Binding of H to Components of C3 and C5 Convertases

The binding characteristics of H to C3b attached to particles appeared to be somewhat unusual. Several groups have shown a curvilinear Scatchard plot for H binding to EAC1423 (CONRAD et al. 1978), zymosan C3b (DiSCIPIO 1981), or EC3b (PARKER et al. 1983). The binding data do not fit a simple two-site model, and the curvature in the Scatchard plot is apparently not a result of the clustered distribution of C3b on the surface, because when the C3b was deposited randomly on the cell surface, Scatchard analysis of H binding was still curvilinear (PARKER et al. 1983). Thus, this curvature in the Scatchard plot may be due to the influence of the microenvironment of the C3b on the affinity of H binding.

The binding of H to C3b was found to be affected by the nature of the surface to which the C3b was attached. The presence of sialic acid on a particle surface renders the particle a nonactivator of the alternative complement pathway. Enzymatic removal or chemical modification of sialic acid converts the particle to an activator of the alternative pathway (FEARON 1978; PANGBURN and MÜLLER-EBERHARD 1978). These relationships could be quantified by determining the affinity constants at equilibrium of H and B for C3b on activators and nonactivators. H bound to C3b-coated sheep erythrocytes (E^{s}C3b) with an affinity of $1 \times 10^7 M^{-1}$ whereas the affinity of B for E^{s}C3b was five fold less. When the sialic acid on the E^{s}C3b was removed or modified, the affinity constant of B for C3b was unaffected. The measured affinity of H for C3b also appeared unchanged; however, although the number of C3b molecules was unaltered, the number of detectable binding sites for H was decreased dramatically, by 80%–85%. This has been interpreted to indicate that a majority of H binding sites on C3b are sensitive to desialation, having in the desialated form a very low residual affinity for H. The remaining 15%–20% of the sites appeared to have retained high affinity for H. Thus, the apparent loss in number of sites on the cell was actually just a result of a drastic decrease in the affinity of H for those sites.

The ability of H to interact with various combinations of C3 and other complement components has been investigated by several groups. H was originally reported to interact with C3b,Bb and C3b,Bb,P (WHALEY and RUDDY 1976a), the alternative pathway C3 convertases. It is also capable of reacting with C1,4b,2a,3b, the classical pathway C5 convertase (WHALEY and RUDDY 1976b; ISENMAN et al. 1980); however, the extent of this reaction was dependent upon the number of C4b,2a,3b sites on the cell (NAGAKI et al. 1978; ITO and TAMURA 1983). H inhibits the interaction of C3b with C5 only if the density of C3b on the cell surface is high. At low density, H has little or no inhibitory activity.

H was also found to interact with the alternative pathway C5 convertase C3b,Bb,C3b,P (FISCHER and KAZATCHKINE 1983). H inhibited C5 cleavage in a dose-dependent manner independent of the number of convertase sites/cell. In this setting, H inhibition was also apparently not affected by the presence of sialic acid, as its inhibition of C5 convertases on desialated E^s more closely resembled normal E^s rather than rabbit erthrocytes. These data suggest that H may be interacting not only with C3b but also with C5 (FISCHER and KAZATCHKINE 1983). When the fluid-phase binding of H to various complement fragments and intermediates was examined, it was found that H had the highest affinity for C3b,B, C3b, and C3b,Bb, with an approximately ten fold lower affinity for C4b and C4b,2a, and a 10000-fold lower affinity for Bb alone (PANGBURN 1986).

1.3 Binding Domains of H and C3

The location of the binding site on C3 for H has been investigated by several groups. Preliminary experiments demonstrated that H bound to both C3b and, more weakly, to iC3b (ROSS et al. 1983). Polyclonal antibodies against C3c were inhibited from binding to C3b in the presence of H (NILSSON and NILSSON 1986), but only antibody specific for the α-chain was able to block H and I-mediated cleavage of C3. In another study, both H and anti-idiotype anti-H antibody bound to C3b and C3d but not C3c, as measured in an enzyme-linked immunosorbent assay (LAMBRIS et al. 1988). This suggested that the primary site of H interaction was in the C3d region. The binding site in C3d was further localized to a 8.6 kDa CNBr fragment representing amino acids 1178–1252 of the C3 sequence (DE BRUIJN and FEY 1985). Three synthetic peptides, representing residues 1192–1249, 1222–1249, and 1234–1249, respectively, were also able to bind to H and partially inhibit H binding to C3b. Because peptide-mediated inhibition of H binding to C3b was not complete, it is likely that there is more than one epitope on C3b for binding factor H, located in another portion of the molecule (LAMBRIS et al. 1988). This conclusion is also supported by the above mentioned data on H-mediated inhibition of anti-C3c antibody binding to C3b. The binding site on C3b for factor H appears to be distinct from that for properdin, as neither protein affects the binding of the other (DISCIPIO 1981). In contrast, C5 and B do cause partial inhibition of H binding to C3b, perhaps indicating that their binding sites are nearby, or that there are conformational changes in C3b induced by these two proteins.

The binding site on H for C3b has been localized by the use of monoclonal anti-H antibodies and proteolytic cleavage fragments of H. Purified H from both humans (CHARLESWORTH et al. 1979; HARRISON and LACHMANN 1979; GARDNER et al. 1980; SIM and DISCIPIO 1982) and guinea pigs (BITTER-SUERMANN et al. 1981) has been reported to contain varying amounts of partially cleaved protein, of M_r 118000–142000. This cleavage can be reproduced *in vitro* with trypsin, yielding fragments of M_r 38000 and 142000 (HONG et al. 1982; SIM and DISCIPIO 1982), and trypsin-cleaved H was more active as a cofactor for I-mediated cleavage of fluid-phase C3b than normal H (HONG et al. 1982). However, its activity on membrane-bound C3b was greatly diminished. When the two trypsin-cleaved fragments were isolated by gel filtration followed by affinity chromatography, it was found that both the binding site for C3b and the cofactor activity for I-mediated cleavage resided in the 38-kDa fragment of H and not the 142-kDa fragment (ALSENZ et al. 1984). N-terminal sequence analysis of the 38-kDa H fragment revealed a sequence identical to that of the N-terminus of H, and thus this fragment comprises the amino terminus of the molecule (ALSENZ et al. 1985).

1.4 H Deficiencies

A number of patients have been reported who are homozygous or heterozygous deficient in H. Clinical manifestations are quite variable. Some patients appear symptom free (THOMPSON and WINTERBORN 1981; BRAI et al. 1988). Others have a clinically benign nephropathy with membranous deposits found on renal biopsy (LEVY et al. 1986). Some patients develop the hemolytic uremic syndrome (THOMPSON and WINTERBORN 1981) or an aggressive, destructive nephropathy (WYATT et al. 1982). This is in contrast to reported cases of I deficiency, where all patients are subject to recurrent bacterial infections (ALPER et al. 1970; THOMPSON and LACHMANN 1977; ROSS and DENSEN 1984). The reason for these differences in clinical expression between the two deficiencies is unclear, since both conditions result in increased complement activation and cleavage of C3. Resolution of this distinction may need to await identification of more patients with these deficiencies.

2 Structure and Variant Forms of Factor H

H is a single-chain glycoprotein, containing 9.3%–18.5% carbohydrate (SIM and DISCIPIO 1982; JOUVIN et al. 1984). Estimates of the molecular weight of H, as determined by sodium dodecyl sulfate (SDS) polyacrylamide gel electrophoresis, range from 150000 to 170000 (WHALEY and RUDDY 1976a; WEILER et al. 1976; FEARON and AUSTEN 1977; SIM and DISCIPIO 1982). The molecular weight determined by equilibrium sedimentation is 150000 to 156500 (WHALEY and RUDDY 1976a; SIM and DISCIPIO 1982). H has an unusual circular dichroism spectrum, in that it apparently has no α-helices or β-strand conformations (DISCIPIO and HUGLI 1982). Based on its unusual circular dichroism spectrum and its high frictional ratio of 2.11 (SIM and DISCIPIO 1982), H was proposed to be rodlike in shape. This was confirmed by electron microscopic studies in which H appeared as an asymmetric, elongated molecule, approximately 28 nm × 3 nm, with one end slightly larger and more rounded (SMITH et al. 1983).

The carbohydrate residues are apparently not important for the function of the molecule, since neither removal of sialic acid nor total deglycosylation affected binding of H to C3b (JOUVIN et al. 1984). However, when the protein is reduced and alkylated (DISCIPIO and HUGLI 1982) or has its lysine residues chemically modified (JOUVIN et al. 1984), it loses most of its activity. Thus, the disulfide bridges are probably providing a structural support for the function of the molecule, and one or more lysine residues may also be important in maintaining secondary structure or may be located in the active site and involved in the binding of H to C3b.

Several variant forms of H, based on physicochemical properties, have been described. Two different forms of H were separated based on their differing abilities to bind to phenyl-Sepharose (RIPOCHE et al. 1984). Seventy percent of the H preparation bound to the column, and the remaining fraction was only retarded. These two fractions appeared identical by other parameters, such as mobility under SDS polyacrylamide gel electrophoresis.

Five allelic variants of H have been identified using isoelectric focusing after neuraminidase treatment (RODRÍGUEZ DE CÓRDOBA and RUBINSTEIN 1984, 1987). Each of the variants consists of one major band and four minor bands, two with slightly higher pI values and two slightly lower. The range of pI for all of the bands is 6.50–6.75. Based on an analysis of 208 unrelated Caucasians, the gene frequencies for the H alleles FH*1, FH*2, FH*3, FH*[4], and FH*5 are 0.685, 0.301, 0.006, 0.002, and 0.006, respectively. Variations in protein sequence have also been observed (DAY et al. 1988). When tryptic peptides of H purified from pooled plasma of 12 donors were subject to sequence analysis, one peptide, 17 residues in length, had both a Tyr and a His in position 15, in a ratio of approximately 2 : 1, which is similar to the ratio of alleles FH*1 to FH*2 (RODRÍGUEZ DE CÓRDOBA and RUBINSTEIN 1987).

3 Sequence Analysis and Genomic Organization

3.1 Characterization of the cDNA

cDNA clones for H have been isolated and sequenced in both mice (KRISTENSEN and TACK 1986) and humans (KRISTENSEN et al. 1986; SCHULZ et al. 1986; RIPOCHE et al. 1988). The derived amino acid sequence indicates that the mature protein consists of 1213 residues in humans and 1216 residues in mice, and the two sequences are 61% homologous (RIPOCHE et al. 1988). A single base pair substitution was found in some human H clones, resulting in the charge from Tyr to His (DAY et al. 1988). This confirms the allelism seen in the peptide sequencing data (see above). Both human and mouse H are comprised of 20 consensus repeating units, each approximately 60 amino acids in length, arranged tandemly. The conserved residues in these repeats include four Cys, two Pro, two Gly, and one Trp (Fig. 2). These consensus repeating units have been identified in a number of complement and noncomplement proteins, including human complement receptors type 1 (CR1; KLICKSTEIN et al. 1987) and 2 (CR2; MOORE et al. 1987) and the related murine homolog mCRY (AEGERTER-SHAW et al. 1987; PAUL et al. 1989), C1r (JOURNET and TOSI 1986; LEYTUS et al. 1986), C1s (TOSI et al. 1987), C2b (BENTLEY 1986), Ba (MORLEY and CAMPBELL 1984; MOLE et al. 1984), C4b binding protein (C4bp; CHUNG et al. 1985; KRISTENSEN et al. 1987a), I (GOLDBERGER et al. 1987; CATTERAL et al. 1987), decay-accelerating factor (DAF; CARAS et al. 1987), membrane cofactor protein (MCP; LUBLIN et al. 1988), β_2-glycoprotein I (LOZIER et al. 1984), haptoglobin (KUROSKY et al. 1980), the interleukin 2 receptor (SHIMUZU et al. 1985), a vaccinia virus secretory polypeptide (KOTWAL and MOSS 1988), and clotting factor XIIIb (ICHINOSE et al. 1986). The function of the 60 amino acid consensus repeating unit is not known, although it is tempting to speculate that these repeats are involved in some manner with the binding of C3b or C4b. However, because of its presence in noncomplement proteins such as the interleukin 2 receptor and β_2-glycoprotein I, this consensus repeat must also function in other capacities. Perhaps conserved residues provide a structural framework for the molecule, and the variable residues provide the specificity for interaction with various other proteins. (For reviews of this family of related proteins, see REID et al. 1986; KRISTENSEN et al. 1987b.)

```
    4      8          26        36       43    49  52   54   57  59
- CYS - PRO ------ PHE  --- CYS --- GLY -- CYS - GLY -TRP - PRO-CYS -
                   TYR
```

Fig. 2. Consensus repeating unit structure in H. These ten residues, with their approximate locations in the 60 amino acid repeating unit indicated, are conserved in at least 17 of the 20 repeats, except for GLY-43, which is present in 13 of 20 repeats

3.2 Genomic Organization and Structure

In order to better understand the nature of this repeating unit structure, its possible role in the function of the protein, and its evolution, we began to examine the genomic organization of H (VIK et al. 1988). A cosmid library of BALB/c genomic liver DNA was screened using the H cDNA as a probe and 17 overlapping clones were isolated, spanning a region of approximately 120 kb. Four clones were selected for further analysis of the intron/exon junction regions and the mapping of the exons in the gene. DNA from these clones was sonicated and shotgun cloned into M13mp8 bacteriophage. Exon-containing subclones were identified and sequenced to determine the intron/exon boundaries. The locations of these exons were then mapped. The gene was found to be composed of 22 exons (Fig. 3). The first exon encoded the 5' untranslated region and the leader sequence peptide. Nineteen of the 20 consensus repeating units were encoded by single exons. The one exception was the second repeat, which was encoded by two exons. These exons spanned approximately 100 kb and were generally grouped in pairs with exons 12–20 clustered together in a region of approximately 13 kb. A dot matrix plot of a comparison of H nucleotide sequence to itself did not reveal any large regions of internal homology, as has been noted for CR1 (KLICKSTEIN et al. 1987) and CR2 (MOORE et al. 1987; WEIS et al. 1988). This may indicate that the gene arose from duplication of individual exons rather than by duplicating longer segments of DNA containing several exons. The dot matrix analysis did indicate, however, that exons 13–20 are more closely related to one another than they are to other exons, suggesting that they were duplicated more recently.

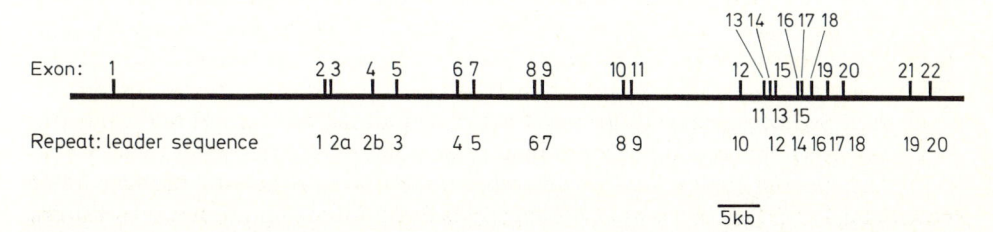

```
                                                      13 14  16 17 18
Exon:    1             2 3  4 5     6 7     8 9   10 11   12  15   19 20  21 22
                                                           11 13 15
Repeat: leader sequence  1 2a 2b 3    4 5    6 7    8 9   10   12 14 16 17 18  19 20
                                                                    5kb
```

Fig. 3. Map of the murine H gene. *Vertical lines,* locations of the 22 exons in this gene, numbered above the live. The repeating units in the protein sequence are identified below the line

A cosmid clone from the 5' region of the H gene was selected for further analysis. A 3.5-kb fragment from this clone that contains the leader peptide exon was fully sequenced, and this sequence was compared to that for other known regulatory elements. Several 5' H sequences were found to share homologies with other known regulatory elements, including the heat-shock consensus (79% homology; PELHAM 1982), the mouse Cytomegalovirus (CMV) enhancer 18 bp consensus (two regions of 60% and 56%; DORSH-HÄSLER et al. 1985), the adenovirus enhancer consensus (91%; HEARING and SHENK 1983), the glucocorticoid responsive element (core) antisense (100%; SCHÜTZ 1988), and the tetradecanoyl phorbol acetate (TPA) responsive element (100%; ANGEL et al. 1987). Although none of these putative regulatory sequences has been shown to function in controlling H expression *in vivo,* their possible effects on H transcription and translation warrant investigation.

This 3.5-kb fragment was also used in S1 nuclease analysis and the RNase protection assay to determine the transcription start site for H. Radiolabeled DNA or RNA, corresponding to the region 5' to the ATG translation start site, was hybridized to liver mRNA and then digested with S1 nuclease or RNase, respectively. Both analyses revealed two possible start sites at positions -105 and -191 (relative to the ATG translation start site). Putative TATA boxes and CAAT boxes were found upstream of each of these start sites. The significance of these alternative start sites and the possibility that their usage may be differentially regulated are under investigation.

3.3 Linkage Studies

The chromosomal location of this gene has been determined in both mice (D'EUSTACHIO et al. 1986) and human (RODRÍGUEZ DE CÓRDOBA et al. 1985). In humans, the H gene is part of a cluster of genes on the long arm of chromosome 1 termed the regulator of complement activation (RCA). This cluster is composed of the genes for C4bp, CR1 (RODRÍGUEZ DE CÓRDOBA et al. 1984), CR2 (WEIS et al. 1987), H (RODRÍGUEZ DE CÓRDOBA et al. 1985), DAF (LUBLIN et al. 1987) and MCP (LUBLIN et al. 1988), whose proteins are comprised partly or entirely of the 60 amino acid consensus repeating unit. In addition, the genes for C4bp, DAF, CR1, and CR2 have been physically mapped using pulse-field gel electrophoresis (CARROLL et al. 1988; REY-CAMPOS et al. 1988). These genes were found to be clustered on a 750-kb segment in the order CR1, CR2, DAF, and C4bp. Linkage analysis of the genes for H and C4bp in humans has determined that they are approximately 6.9 cM apart (RODRÍGUEZ DE CÓRDOBA and RUBINSTEIN 1987; Fig. 4A).

In mice, we have examined the gene linkage relationship between H and a number of other known markers on chromosome 1, including C4bp, renin, CRY and CD45 (Ly5, T200; KINGSMORE et al., 1989). Based on over 200 back-crosses, the gene for H was found to be 9 cM from the C4bp gene, 7.5 cM from the renin gene, 30 cM from the CRY gene, and 0.5 cM from the CD45 gene. The order of these genes was C4bp, renin, CD45, H, CRY (Fig. 4B).

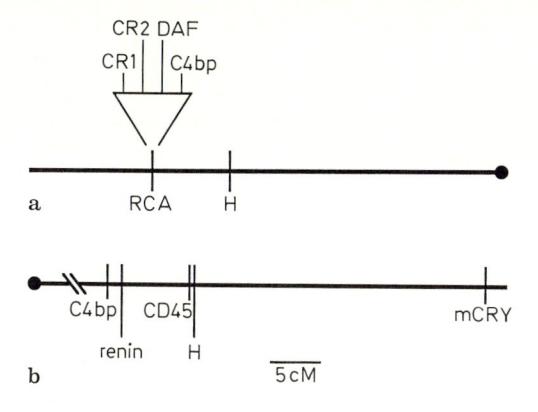

Fig. 4a, b. Linkage map of human chromosome 1q (**a**) and murine chromosome 1 (**b**). The relative positions of the regulator of complement activation (*RCA*) locus and of the genes for CR1, CR2, decay-accelerating factor (*DAF*), C4b binding protein (*C4bp*), renin, CD45 (Ly5, T200), H, and murine complement receptor Y (*mCRY*) are shown in relation to the centromere

4 H-Related Molecules

4.1 A Truncated Form of H in Humans

In both humans and mice, several H-related proteins and/or mRNA transcripts have been identified. In humans, two partial H cDNA sequences were published that were recognized as anomalous. The first report (KRISTENSEN et al. 1986) identified a cDNA clone of 1050 bp that was polyadenylated at its 3′ end. However, when this sequence was compared to that of a full-length murine cDNA sequence, it was found to be most homologous with the 5′ region of the sequence, not its 3′ end. Similarly, another cDNA clone was sequenced that also contained a 3′ poly(A) tail (SCHULZ et al. 1986). However, when the derived amino acid sequence was compared to known H protein sequence, it was found to be identical to the N-terminal region of the molecule (which had been previously sequenced at the protein level) and not the C-terminus (Fig. 5). Thus, these truncated forms of H were proposed to arise from alternative splicing and usage of different polyadenylation sequences. The alternative transcripts can be easily detected in Northern blots (RIPOCHE et al. 1987). Human acute-phase liver poly(A)$^+$ mRNA that was hybridized with an H cDNA probe revealed two transcripts of 4.4 kb and 1.8 kb, respectively. This smaller mRNA was subsequently shown to be transcribed in both acute-phase and non-acute-phase livers (SCHWÄBLE et al. 1987), and a novel, shorter form of H of 43 kDa was detected in human serum.

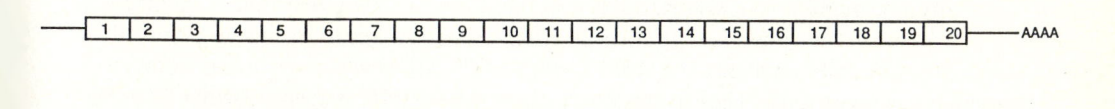

Fig. 5. Schematic representation of the two forms of human H. *Above*, full-length H, with 20 repeats; *below*, the truncated form, with 7 repeats

H and its related truncated form were examined in biosynthetic studies (KATZ and STRUNK 1988). Both proteins are expressed in human skin fibroblasts, in a ration of approximately 10:1, but neither was expressed in HepG2 cells (a hepatocyte cell line) or human peripheral blood monocytes. Pulse-chase experiments indicated that the smaller protein was not a degradation product of H, and Northern blot experiments demonstrated two different H-related transcripts of 4.3 kb and 1.8 kb, respectively, corresponding to these two protein products. Interferon-γ, but not lipopolysaccharide, increased mRNA levels and protein synthesis for both of these molecules.

4.2 mRNA, cDNA, and Genomic DNA Homologous to H in Mice

In mice, the situation is different. When we probe Northern blots of murine liver poly(A)$^+$ mRNA with a full-length H cDNA probe, four different transcripts are identified, with lengths of 4.4 kb, 3.5 kb, 2.8 kb, and 1.8 kb, respectively. However, in contrast to the human, only the full-length 4.4-kb transcript is reactive with a 5' H probe containing the 5' untranslated region. Both the 1.8-kb and 4.4-kb H transcripts hybridize to a 3' H probe containing sequence from repeats 18–20 and the 3' untranslated region, and the 3.5-kb and the 2.8-kb mRNAs are weakly reactive with this probe. Thus, in the mouse, these three H-related transcripts are less likely to have arisen by alternative splicing because of their differing 5' regions. We propose that they are products of one or more different loci related to H.

We have also probed Northern blots of other murine tissues and cell lines with H cDNA, The 4.4-kb transcript was expressed by kidney and L929 fibroblasts at approximately 50% of the level of liver and by thymus, spleen, and the hepatic cell line 1469 at approximately 10% of the level of liver. All of these cells also expressed small amounts of the 1.8-kb transcript at levels approximately 5% of those in the liver. mRNA from the macrophage cell lines WR19.1M and P388.D1 did not hybridize with H cDNA.

To analyze further the nature of these mRNA transcripts, we screened a non-size-selected C57B10.WR liver cDNA library with a full-length H cDNA probe in order to isolate clones corresponding to these transcripts. Eight distinct clones, different from H cDNA by restriction mapping, were isolated, and thus far seven of them have been sequenced. They appear to represent three separate types of mRNA. The first type, defined by five clones, has a unique 5' untranslated region and leader sequence, and then has over 90% homology with H sequence. Some of the clones appear to be products of alternative splicing of a novel H-related gene; whereas all of the clones contain regions highly homologous to but distinct from H repeats 5–7 and 19, some also contain regions highly homologous to H repeats 8, 9, and 20 and the 3' untranslated region. The second type is defined by a single clone that is incomplete at its 5' end. It is comprised of 13 repeating units that share extensive (over 90%) homology with H repeats 6–14, 16, 17, 19, and 20. However, there are a number of unique base pair differences in this clone that demonstrate it to be distinct from the first type of mRNA. The third type is also defined by only one clone. It has a unique 5' untranslated region and putative leader sequence region, two consensus repeating units which share approximately 56% homology with H repeats 6 and 7, and a region with 80%–95% homology with H repeats 19 and 20 and the 3' untranslated region.

Although all of these clones contain regions highly homologous with H sequence, we believe that they are distinct from H, rather than allelic differences. The full-length H cDNA clone, derived from a C57B10.WR library, and the exons of the H genomic clones, derived from a BALB/c library, shared sequence homology of 99.93% despite coming from two different strains. Thus, even the difference in sequence of 5%–10% between the H and H-related cDNAs that we see is significant, especially since the clones were derived from the same strain of mice.

We have attempted to relate the isolated cDNA clones to the different H-related transcripts that we have observed. Unique 5' cDNA probes from mRNA types 1 and 3 were used to probe Northern blots of poly(A)$^+$ mRNA from murine liver. Both probes hybridized to a 1.8-kb transcript in liver, indicating that at least two distinct 1.8-kb H-related mRNAs are transcribed in this tissue. In addition, the type 1 probe also hybridized to transcripts of 3.5 and 2.8 kb. The L929 mRNA was unreactive with the probe from type 3 cDNA in Northern blot analysis.

In the course of screening the murine genomic cosmid library for H clones, 33 other clones had been isolated that showed restriction maps and hybridization pattern distinct from the H gene cluster. DNA from each of these cosmids has now been probed in Southern blots with both unique H-related cDNA fragments and oligonucleotides specific for each of the 3 cDNA groups. Nine of the cosmids hybridized with a unique 3' probe from group 1. A unique 5' probe from group 3 hybridized to two other cosmids. No cosmid clones have yet been identified with sequence from group 2. Thus, probes from the H-related cDNA clones have identified two new H-related genes, to which 11 of the 33 cosmid clones have been assigned. The remaining 22 cosmid clones may be part of these two genes already defined, or they may represent other genes, one of which may include the type 2 cDNA clone.

The same probe from group 1 that was used in the cosmid Southern blots was used in linkage analysis of recombinant inbred mice (S. F. KINGSMORE, D. P. VIK, B. F. TACK, M. F. SELDIN, unpublished results). Genomic DNA from back-crossed mice was probed with cDNA from both H and group 1 clones. No crossover events were observed in over 200 mice, indicating that the group 1 locus is ≤ 0.5 cM from the H gene locus. We do not know at this time whether these newly defined H-related genes direct the synthesis of protein products, and whether these anticipated proteins have complement regulatory activity. These issues are currently being investigated in our laboratories.

5 Summary

While the mouse and human H proteins are structurally and functionally similar, they differ in their genetics. Whereas there is no evidence in humans for more than one gene; in mice the H locus is complex. Based on cDNA sequence and hybridization analysis of genomic cosmid clones, there are at least three distinct genes, all highly related to one another. The consensus repeating unit that comprises this molecule has obviously been duplicated numerous times, since it is present in many other molecules. Thus, it is not surprising to discover that there are several genes related to H in the mouse. A similar case has been described for two other members of this family. In humans, CR1 cDNA hybridizes to two distinct genomic clusters in the CR1 locus (Wong et al. 1989), and in mice, mCRY hybridizes to two regions in the genome, one on chromosome 1 and another on chromosome 8 (Aegerter-Shaw et al. 1987). It will be of interest to see if any other members of this family display as complex a genetic locus as murine H.

Acknowledgments. We greatly acknowledge the secretarial assistance of Bonnie Towle. This is publication number 5803IMM from The Research Institute of Scripps Clinic.

References

Aegerter-Shaw M, Cole JL, Klickstein LB, Wong WW, Fearon DT, Lalley PA, Weis JH (1987) Expansion of the complement receptor gene family. Identification in the mouse of two new genes related to the CR1 and CR2 gene family. J Immunol 138: 3488–3493

Angel P, Imagawa M, Chiu R, Stein B, Imbra RJ, Rahmsdorf HJ, Jonat C, Herrlich P, Karin M (1987) Phorbol ester-inducible genes contain a common *cis* element recognized by a TPA-modulated trans-acting factor. Cell 49: 729–739

Alper CA, Abramson N, Johnston RB, Jandl JH, Rosen FS (1970) Increased susceptibility to infection associated with abnormalities of complement-mediated functions and of the third component of complement (C3). N Engl J Med 282: 349–353

Alsenz J, Lambris JD, Schulz TF, Dierich MP (1984) Localization of the complement-component-C3b-binding site and the cofactor activity for factor I in the 38 kDa tryptic fragment of factor H. Biochem J 224: 389–398

Alsenz J, Schulz TF, Lambris JD, Sim RB, Dierich MP (1985) Structural and functional analysis of the complement component factor H with the use of different enzymes and monoclonal antibodies to factor H. Biochem J 232: 841–850

Bentley DR (1986) Primary structure of human complement component C2. Homology to two unrelated protein families. Biochem J 239: 339–345

Bitter-Suermann D, Burger R, Hadding U (1981) Activation of the alternative pathway of complement: efficient fluid-phase amplification by blockade of the regulatory complement protein β_1H through sulfated polyanions. Eur J Immunol 11: 291–295

Brai M, Misiano G, Maringhini S, Cutaja I, Hauptmann G (1988) Combined homozygous factor H and heterozygous C2 deficiency in an Italian family. J Clin Immunol 8: 50–56

Caras IW, Davitz MA, Rhee L, Weddell G, Martin DW, Nussenzweig V (1987) Cloning of decay-accelerating factor suggests novel use of splicing to generate two proteins. Nature 325: 545–549

Carroll MC, Alicot EM, Katzman PJ, Klickstein LB, Smith JA, Fearon DT (1988) Organization of the genes encoding complement receptors type 1 and 2, decay-accelerating factor, and C4-binding protein in the RCA locus on human chromosome 1. J Exp Med 167: 1271–1280

Catteral CF, Lyons A, Sim RB, Day AJ, Harris TJR (1987) Characterization of the primary amino acid sequence of human complement control protein factor I from an analysis of cDNA clones. Biochem J 242: 849–856

Charlesworth JA, Scott DM, Pussell BA, Peters DK (1979) Metabolism of human β_1H: studies in man and experimental animals. Clin Exp Immunol 38: 397–404

Chung LP, Bentley DR, Reid KBM (1985) Molecular cloning and characterization of the cDNA coding for C4b-binding protein, a regulatory protein of the classical pathway of the human complement system. Biochem J 230: 133–141

Conrad DH, Carlo JR, Ruddy S (1978) Interaction of β1H globulin with cell-bound C3b: quantitative analysis of binding and influence of alternative pathway components on binding. J Exp Med 147: 1792–1805

Day AJ, Willis AC, Ripoche J, Sim RB (1988) Sequence polymorphism of human complement factor H. Immunogenetics 27: 211–214 de Bruijn MHL, Fey GH (1985) Human complement component C3: cDNA coding sequence and derived primary structure. Proc Natl Acad Sci USA 82: 708–712

D'Eustachio P, Kristensen T, Wetsel RA, Riblet R, Taylor BA, Tack BF (1986) Chromosomal location of the genes encoding complement components C5 and factor H in the mouse. J Immunol 137: 3990–3995

DiScipio RG (1981) The binding of human complement proteins C5, factor B, β_1H and properdin to complement fragment C3b on zymosan. Biochem J 199: 485–496

DiScipio RG, Hugli TE (1982) Circular dichroism studies of human factor H a regulatory component of the complement system. Biochim Biophys Acta 709: 58–64

Dorsch-Häsler KDA, Keil GM, Weker F, Jasin M, Schaffner W, Koszinowski UH (1985) A long and complex enhancer activates transcription of the gene coding for the highly abundant immediate early mRNA in murine cytomegalovirus. Proc Natl Acad Sci USA 82: 8325–8329

Fearon DT (1978) Regulation by membrane sialic acid of β_1H-dependent decay-dissociation of amplification C3 convertase of the alternative complement pathway. Proc Natl Acad Sci USA 75: 1971–1975

Fearon DT, Austen KF (1977) Activation of the alternative complement pathway due to resistance of zymosan-bound amplification convertase to endogenous regulatory mechanisms. Proc Natl Acad Sci USA 74: 1683–1687

Fischer E, Kazatchkine MD (1983) Surface-dependent modulation by H of C5 cleavage by the cell-bound alternative pathway C5 convertase of human complement. J Immunol 130: 2821–2824

Gardner WD, White PJ, Hoch SO (1980) Identification of a major human serum DNA-binding protein as β_1H of the alternative pathway of complement activation. Biochem Biophys Res Commun 94: 61–67

Goldberger G, Bruns GAP, Rits M, Edge MD, Kwiatkowski DJ (1987) Human complement factor I: analysis of cDNA-derived primary structure and assignment of its gene to chromosome 4. J Biol Chem 262: 10065–10071

Harrison RA, Lachmann PJ (1979) An improved purification procedure for the third component of complement and β_1H globulin from human serum. Mol Immunol 16: 767–776

Hearing P, Shenk T (1983) The adenovirus type 5 E1A transcriptional control region contains a duplicated enhancer element. Cell 33: 695–703

Hong K, Kinoshita T, Dohi Y, Inoue K (1982) Effect of trysinization on the activity of human factor H. J Immunol 129: 647–652

Ichinose A, McMullen BA, Fujikawa K, Davie EW (1986) Amino acid sequence of the b subunit of human factor XIII, a protein composed of ten repetitive segments. Biochemistry 25: 4633–4638

Isenman DE, Podack ER, Cooper NR (1980) The interaction of C5 with C3b in free solution: a sufficient condition for cleavage by a fluid phase C3/C5 convertase. J Immunol 124: 326–331

Ito S, Tamura N (1983) Inhibition of classical C5 convertase in the complement system by factor H. Immunology 50: 631–635

Journet A, Tosi M (1986) Cloning and sequencing of full-length cDNA encoding the precursor of human complement component C1r. Biochem J 240: 783–787

Jouvin M-H, Kazatchkine MD, Cahour A, Bernard N (1984) Lysine residues, but not carbohydrates, are required for the regulatory function of H on the amplification C3 convertase of complement. J Immunol 133: 3250–3254

Katz Y, Strunk RC (1988) Synthesis and regulation of complement protein factor H in human skin fibroblasts. J Immunol 141: 559–563

Kazatchkine MD, Fearon DT, Austen KF (1979) Human alternative complement pathway: membrane-associated sialic acid regulates the competition between B and β_1H for cell-bound C3b. J Immunol 122: 75–81

Kingsmore SF, Vik DP, Kurtz CB, Leroy P, Tack BF, Weis JH, Seldin MF (1989) Genetic organization of complement receptor related genes in the mouse. J Exp Med 169: 1479–1484

Klickstein LB, Wong WW, Smith JA, Weis JH, Wilson JG, Fearon DT (1987) Human C3b/C4b receptor (CR1). Demonstration of long homologous repeating domains that are composed of the short consensus repeats characteristic of C3/C4 binding proteins. J Exp Med 165: 1095–1112

Kotwal GJ, Moss B (1988) Vaccinia virus encodes a secretory polypeptide structurally related to complement control proteins. Nature 335: 176–178

Kristensen T, Tack BF (1986) Murine protein H is comprised of 20 repeating units, 61 amino acids in length. Proc Natl Acad Sci USA 83: 3963–3967

Kristensen T, Wetsel RA, Tack BF (1986) Structural analysis of human complement protein H: homology with C4b binding protein, β_2-glycoprotein I, and the Ba fragment of B. J Immunol 136: 3407–3411

Kristensen T, Ogata RT, Chung LP, Reid KBM, Tack BF (1987a) cDNA structure of murine C4b-binding protein, a regulatory component of the serum complement system. Biochemistry 26: 4668–4674

Kristensen T, D'Eustachio P, Ogata RT, Chung LP, Reid KBM, Tack BF (1987b) The superfamily of C3b/C4b-binding proteins. Fed Proc 46: 2463–2469

Kurosky A, Barnett DR, Lee T-H, Touchstone B, Hay RE, Arnott MS, Bowman BH, Fitch WM (1980) Covalent structure of human haptoglobin: a serine protease homolog. Proc Natl Acad Sci USA 77: 3388–3392

Lambris JD, Avila D, Becherer JD, Müller-Eberhard HJ (1988) A discontinuous factor H binding site in the third component of complement as delineated by synthetic peptides. J Biol Chem 263: 12147–12150

Levy M, Halbwachs-Mecarelli L, Gubler M-C, Kohout G, Bensenouci A, Niaudet P, Hauptmann G, Lesavre P (1986) H deficiency in two brothers with atypical dense intramembranous deposit disease. Kidney Int 30: 949–956

Leytus SP, Kurachi K, Sakariassen KS, Davie EW (1986) Nucleotide sequence of the cDNA coding for human complement C1r. Biochemistry 25: 4855–4863

Lozier J, Takahashi N, Putnam FW (1984) Complete amino acid sequence of human plasma β_2-glycoprotein I. Proc Natl Acad Sci USA 81: 3640–3644

Lublin DM, Lemons RS, LeBeau MM, Holers VM, Tykocinski ML, Medof ME, Atkinson JP (1987) The gene encoding decay-accelerating factor (DAF) is located in the complement-regulatory locus on the long arm of chromosome 1. J Exp Med 165: 1731–1736

Lublin DM, Liszewski MK, Post TW, Arce MA, LeBeau MM, Rebentisch MB, Lemons RS, Seya T, Atkinson JP (1988) Molecular cloning and chromosomal localization of human membrane cofactor protein (MCP). Evidence for inclusion in the multigene family of complement-regulatory proteins. J Exp Med 168: 181–194

Mole JE, Anderson JK, Davison EA, Woods DE (1984) Complete primary structure for the zymogen of human complement factor B. J Biol Chem 259: 3407–3412

Moore MD, Cooper NR, Tack BF, Nemerow GR (1987) Molecular cloning of the cDNA encoding the Epstein-Barr virus/C3d receptor (complement receptor type 2) of human B lymphocytes. Proc Natl Acad Sci USA 84: 9194–9198

Morley BJ, Campbell RD (1984) Internal homologies of the Ba fragment from human complement component factor B, a class III MHC antigen. EMBO J 3: 153–157

Nagaki K, Iida K, Okubo M, Inai S (1978) Reaction mechanisms of β1H globulin. Int Arch Allergy Appl Immunol 87: 221–232

Nilsson B, Nilsson UR (1986) Antigens of complement factor C3 involved in the interactions with factors I and H. Scand J Immunol 23: 357–363

Nilsson UR, Müller-Eberhard HJ (1965) Isolation of β_{IF}-globulin from human serum and its characterization as the fifth component of complement. J Exp Med 122: 277–298

Pangburn MK (1986) Differences between the binding sites of the complement regulatory proteins DAF, CR1 and factor H on C3 convertases. J Immunol 136: 2216–2221

Pangburn MK, Schreiber RD, Müller-Eberhard HJ (1977) Human complement C3b inactivator: isolation, characterization, and demonstration of an absolute requirement for the serum protein β1H for cleavage of C3b and C4b in solution. J Exp Med 146: 257–270

Pangburn MK, Müller-Eberhard HJ (1978) Complement C3 convertase: cell surface restriction of $\beta_1 H$ control and generation of restriction on neuraminidase-treated cells. Proc Natl Acad Sci USA 75: 2416–2420

Parker CJ, Baker PJ, Rosse WF (1983) Comparison of binding characteristics of factors B and H to C3b on normal and paroxysmal nocturnal hemoglobinuria erythrocytes. J Immunol 131: 2484–2489

Paul MS, Aegerter M, O'Brien SE, Lurtz CB, Weis JH (1989) The murine complement receptor gene family. Analysis of mCRY gene products and their homology to human CR1. J Immunol 142: 582–589

Pelham HRB (1982) A regulatory upstream promoter element in the *Drosophila* hsp 70 heat-shock gene. Cell 30: 517–528

Reid KBM, Bentley DR, Campbell RD, Chung LP, Sim RB, Kristensen T, Tack BF (1986) Complement system proteins which interact with C3b or C4b. Immunol Today 7: 230–234

Rey-Campos J, Rubinstein P, Rodríguez de Córdoba S (1988) A physical map of the human regulator of complement activation gene cluster linking the complement genes CR1, CR2, DAF, and C4BP. J Exp Med 167: 664–669

Ripoche J, Al Salihi A, Rousseaux J, Fontaine M (1984) Isolation of two molecular populations of human complement factor H by hydrophobic affinity chromatography. Biochem J 221: 89–96

Ripoche J, Day AJ, Moffatt B, Sim RB (1987) mRNA coding for a truncated form of human complement factor H. Biochem Soc Trans 15: 651–652

Ripoche J, Day AJ, Harris TJR, Sim RB (1988) The complete amino acid sequence of human complement factor H. Biochem J 249: 593–602

Rodríguez de Córdoba S, Rubinstein P (1984) Genetic polymorphism of human factor H (β1H). J Immunol 132: 1906–1908

Rodríguez de Córdoba S, Rubinstein P (1987) New alleles of C4-binding protein and factor H and further linkage data in the regulator of complement activation (RCA) gene cluster in man. Immunogenetics 25: 267–268

Rodríguez de Córdoba S, Dykman TR, Ginsberg-Fellner F, Ercilla G, Agua M, Atkinson JP, Rubinstein P (1984) Evidence for linkage between the loci coding for the binding protein for the fourth component of human complement (C4BP) and for the C3b/C4b receptor. Proc Natl Acad Sci USA 81: 7890–7892

Rodríguez de Córdoba S, Lublin DM, Rubinstein P, Atkinson JP (1985) Human genes for three complement components that regulate the activation of C3 are tightly linked. J Exp Med 161: 1189–1195

Ross GD, Newman SL, Lambris JD, Devery-Pocius JE, Cain JA, Lachmann PJ (1983) Generation of three different fragments of bound C3 with purified factor I or serum. II. Location of binding sites in the C3 fragments for factors B and H, complement receptors, and bovine conglutinin. J Exp Med 158: 334–352

Ross SC, Densen P (1984) Complement deficiency states and infection: epidemiology, pathogenesis and consequences of neisserial and other infections in an immune deficiency. Medicine 63: 243–273

Schulz TF, Schwäble W, Stanley KK, Weiß E, Dierich MF (1986) Human complement factor H: isolation of cDNA clones and partial cDNA sequence of the 38-kDa tryptic fragment containing the binding site for C3b. Eur J Immunol 16: 1351–1355

Schütz G (1988) Control of gene expression by steroid hormones. Biol Chem Hoppe Seyler 369: 77–86

Schwäble W, Zwirner J, Schulz TF, Linke RP, Dierich MP, Weiß EH (1987) Human complement factor H: expression of an additional truncated gene product of 43 kDa in human liver. Eur J Immunol 17: 1485–1489

Shimuzu A, Kondo S, Takeda S, Yodoi J, Ishida N, Sabe H, Osawa H, Diamantstein T, Nikaido T, Honjo T (1985) Nucleotide sequence of mouse IL-2 receptor cDNA and its comparison with the human IL-2 receptor sequence. Nucleic Acids Res 5: 1505–1516

Sim RB, DiScipio RG (1982) Purification and structural studies on the complement-system control protein β_1H (factor H). Biochem J 205: 285–293

Smith CA, Pangburn MK, Vogel C-W, Müller-Eberhard HJ (1983) Structural investigations of properdin and factor H of human complement. Immunobiology 164: 298

Thompson RA, Lachmann PJ (1977) A second case of human C3b inhibitor (KAF) deficiency. Clin Exp Immunol 27: 23–39

Thompson RA, Winterborn MH (1981) Hypocomplementaemia due to a genetic deficiency of β1H globulin. Clin Exp Immunol 46: 110–119

Tosi M, Duponchel C, Meo T, Julier C (1987) Complete cDNA sequence of human complement C1s and close physical linkage of the homologous genes C1s and C1r. Biochemistry 26: 8516–8524

Vik DP, Keeney JB, Muñoz-Cánoves P, Chaplin DD, Tack BF (1988) Structure of the murine complement factor H gene. J Biol Chem 263: 16720–16724

Weiler JM, Daha MR, Austen KF, Fearon DT (1976) Control of the amplification convertase of complement by the plasma protein β1H. Proc Natl Acad Sci USA 73: 3268–3272

Weis JH, Morton CC, Bruns GP, Weis JJ, Klickstein LB, Wong WW, Fearon DT (1987) A complement receptor locus: genes encoding C3b/C4b receptor and C3d/Epstein-Barr virus receptor map to 1q32. J Immunol 138: 312–315

Weis JJ, Toothaker LE, Smith JA, Weis JH, Fearon DT (1988) Structure of the human B lymphocyte receptor for C3d and the Epstein-Barr virus and relatedness to other members of the family of C3/C4 binding proteins. J Exp Med 167: 1047–1065

Whaley K, Ruddy S (1976a) Modulation of the alternative complement pathway by β1H globulin. J Exp Med 144: 1147–1163

Whaley K, Ruddy S (1976b) Modulation of C3b hemolytic activity by a plasma protein distinct from C3b inactivator. Science 193: 1011–1013

Wong WW, Cahill JM, Rosen MD, Kennedy CA, Bonaccio ET, Morris MJ, Wilson JG, Klickstein LB, Fearon DT (1989) Structure of the CR1 gene: molecular basis of the structural and quantitative polymorphisms and identification of a new CR1-like allele. J Exp Med 169: 847–863

Wyatt RJ, Julian BA, Weistein A, Rothfield NF, McLean RH (1982) Partial H (β1H) deficiency and glomerulonephritis in two families. J Clin Immunol 2: 110–117

C3 Binding Proteins of Foreign Origin*

M. P. Dierich, H. P. Huemer, and W. M. Prodinger

1 Introduction   163
2 Herpes Simplex Virus   164
3 Epstein-Barr Virus   168
4 Vaccinia Virus   169
5 *Trypanosoma cruzi*   170
6 Malaria Parasites   171
7 *Schistosoma mansoni*   172
8 *Candida albicans*   173
9 Conclusion   175
References   176

1 Introduction

The generation of C3b from C3 releases the internal thioester and allows C3 to bind covalently to OH and NH_2 groups, using the carbonyl group and leaving the SH group unoccupied. The molecules to which C3b binds in this way are commonly referred to as C3 acceptors and range from H_2O to large fluid-phase and membrane molecules. C3b and its derivatives — iC3b, C3d,g, and C3d, covalently bound to C3 acceptors — have the potential to interact noncovalently with a number of fluid-phase and membrane-associated molecules (Lambris 1988). Such molecules on pathogens are the subject of this chapter, particularly whether the presence of such C3 binding molecules constitute advantages for the pathogen in coping with the control mechanisms of the host, i.e., whether they can be considered as pathogenicity factors.

The complement system consists of two independent activation pathways, the central component C3, the terminal sequence (C5–C9), regulatory molecules, and receptors for the biologically active fragments. The antibody-independent, alternative activation pathway (C3b, B, D, P) generates the enzyme C3b,Bb. The antibody-dependent, classical pathway (C1, C4, C2) generates the enzyme C4b,C2a. These alternative and classical pathway convertases activate C3 and generate C3b and C3a. C3b, while remaining covalently bound on the acceptor, may be modulated by several

---

Institut für Hygiene, University of Innsbruck, Fritz-Pregl-Straße 3, A-6010 Innsbruck, Austria
* The authors' own work cited in this study was supported by grants from the FWF (P6920 and P6923).

cleavages: first to form iC3b, then by releasing C3c to form C3d,g, and finally C3d. The function of C3b and its derivatives is to mediate adherence to cells carrying receptors for these ligands. It is most interesting that depending on whether C3 and its derivatives are located on tumor cells, on bacteria, viruses, or other pathogens, or even on the organism's own cells, the complement receptor carrying cells by adhering to the C3 ligand may lead to destruction of the objects carrying covalently attached C3. But it is not only adherence that may be mediated by covalently bound C3. C3b also triggers the terminal sequence of C5–C9, thus generating a complex of C5b–C9, the membrane attack complex. This complex becomes deposited on target structures and may cause lysis of cells or destruction of other targets.

These killing processes operating via cells (phagocytes, natural killer cells) or via C5b–C9 are beneficial as long as they act against pathogens or, for example, tumors cells. On the other hand, they could be detrimental to the organism if they were to act on the organism's own cells, that is, if they acted, against self.

Given this central role of C3 in destructive action, it is not surprising that the human organism has developed several control mechanisms. Some of these mechanisms act by reducing the activity of the two convertases by accelerating their decay, while others operate by blocking the interaction of C3b with C5 or by helping to inactivate C3b. These various factors (dealt with in detail in other chapters of this volume) are of critical importance to protect "self" cells from being killed by complement action. It has also been demonstrated that these factors show species specificity (ATKINSON and FARRIES 1987).

It is reasonable to assume that pathogens also make use of such self-protection mechanisms by the organism as a shelter against the adverse effects upon them by the host's complement system, and in recent years this has been shown indeed to be the case. These adverse effects do not necessarily consist of destruction of the pathogen; merely the neutralization of a virus would provide enough defense for the host. Such mechanisms have been observed on viruses such as herpes simplex virus (HSV) types 1 and 2 and Epstein-Barr virus (EBV) and in association with vaccinia virus. They have also been observed on certain protozoa such as *Trypanosoma cruzi* and on worms such as *Schistosoma mansoni*. While to our knowledge they have not been described on bacteria or on fungi, iC3b and C3d binding molecules have been described on fungi. These various C3 binding structures (Table 1) are discussed below.

2 Herpes Simplex Viruses

HSV types 1 and 2 are members of the α-Herpesviridae. These are fast growing cytolytic viruses with a broad host range of in vitro cultivated cells and animal systems, in contrast to the β- (cytomegalovirus) and γ-Herpesviridae (EBV). The Herpesviridae are enveloped DNA viruses with a diameter of 120–280 nm. Their DNA is embedded in the capsid (diameter 90–110 nm; icosaedric structure) which is formed by 162 so-called capsomeres. The capsid is surrounded by an envelope consisting of a lipid membrane derived from the host cell. In the electron microscope the glycoproteins appear as spikes upon this.

Table 1. C3 binding structures on pathogenes*

Pathogen	Structure on pathogen	Interaction with complement	Decay acceleration of C3 convertase	Cofactor for I	Relationship to known human structures
HSV-1	gC-1	Noncovalent binding of C3b	+ APCA − CPCA	−	Functional to CR1
HSV-2	gC-2	Noncovalent binding of C3b by isolated gC-2	− APCA − CPCA	?	
EBV	gp350	Binding to CR2 acceptor for nascent C3b (covalent binding)			C3d.g (partial sequence homology)
	Undefined molecule(s)	?	+ APCA − CPCA	+	Functional to CR1
Vaccinia virus	p35 (secreted by infect. cells)	?	+ CPCA		C4bp (partial sequence homology)
Trypanosoma cruzi	CMTp87/93	Binding to factor B	+ APCA + CPCA	−	Functional to DAF
	CMTgp58/68	Inhibition of APCA C3 convertase formation	− APCA − CPCA	−	
	EPIgp72	Acceptor of nascent C3b (covalent binding)			
Schistosoma mansoni	Undefined molecules on immature formes	Noncovalent binding of C3b(?)			
Candida albicans	Partially defined on pseudohyphae	Binding of iC3b, C3d			Functional and antigenic relation to CR3
	gp60	Binding of C3d			

*For references see the text. gC-1, Glycoprotein C of HSV-1; gC-2, glycoprotein C of HSV-2; CMT, culture-derived metacyclic trypomastigotes; EPI, epimastigotes; APCA, alternative pathway of complement activation; CPCA, classical pathway of complement activation

The HSV genome consists of a linear double-stranded DNA of approximately 150 kb with 50% sequence homology between the two serotypes. From the occurrence of intertypic recombinants following a mixed infection with serotypes 1 and 2 one can assume that the sequences in common, which are distributed all over the genome, are collinear in the two genomes (Morse et al. 1978). HSV-1 encodes at least seven viral glycoproteins, designated gB, gC, gD, gE (Spear 1984, 1985), gG (Richman et al. 1986), gH (Buckmaster et al. 1984), and gI (Johnson and Feenstra 1987). These glycoproteins are incorporated into the virion and are expressed on the cell membrane of infected cells. Because of their location on the surface of infected cells the HSV glycoproteins act as major antigenic determinants for cellular and humoral immune attack by the host (Norrild 1985; Spear 1985).

Although their functions are not completely defined, certain specific functions have been assigned to the various glycoproteins. The envelope protein gB is involved in virus penetration and adhesion to cell membranes (Little et al. 1981) as well as to serum lipoproteins (Huemer et al. 1988), gD plays a role in inducing cell fusion (Noble et al. 1983). The envelope proteins gE/gI and glycoprotein C have functions which probably modulate the immune response; gE and gI function as a complex in binding the Fc portion of IgG (Johnson et al. 1988; Dowler and Veltri 1984), while glycoprotein C has been shown to act as a receptor for the C3b fragment of the third component of complement (Friedman et al. 1984).

The gene for gC-1 has been mapped on the *hin*dIII L fragment of the viral genome, and the DNA has been sequenced (Frink et al. 1983; Draper et al. 1984). As deduced from these data, gC-1 has an open reading frame of 511 amino acids, with a hydrophobic putative signal sequence and a hydrophobic membrane anchoring sequence and exhibits nine potential N-linked glycosylation sites. The relative molecular weight of gC-1 measured by sodium dodecyl sulfate polycrylamide gel electrophoresis (SDS-PAGE) analysis is approximately 130 kDa, but the molecular mass calculated from the amino acid sequence is only approximately 55 kDa. This is evidence that gC-1 is posttranslationally modified, as has been demonstrated for attachment of N-linked oligosaccharides (Campadelli-Fiume et al. 1982; Wenske et al. 1982) and O-linked oligosaccharides (Johnson and Spear 1983; Olofsson et al. 1983) and in sulfatation (Hope et al. 1982). The heavy glycosylation has been shown to influence the immunoreactivity with monoclonal antibodies (Sjöblom et al. 1987) and seems to be crucial for C3b receptor activity (Smiley et al. 1985; Smiley and Friedman 1985). There is also evidence that host cell differences in the extent of glycosylation of gC are responsible for variations in C3b receptor expression on various cell lines (Smiley et al. 1985; Smiley and Friedman 1985). Neuraminidase treatment of HSV-infected cells enhances receptor activity, suggesting that sialic acid residues on gC interfere with C3b binding (Smiley and Friedman 1985; Eisenberg et al. 1987).

Using monoclonal antibodies against gC-1 two major antigenic sites, designed I and II, have been identified (Marlin et al. 1985). Antigenic site II, which is composed of three subsites, is probably located on the amino-terminal half of gC-1 (Holland et al. 1984; Marlin et al. 1985) and contains also eight of the nine potential glycosylation sites. Antigen site I is probably located between amino acids 297 and 359 on the carboxy-terminal half (Homa et al. 1986). All antibodies tested so far, directed against these two sites on gC-1, have been shown to block C3b binding (Friedman et al. 1986) whereas we could show that antibodies directed against the middle third of gC-1 had no inhibitory effect (HUEMER et al. 1989). This indicates that the antigenic structure of gC-1 does not define one particular region of gC which binds C3b.

Although the expression of glycoprotein C is not required for infectivity in cell culture (HOLLAND et al. 1984), and gC-deficient mutants are still pathogenic in animal systems (JOHNSON et al. 1986), gC expression seems to be a conserved function in vivo (FIEDMAN et al. 1986), and the presence of gC in clinical isolates (PEREIRA et al. 1982) suggests that C3b receptor activity of gC-1 is important for viral pathogenesis. HSV-1 and equine herpesvirus type 1 (BIELEFELDT-OHMANN and BABIUK 1988) are the only viral agents known to induce C3b receptor activity on a variety of cell types following infection (BIELEFELDT-OHMANN and BABIUK 1988; FRIEDMAN et al. 1984; KUBOTA et al. 1987; SMILEY et al. 1985); other herpesviruses tested do not induce detectable expression of the receptor on the surface of infected cells (BIELEFELDT-OHMANN and BABIUK 1988; SMILEY et al. 1985; FRIEDMAN et al. 1984).

HSV-2 does not express a C3 receptor on infected cells when tested for the ability of rosette formation between HSV-2 infected cells and C3b-coated erythrocytes (FRIEDMAN et al. 1984). Recently it could be shown that purified glycoprotein C from HSV-2 infected cells (gC-2) also binds C3b (EISENBERG et al. 1987; MCNEARNEY et al. 1987) even though no receptor activity could be found on the surfaces of HSV-2 infected cells. While at present it is not clear why HSV-2 infected cells do not express C3b receptor activity on their surfaces, the simplest explanation would be that there is less gC-2 than gC-1 expressed on infected cell membranes (SEIDEL-DUGAN et al. 1988).

HSV-1 glycoprotein C and that of HSV-2 show amino acid homology (DOWBENKO and LASKY 1984; FRINK et al. 1983; SWAIN et al. 1985) and are antigenetically related (ZWEIG et al. 1983), but there are a number of differences between them. Although the genes for gC-1 and gC-2 occupy collinear positions in their respective viral genomes (ZEZULAK and SPEAR 1984) and are homologous in sequence (DOWBENKO and LASKY 1984; SWAIN et al. 1985), the glycoproteins specified by these genes are quite different in size. This difference is in part due to the fact that gC-1 contains a stretch of 27 amino acids near its NH_2-terminus that are not present in gC-2 (DOWBENKO and LASKY 1984; SWAIN et al. 1985). Moreover, the mechanism of C3b binding to gC-1 and gC-2 appears different since treatment with endoglycosidase (F and H, respectively) of gC-1 and gC-2 inhibited C3b binding to gC-2 but had no effect on its binding to gC-1, thus indicating that N-linked oligosaccharides are involved in gC-2 but not in gC-1.

GC-1 shows binding specificities and functional properties very similar to known human complement receptors (KUBOTA et al. 1987; FRIES et al. 1986). Like two human complement regulatory proteins, decay-accelerating factor (DAF) and the C3b/C4b receptor (CR1), it possesses decay-accelerating activity for the alternative pathway C3 convertase (FRIES et al. 1986) while gC-2 shows no decay acceleration (EISENBERG et al. 1987). In addition, gC-1 is thought to impair the overall lytic efficiency of such convertases by interfering with the interaction of C5/C5b with C3b much more efficiently than gC-2 (EISENBERG et al. 1987). Thus, gC-1 is at least 100-fold more potent than factor H in disrupting the C3b–C5b interaction. But unlike factor H and CR1, it does not promote the factor I mediated cleavage of C3b (FRIES et al. 1986).

In spite of these differences gC-2 has been shown recently to provide protection against complement-mediated viral neutralization (MCNEARNEY et al. 1987). In these neutralization assays, a gC-deficient mutant strain and recombinants of the same strain expressing gC-1 or gC-2, were compared with respect to sensitivity to neutralization by complement in the presence and absence of anti-HSV antibody. This would indicate that both the classical and the alternative pathway could be important in viral neutralization. This is new information, because the role of complement-mediated lysis in viral immunity is still unclear, although some viruses and virus-infected cells are susceptible to complement action (COOPER and NEMEROW 1983). Interestingly, lysis of such targets generally proceeds via the alternative pathway in homologous systems, even in presence of antibody (COOPER and NEMEROW 1983). Thus the specificity of gC-1 for the alternative pathway convertase may be quite appropriate. In addition to causing lysis, complement activation by viruses or infected cells may lead directly to viral neutralization or provide binding sites for complement receptors on a variety of host immune effector cells, including polymorphonuclear leukocytes, mononuclear phagocytes, and lymphocytes with killer and/or natural killer activity. Glycoprotein C might also interfere with the binding of target-bound C3b or its derivative fragments to receptors on such effector cells.

3 Epstein-Barr Virus

EBV is a human γ-herpesvirus. As in the case of the other herpesviruses, its genome consists of double-stranded DNA; it has an envelope derived from the nuclear membrane of the EBV-producing cells. The envelope gp350 shows some amino acid sequence homology with that of human C3d,g (NEMEROW et al. 1987). This offers an explanation for the binding of EBV to CR2, the C3d/EBV receptor, on human B-lymphocytes (for a recent review see N. R. COOPER et al. 1988) and human epithelial cells during the early phase of their ontogeny (SIXBEY et al. 1987). This gp350, whole EBV, and latently EBV-infected and EBV-producing cells, stimulate the alternative pathway activation of complement (MAYES et al. 1984; MOLD et al. 1988b; SCHULZ et al. 1980; MCCONNELL et al. 1978), in the course of which gp350 may covalently bind C3b and serve as an acceptor for C3b.

Most recently MOLD et al. (1988a) have detected cofactor activity for factor I dependent cleavage of C3b, iC3b, C4b, and iC4b in purified preparations of EBV obtained from marmoset and human B lymphoblastoid cells. Furthermore, EBV enhanced decay of the alternative pathway C3 convertase, but not that of the classical pathway C3 convertase. Although such activities would suggest binding to C3b and C4b, the authors state that EBV does not bind preformed C3b or C4b. These characteristics differ from those of other cofactor molecules such as H, CR1, and CR2. Since the structures responsible for this cofactor activity on EBV are functionally similar to CR1 but are not yet isolated, MOLD et al. in describing this activity consider a long list of control experiments. The CR1-like molecules have no protease activity; they are not inhibited by antibodies against CR1. A strong argument for the hypothesis that the molecules are EBV coded is the fact that both marmoset and human cells give rise to EBV endowed with such activity. MOLD et al. provide a calculation concerning the strength of the EBV cofactor activity: 2×10^9 EBV particles have about the same cofactor activity as 10^7 human erythrocytes, carrying roughly 10^3 molecules of CR1 per cell. Since EBV would certainly carry more than five copies of the cofactor molecule, the EBV cofactor must be less active than CR1, the erythrocyte cofactor.

The biological function of these as yet unisolated CR1-like molecules on EBV can only be the subject of speculation. C3b molecules that become covalently bound to EBV in the course of alternative pathway activation are quickly degraded to iC3b and further to C3d,g. This has two consequences: firstly, covalently attached C3b cannot induce further alternative pathway activation; thus, further deposition of C3b and possible uptake by phagocytes are less likely. Secondly, C3d,g covalently fixed on EBV would enhance binding of EBV to its own receptor, CR2, on B cells and epithelial cells. C3, which otherwise might cover the CR2-recognizing epitope on gp350 and effect virus neutralization, might thus support EBV infection of CR2-carrying cells. Activation of the alternative pathway of complement, resulting in deposition of C3b and consecutive degradation of C3b to C3d,g and ensuring binding to CR2, may be viewed from the position of the virus as a very effective way to secure continuation of infection.

4 Vaccinia Virus

Vaccinia virus belongs to the family of Poxviridae, genus *Orthopox*. It is a brick-shaped particle with rounded corners. The DNA is contained in a central nucleoid surrounded by lipoprotein membranes, an ellipsoidal body, and the outer viral coat. Cells infected with vaccinia virus secrete several polypeptides into the culture medium. Among these are a 35-kDa protein (MCCRAE and PENNINGTON 1978), a 12-kDa protein (MCCRAE and PENNINGTON 1978), and a 19-kDa protein (VENKATESAN et al. 1982). Studying the molecular genetics of the 35-kDa protein KOTWAL and MOSS (1988) found the coding DNA to have an open reading frame for a protein of 263 amino acids and a calculated mass of 28.6 kDa. This protein contains no glycosylation sites. In addition to internal repetitions the authors detected conserved elements at four sites. These elements resemble the short consensus repeats consisting of 60 amino acids that are found in a series of molecules, of which many are complement receptors (CR1, CR2) or do react with C3b or C4b (H, DAF, C2, B, C4 binding protein; REID et al. 1986; DIERICH et al. 1988). The 35-kDa protein bears the closest resemblance to the human C4 binding protein, displaying 38% identity with the first half and 28% identity with the second half (KOTWAL and MOSS 1988).

Preliminary data by KOTWAL and MOSS (1988) indicate that the medium containing this protein may inhibit the classical pathway of complement activation. Thus, one may speculate that vaccinia virus protects virus production by the infected cell against complement attack by means of such a 35-kDa protein. Before such a speculation can be substantiated, it must be clarified whether the occurrence of the 60 amino acid units endows the 35-kDa protein with a capacity to react with C4b or C3b, and whether the presumable consequences for the action on complement activation actually ensue. Furthermore, one wonders whether the 35-kDa protein is also associated with the cell membrane.

5 Trypanosoma cruzi

T. cruzi, a hemoflagellate protozoan causing Chagas' disease in humans, passes through a complex life-cycle. Infection of vertebrate hosts is initiated by metacyclic trypomastigote forms which differentiate from noninfectious epimastigotes in the gut of blood-feeding reduviids, the insect vector. Whereas epimastigotes are lysed in nonimmune normal human serum through activation of the alternative pathway of complement (NOGUEIRA et al. 1975; RUBIO 1956), metacyclic trypomastigotes – as well as culture-derived metacyclic trypomastigotes (CMT), their in vitro equivalent – are resistant to complement lysis. Similarly, the two stages developing in the vertebrate host, the intracellularly replicating amastigote and the bloodstream trypomastigote form, are not susceptible to complement (NOGUEIRA et al. 1975; KIPNIS et al. 1981). The same is true for tissue culture derived trypomastigotes (TCT), as the representatives of the latter. Treatment of CMT with pronase or neuraminidase resulted in loss of or decrease in resistance to complement (KIPNIS et al. 1981; SHER et al. 1986), but allowed to redevelop for 10 h in culture medium, these pretreated CMT or TCT regained full insusceptibility. Inhibitors of protein synthesis or N-glycosylation (e.g., puromycin and tunicamycin, respectively) abrogated this reconstitution, thus confirming glycoconjugates on CMT and TCT as being the responsible structures for lysis escape (SHER et al. 1986).

The deposition of C3 cleavage products on *T. cruzi* manifests itself in two ways. Whereas the membrane glycoprotein gp72 on epimastigotes is a potent membrane acceptor for C3b (JOINER et al. 1985) and allows binding and activation of factor B to a great extent (JOINER et al. 1986), the related CMT gp 72 is an inefficient acceptor for C3b (JOINER et al. 1985), most of the fivefold lesser amount of C3b being bound to 25- to 30-kDa molecules on CMT (JOINER et al. 1986).

The evading of alternative pathway activation for CMT and TCT seems to be due mainly to the fact that factor B is bound with lower affinity and is therefore activated to a lesser extent than in epimastigotes; in addition, factor H is bound with high affinity (JOINER et al. 1986). This results in inactivation of C3b so that iC3b constitutes the major form of C3 on CMT and TCT (JOINER et al. 1986).

Moreover, not only the assembly of C3b,Bb, the alternative pathway C3 convertase, is inhibited on CMT and TCT, but they also possess decay-accelerating activity for preformed alternative and classical complement pathway C3 convertases (RIMOLDI et al. 1988). As cofactor function for the factor I mediated cleavage of C3b is lacking, CMT and TCT seem to possess factors that could somewhat be compared to human DAF (MEDOF et al.1984; NICHOLSON-WELLER et al. 1982). The molecules obviously responsible for this phenomenon are not yet definitely demonstrated. Nevertheless, JOINER et al. (1988) have recently characterized molecules produced and spontaneously shed by these forms that might account for this developmentally regulated escape from complement-mediated lysis. The molecules were copurified on fast protein liquid chromatography with a glycoprotein fraction of 87–93 kDa from CMT and TCT lysates and culture supernatants. Antisera against human DAF or other human regulatory proteins of complement activation did not recognize these glycoproteins, which could however be immunoprecipitated by sera from Chagas' patients. The responsibility of the glycoproteins for the postulated DAF-like activity remains to be substantiated further.

Other groups assigned inhibition of classical pathway C3 convertase formation to, as not yet further defined, molecules of 60 kDa (KIPNIS et al. 1986) or demonstrated that a glycoprotein present only on the trypomastigote forms was able to counteract cell-bound or fluid-phase alternative pathway C3-convertase formation (FISCHER et al. 1988). This glycoprotein showed a relative molecular mass of 58 kDa on unreduced and 68 kDa on reduced SDS-PAGE. However, this glycoprotein could not enhance decay of alternative or classical pathway C3 convertases or exert cofactor activity for factor I. Additionally, it showed specificity for wheatgerm agglutinin (FISCHER et al. 1988), whereas gp87-93 bound to concanavalin A and was negative for wheatgerm agglutinin. This may indicate that there is more than one clearly distinguishable molecule responsible for complement resistance.

6 Malaria Parasites

The alternative pathway C3 convertase C3b,Bb has a short half-life. Properdin is capable of stabilizing this enzyme by binding to C3b and forming the complex C3b,Bb,P. Most recently GOUNDIS and REID (1988) reported that most of the amino acid sequence of properdin is composed of six copies of a 60 amino acid motif. Each of these motifs is characterized by conservation of six cysteines and three tryptophans. Obviously these motifs are each encoded by a separate exon comparable to the short consensus repeats in other C3b and C4b binding proteins (REID et al. 1986; DIERICH et al. 1988). It is most telling that similar sequences can be found not only in membrane attack components of complement (DISCIPIO et al. 1988) and thrombospondin (LAWLER and HYNES 1986) but also in the circumsporozoite protein (CSP) of *Plasmodium falciparum* between residues 340 and 395 (DAME et al. 1984), of *P. vivax* (ARNOT et al. 1985) and *P. knowlesi* (OZAKI et al. 1983; GOUNDIS and REID 1988).

It has not yet been reported whether the sequence homology of properdin and CSP also reflects a potential of the parasites to bind C3b as properdin does, or whether this is of biological importance for survival of the parasites, for the entry into host cells, or for other purposes. Whether there is a functional connection to the fact that *P. falciparum* infected human erythrocytes activate the alternative pathway and C3 is deposited on their surface also (STANLEY et al. 1984) remains unclear.

The molecular mass of the CSP in various *Plasmodium* species ranges from 30 to 60 kDa. It is the major protein in sporozoites, covering their complete surface. Its amino acid sequence was analyzed in detail in the case of *P. knowlesi* (NUSSENZWEIG and NUSSENZWEIG 1985; GODSON 1985). The gene for CSP has unique features in that introns are absent (DAME et al. 1984). The CSP of *P. falciparum* contains a central area composed of 41 tandem repeats of a tetrapeptide. It is essential for the sporozoite's invasion of the host cell.

7 Schistosoma mansoni

Of *S. mansoni*, a widespread trematode causing bilharziasis in man, only the very early stages — those of cercariae and immature schistosomula — seem to be susceptible to killing by complement. There is evidence for activation of the alternative (MACHADO et al. 1975; OUAISSI et al. 1980a) as well as the classical pathway (CAPRON et al. 1974; TAVARES et al. 1978) of complement and for C3-dependent adherence of eosinophilic leukocytes (OTTESEN et al. 1977; RAMALHO-PINTO et al. 1978). The sophisticated life-cycle of *S. mansoni* starts with the development in an intermediate host (the snail *Biomphalaria glabrata*) of cercariae, the infectious stage of *S. mansoni*. After penetrating the skin of the human host cercariae start to transform to schistosomula. Schistosomula immediately effect intensive remodeling of their exterior, and after a complicated route through the body the further differentiation into adult worms takes place in the intrahepatic branches of the portal vein.

A distinctive feature of immature schistosomula seems to be the shedding of their — cercarial — glycocalyx, the coat of their tegumental membrane (SAMUELSON and CAULFIELD 1985) in the first few hours after infection (MARIKOVSKY et al. 1988a), thus rendering them insusceptible to destruction by activation of the alternative pathway (SAMUELSON and CAULFIELD, 1986; SANTORO et al. 1979; MARIKOVSKY et al. 1986) or eosinophils (RAMALHO-PINTO et al. 1978) when found in the lungs after about 4 days. To this end they seem to employ glandular secretion of enzymes, the inhibition of which blocks shedding of surface material (SAMUELSON and CAULFIELD 1985) as well as conversion to complement resistance (SAMUELSON and CAULFIELD 1986).

Two proteases from the secreted material have been purified (MARIKOVSKY et al. 1988b) and shown to cleave off the same surface material from schistosomula as is shed spontaneously by transforming schistosomula (MARIKOVSKY et al. 1988a). The carbohydrate-rich (and highly antigenic) glycocalyx has been shown to be a potent activator of the alternative pathway of complement activation (SAMUELSON and CAULFIELD 1986; MARIKOVSKY et al. 1986) and also to possess IgG-Fc receptor function (TORPIER et al. 1979; OUAISSI et al. 1981). Fc-bound IgG as well as the Fc trunk of the immunoglobulin remaining surface bound after cleavage by parasite enzymes (AURIAULT et al. 1981) were able to fix C1q (presumably via the CH2 domain) and hence activate the classical complement pathway independent of specific antibodies (SANTORO et al. 1979, 1980; OUAISSI et al. 1981).

Quaissi et al. postulated a C3b receptor activity on cercariae and schistosomula showing adherence of EAC1-3b which could be inhibited by purified human C3 (OUAISSI et al. 1980b). Yet the actual presence of receptors and their localization must be further clarified. Molecular characteristics are lacking. Complement activation as well as IgG-Fc or C3b receptor activity are features mainly of immature and not of adult schistosomes. TARLETON and KEMP (1981) suggested that this could be due to continuous shedding of surface-bound "proteins" on adult schistosomes, as they observed to be true for surface-attached immune complexes. In addition, they postulated receptor activity for a not further specified "activated form of C3" as concluded from binding of immune complexes activated with complement in native human serum. However, the need to use an amplification technique to obtain reproducibly visible fluorescence perhaps indicates the low amount of receptors on adult schistosomes.

8 *Candida albicans*

In this section we discuss C3 binding proteins whose function is probably not to lessen damage arising from activation of the complement cascade but rather to serve other purposes. *Candida albicans* is a dimorphic fungus accounting for the greatest part of yeast infections in man.

We could demonstrate that *C. albicans* pseudohyphae (pseudomycelial forms), but not yeast forms, have receptor activity for both iC3b and C3d (HEIDENREICH and DIERICH 1985; Fig. 1); this has subsequently been confirmed by others (GILMORE et al. 1988; EDWARDS et al. 1986; CALDERONE et al. 1988). The binding of C3 degradation products occurs with high affinity by pseudohyphae and is saturable (GILMORE et al. 1988). This may be a factor contributing to pathogenicity, since in contrast to yeast forms the pseudohyphae are considered to be the tissue-invading forms in human candidiasis. In addition, evidence that only *C. albicans* and its close relative *C. stellatoidea* exhibit C3bi and C3d binding while the other, less pathogenic species such as *C. tropicalis, C. parapsilosis,* and *C. krusei* do not, suits well in this connection and strengthens the hypothesis that the C3 binding structures on pseudohyphae serve as a pathogenicity factor.

The iC3b binding structure on *Candida* was demonstrated to be specifically stained by some of the monoclonal antibodies (OKMI and M522) directed against the human CR3 α-chain (GILMORE et al. 1988; EDWARDS et al. 1986; EIGENTLER et al. 1989). Other antibodies against the α-chain of CR3 (MN41, Leu15) and an antibody against the β-chain (MHM23) did not stain. However, blocking of adherence of EAC3bi was not observed with all of these antibodies (EIGENTLER et al. 1989; EDWARDS et al. 1986) except with OKM1, when applied at the relatively high concentration of 300 µg/ml (EIGENTLER et al. 1989). Nevertheless, both EAC3bi rosette formation – taken as an indication of "CR3" expression – and antibody staining were similar in their dependence on the growth temperature of the fungi: pseudohyphae grown at 30 °C showed an optimal, at 37 °C a reduced, and at 38 °C no expression of the CR3-like structures (EIGENTLER et al. 1989). The fact that the antibodies were poor inhibitors of rosette formation may be explained by assuming that the epitopes recognized by the antibodies are not directly involved in the binding site. It is of interest to note that one of the anti-CR3 antibodies, OKM1, could precipitate mainly a 130-kDa molecule in addition to 50-, 100-, and 500-kDa molecules from pseudohyphal extracts following ^{125}I labeling of intact pseudomycelia (EIGENTLER et al. 1989).

However, these and other data – and the fact that, in contrast to human CR3, the corresponding candidal molecule is functionally independent of divalent cations (EIGENTLER et al. 1989) – indicate that there is only a limited antigenic and functional relationship.

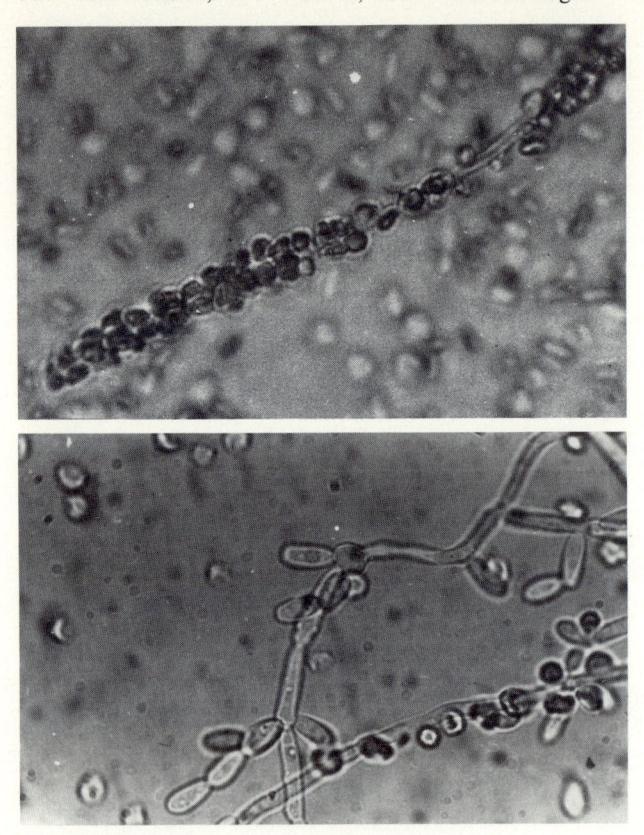

Fig. 1. The strong binding of EAC1423bi to *Candida albicans* pseudohyphae (upper half of figure) is in contrast to only little EAC1423b adherence (lower half). This finding led to further investigations on CR3-like structures on *Candida albicans*

CALDERONE et al. have recently elucidated the structure probably responsible for C3d binding to *C. albicans* (CALDERONE et al. 1988; LINEHAN et al. 1988). They could purify a 62-kDa molecule out of pseudohyphal extract, which was suggested to be a mannoprotein from its affinity to concanavalin A, although its carbohydrate moiety does not seem to be involved in the receptor-ligand interaction (CALDERONE et al. 1988). This 62-kDa receptor could exclusively block adherence of EAC3d to pseudohyphae. If the same receptor is also involved in iC3b binding, remains to be investigated.

A monoclonal antibody directed against this C3d receptor (and shown to have its epitope on the outer fibrillar layer of the cell wall) did not react with human cells carrying CR2 or CR3 (LINEHAN et al. 1988). Indeed, an antibody against human CR2 also failed to stain pseudomycelia (GILMORE et al. 1988; EIGENTLER et al. 1989) and could block C3d rosette formation only at high concentration (EDWARDS et al. 1986).

One function of these complement receptor-like structures might be that covalently bound iC3b on the candidal surface may interact with fungal receptors on the same organism, thus becoming inaccessible for the phagocytic cells. The clumping of *C.*

albicans in human serum (CHILGREN et al. 1968) might also be due to these molecules, the C3-fragment of the one being bound by the receptor of the neighboring fungus. However, other functions might be assigned to the CR3-like structure on the pseudohyphae.

CR3 on macrophages has been established as one of the molecules forming the Mac-1 family, or β_2-integrins (DIERICH et al. 1988). These molecules have important adherence functions. Binding of iC3b to CR3 is only one — possibly not even the most important — characteristic. Thus, CR3 also binds β-glucan, various bacteria, lipopolysaccharide, fibrinogen (WRIGHT et al. 1988), and other substances (for reference see DIERICH et al. 1988). Most recently it was demonstrated that *Leishmania* promastigotes, by an Arg-Gly-Asp containing region of their major surface glycoprotein gp63, bind to CR3 directly without involvement of complement (RUSSELL and WRIGHT 1988). Others have suggested that C3 secreted by macrophages themselves is important for the uptake of *Leishmania* (A. COOPER et al. 1988). Aside from gp63, the promastigote lipophosphoglycan has been demonstrated to be involved in the uptake (HANDMAN and GODING 1985). Based on these findings it is well conceivable that the *C. albicans* CR3-like structure acts as an adherence-mediating molecule that is instrumental in the tissue invasion by pseudohyphae. Understanding its interaction with iC3b may help to characterize this structure.

9 Conclusion

C3 binding proteins on pathogens is a gradually expanding area of interest. The examples known to date and discussed here have only in some cases reached a considerable degree of biochemical, moleculargenetic, and functional characterization. In several instances it may still be considered a case of "more fantasy than facts." Nevertheless, the data thus far known are very encouraging in that the detailed characterization of the known and assumed C3 binding proteins of pathogens may be detected using complement-related mechanisms to defend themselves against complement of the host or to promote their development in other ways. Dealing with these aspects will certainly lead to better understanding of host-parasite relationships. Whether this will have therapeutic consequences is unclear.

Acknowledgement: The secretarial help of Mrs. S. Wucherer is gratefully recognized.

Note added in proof: During the XIIIth International Complement Workshop, September 1989, NORRIS et al. have reported on the isolation of a 160 kD protein from *Trypanosoma cruzi* which binds to C3b-Sepharose anchored in the membrane via a glycosylphosphatidylinositol linkage and a λgt11 clone expressing the 160 kD protein hybridizes to human DAF cDNA under moderate stringency conditions.
(Norris KA, Bradt B, Flynn J, So M, Cooper NR (1989) Purification, characterization and cloning of a *Trypanosoma cruzi* membrane protein with functional and genetic similarities to C3 binding proteins. Compl. Inflamm 6(5): 378)

References

Arnot DE, Barnwell JW, Tam JP, Nussenzweig V, Nussenzweig RS, Enea V (1985) Circumsporozoite protein of *Plasmodium vivax*: gene cloning and characterization of the immunodominant epitope. Science 230: 815–818

Atkinson JP, Farries T (1987) Separation of self from non-self in the complement system. Immunol Today 8: 212–215

Auriault C, Ouaissi MA, Torpier G, Eisen H, Capron A (1981) Proteolytic cleavage of IgG bound to the Fc receptor of *Schistosoma mansoni* schistosomula. Parasite Immunol 3: 33

Bielefeldt-Ohmann H, Babiuk LA (1988) Induction of receptors for complement and immunoglobulins by herpesviruses of various species. Virus Res 9: 335–342

Buckmaster EA, Gompels U, Minson AC (1984) Characterisation and physical mapping of an HSV-1 glycoprotein of approximately 115 × 1000 molecular weight. Virology 139: 408–413

Calderone RA, Linehan L, Wadsworth E, Sandberg AL (1988) Identification of C3d receptors on *Candida albicans*. Infect Immun 56: 252–258

Campadelli-Fiume G, Poletti L, Dall'olio F, Serafini-Cessi F (1982) Infectivity and glycoprotein processing of herpes simplex virus type 1 grown in a ricin-resistant cell line deficient in N-acetylglycosaminyl transferase I. J Virol 43: 1061–1071

Capron A, Capron M, Dupas H, Bout D, Petitprez A (1974) Etude in vitro des phénomènes immunologiques dans la schistosomiase humaine et expérimentale. I. Etude comparative in vitro de l'activité léthal d'immunosérums sur les formes immatures et sur les adultes de S. mansoni. Int J Parasitol 4: 613–623

Chilgren RA, Hong R, Quie PG (1968) Human serum interaction with *Candida albicans*. J Immunol 101: 128–132

Cooper A, Rosen H, Blackwell JM (1988) Monoclonal antibodies that recognize distinct epitopes of the macrophage type three complement receptor differ in their ability to inhibit binding of *Leishmania* promastigotes harvested at different phases of their groth cycle. Immunology 65: 511–514

Cooper NR, Nemerow GR (1983) Complement, viruses, and virus-infected cells. Springer Sem Immunopathol 6: 327–347

Cooper NR, Moore MD, Nemerow GR (1988) Immunobiology of CR2, the B lymphocyte receptor for Epstein-Barr virus and the C3d complement fragment. Annu Rev Immunol 6: 85–113

Dame JB, Williams JL, McCutchan TF, Weber JL, Wirtz RA, Hockmeyer WT, Maloy WL, Haynes JD, Schneider I, Roberts D, Sanders GS, Reddy EP, Diggs CL, Miller LH (1984) Structure of the gene encoding the immunodominant surface antigen on the sporozoite of the human malaria parasite *Plasmodium falciparum*. Science 225: 593–599

Dierich MP, Schulz TF, Eigentler A, Huemer H, Schwaeble W (1988) Structural and functional relationships among receptors and regulators of the complement system. Mol Immunol 25: 1043–1051

DiScipio RG, Chakravarti DN, Müller-Eberhard HJ, Fey GH (1988) The structure of human complement component C7 and the C5b-7 complex. J Biol Chem 263: 549–560

Dowbenko DJ, Lasky LA (1984) Extensive homology between herpes simplex virus type 2 glycoprotein F gene and the herpes simplex virus type 1 glycoprotein C gene. J Virol 52: 154–163

Dowler KW, Veltri RW (1984) In vitro neutralization of HSV-2: inhibition by binding of normal IgG and purified Fc to virion Fc receptor (FcR). J Med Virol 13: 251–259

Draper KG, Costa RH, Lee G-Y, Spear PG, Wagner EK (1984) Molecular basis of the glycoprotein-C-negative phenotype of herpes simplex virus type 1 macroplaque strain. J Virol 51: 578–585

Edwards JE, Gaither TA, O'Shea JJ, Rotrosen D, Lawley TJ, Wright SA, Frank MM, Green I (1986) Expression of specific binding sites on *Candida* with functional and antigenic characteristics of human complement receptors. J Immunol 137: 3577–3583

Eigentler A, Schulz TF, Larcher C, Breitwieser EM, Myones BL, Petzer AL, Dierich MP (1989) Temperature dependent expression of a C3bi binding protein on *Candida albicans* and its relationship to human CR3. Infect Immun 57/2: 616–622

Eisenberg RJ, Ponce de Leon M, Friedman HM, Fries LF, Frank MM, Hastings JC, Cohen GH (1987) Complement component C3b binds directly to purified glycoprotein C of herpes simplex virus types 1 and 2. Microbiol Pathol 3: 423–435

Fischer E, Ouaissi MA, Velge P, Cornette J, Kazatchkine MD (1988) Gp58/68, a parasite component that contributes to the escape of the trypomastigote form of *Trypanosoma cruzi* from damage by the human alternative complement pathway. Immunology 65(2): 299–305

Friedman HM, Cohen GH, Eisenberg RJ, Seidel CA, Cines DB (1984) Glycoprotein C of herpes simplex virus 1 acts as a receptor for the C3b complement component on infected cells. Nature 309: 633–635

Friedman HM, Glorioso JC, Cohen GH, Hastings JC, Harris SL, Eisenberg RJ (1986) Binding of complement component C3b to glycoprotein C of herpes simplex virus type 1: mapping of gC-binding sites and demonstration of conserved C3b binding in low passage clinical isolates. J Virol 60: 470–475

Fries LF, Friedman HM, Cohen GH, Eisenberg RJ, Hammer CH, Frank MM (1986) Glycoprotein C of herpes simplex virus 1 is an inhibitor of the complement cascade. J Immunol 137: 13636–1641

Frink RJ, Eisenberg R, Cohen G, Wagner EK (1983) Detailed analysis of the portion of the herpes simplex virus type 1 genome encoding glycoprotein C. J Virol 45: 634–647

Gilmore BJ, Retsinas EM, Lorenz JS, Hostetter MK (1988) An iC3b-receptor on *Candida albicans*: structure, function, and correlates for pathogenicity. J Infect Dis 157(1): 38–46

Godson GN (1985) Molecular approach of malaria vaccines. Sci Am 252: 32–39

Goundis D, Reid KBM (1988) Properdin, the terminal complement components, thrombospondin and the circumsporozoite protein of malaria parasites contain similar sequence motifs. Nature 335: 82–85

Handman E, Goding JW (1985) The *Leishmania* receptor for macrophages is a lipid-containing glycoconjugate. EMBO J 4: 329–336

Heidenreich F, Dierich MP (1985) *Candida albicans* and *Candida stellatoidea*, in contrast to other *Candida* species, bind iC3B and C3d but not C3b. Infect Immun 50: 598–600

Holland TC, Homa FL, Marlin SD, Levine M, Glorioso J (1984) Herpes simplex virus type 1 glycoprotein C-negative mutants exhibit multiple phenotypes, including secretion of truncated glycoproteins. J Virol 52: 566–574

Homa FL, Purifoy DJM, Glorioso JC, Levine M (1986) Molecular basis of the glycoprotein C-negative phenotypes of herpes simplex virus type 1 mutants selected with a virus-neutralizing monoclonal antibody. J Virol 58: 281–289

Hope RG, Palfreyman J, Suh M, Marsden HS (1982) Sulphated glycoproteins induced by herpes simplex virus. J Gen Virol 58: 399–415

Huemer HP, Bröker M, Larcher C, Lambris JD, Dierich MP (1989) The central segment of Herpes simplex virus type 1 glycoprotein C (gC) is not involved in C3b binding: Demonstration by using monoclonal antibodies and recominant gC expressed in Escherichia coli. J Gen Virol 70: 1571–1578

Huemer HP, Menzel HJ, Potratz D, Brake B, Falke D, Utermann G, Dierich MP (1988) Herpes simplex virus binds to human serum lipoprotein. Intervirology 29: 68–76

Johnson DC, Feenstra V (1987) Identification of a novel herpes simplex virus type 1-induced glycoprotein which complexes with gE and binds immunoglobulin. J Virol 61: 2208–2216

Johnson DC, Spear PG (1983) O-linked oligosaccharides are acquired by herpes simplex virus glycoproteins in the Golgi apparatus. Cell 32: 987–997

Johnson DC, McDermott MR, Chrisp C, Glorioso JC (1986) Pathogenicity in mice of herpes simplex virus type 2 mutants unable to express glycoprotein C. J Virol 58: 36–42

Johnson DC, Frame MC, Ligas MW, Cross AM, Stow ND (1988) Herpes simplex virus immunoglobulin G Fc receptor activity depends on a complex of two viral glycoproteins, gE and gI. J Virol 62: 1347–1354

Joiner K, Hieny S, von Kirchhoff L, Sher A (1985) gp72, the 72 kilodalton glycoprotein, is the membrane acceptor site for C3 on *Trypanosoma cruzi* epimastigotes. J Exp Med 161: 1196–1212

Joiner K, Sher A, Gaither T, Hammer C (1986) Evasion of alternative complement pathway by *Trypanosoma cruzi* results from inefficient binding of factor B. Proc Natl Acad Sci USA 83: 6593–6597

Joiner KA, Dias da Silva W, Rimoldi MT, Hammer CH, Sher A, Kipnis TL (1988) Biochemical characterization of a factor produced by trypomastigotes of *Trypanosoma cruzi* that accelerates the decay of complement C3 convertases. J Biol Chem 263(23): 11327–11336

Kipnis TL, David JD, Alper CA, Sher A, Dias da Silva W (1981) Enzymatic treatment transforms trypomastigotes of *Trypanosoma cruzi* into activators of the alternative complement pathway and potentiates their uptake by macrophages. Proc Natl Acad Sci USA 78(1): 602–605

Kipnis TL, Tambourgi DV, Sucupira M, Dias da Silva W (1986) Effect of *Trypanosoma cruzi* membrane components on the formation of the classical pathway C3 convertase. Braz J Med Biol Res 19(2): 271–278

Kotwal GJ, Moss B (1988) Vaccinia virus encodes a secretory polypeptide structurally related to complement control proteins. Nature 335: 176–178

Kubota Y, Gaither TA, Cason J, O'Shea JJ, Lawley TL (1987) Characterization of the C3 receptor induced by herpes simplex virus type 1 infection of human epidermal, endothelial, and A431 cells. J Immunol 138: 1137–1142

Lambris JD (1988) The multifunctional role of C3, the third component of complement. Immunol Today 9: 387–393

Lawler J, Hynes RO (1986) The structure of human thrombospondin, an adhesive glycoprotein with multiple calcium-binding sites and homologies with several different proteins. J Cell Biol 103: 1635–1648

Linehan L, Wadsworth E, Calderone RA (1988) *Candida albicans* C3d-receptor, isolated by using a monoclonal antibody. Infect Immun 56(8): 1981–1986

Little SP, Jofre JT, Courtney RJ, Schaffer PA (1981) A virion-associated glycoprotein essential for infectivity of herpes simplex virus type 1. Virology 115: 149–160

Machado AJ, Gazzinelli G, Pellegrino J, Dias da Silva W (1975) *Schistosoma mansoni*: the role of the complement C3 activating system in the cercaricidal action of normal serum. Exp Parasitol 38: 20

Marikovsky M, Levi-Schaffer F, Arnon A, Fishelson Z (1986) *Schistosoma mansoni*: killing of transformed schistosomula by the alternative pathway of human complement. Exp Parasitol 61/1: 86–90

Marikosvky M, Arnon R, Fishelson Z (1988a) Proteases secreted by transforming schistosomula of *Schistosoma mansoni* promote resistance to killing by complement. J Immunol 141(1): 273–278

Marikovsky M, Fishelson Z, Arnon R (1988b) Purification and characterization of proteases secreted by transformed schistosomula of *Schistosoma mansoni*. Mol Biochem Parasitol 1: 45–54

Marlin SD, Holland TC, Levine M, Glorioso JC (1985) Epitopes of herpes simplex virus type 1 glycoprotein gC are clustered in two distinct antigenic sites. J Virol 53: 128–136

Mayes JT, Schreiber RD, Cooper Nr (1984) Development and application of an enzyme-linked immunosorbent assay for the quantitation of alternative complement pathway activation in human serum. J Clin Invest 73: 160–170

McConnell I, Klein G, Lint TF, Lachmann PJ (1978) Activation of the alternative complement pathway by human B cell lymphoma lines is associated with Epstein Barr virus transformation of the cells. Eur J Immunol 8: 453–458

McCrae MA, Pennington TH (1978) Specific secretion of polypeptides from cells infected with vaccinia virus. J Virol 28: 828–834

McNearney TA, Odell C, Holers VM, Spear PG, Atkinson JP (1987) Herpes simplex virus glycoprotein gC-1 and gC-2 bind to the third component of complement and provide protection against complement-mediated neutralization of viral infectivity. J Exp Med 166: 1525–1535

Medof ME, Kinoshita T, Nussenzweig V (1984) Inhibition of complement activation on the surface of cells after incorporation of decay-accelerating factor (DAF) into their membranes. J Exp Med 160: 1558–1578

Mold C, Bradt BM, Nemerow GR, Cooper NR (1988a) Epstein-Barr virus regulates activation and processing of the third component of complement. J Exp Med 168: 949–969

Mold C, Bradt BM, Nemerow GR, Cooper NR (1988b) Activation of the alternative complement pathway by EBV and the viral envelope glycoprotein, gp350. J Immunol 140: 3867–3874

Morse LS et al. (1978) Anatomy of HSV DNA. XI. mapping of viral genes by analysis of polypeptides and functions specified by HSV-1 × HSV-2 recombinants. J Virol 26: 389–410

Nemerow GR, Mold C, Keivens Schwend V, Tollefson V, Cooper NR (1978) Identification of gp350 as the viral glycoprotein mediating attachment of Epstein-Barr virus (EBV) to the EBV/C3d receptor of B cells: sequence homology of gp350 and C3 complement fragment C3d. J Virol 61: 1416–1420

Nicholson-Weller A, Burge J, Fearon DT, Weller PF, Austen KF (1982) Isolation of a human erythrocyte membrane glycoprotein with decay-accelerating activity of C3 convertases of the complement system. J Immunol 129: 184–189

Noble AG, Lee G-Y, Sprague R, Parish ML, Spear PG (1983) Anti-gD monoclonal antibodies inhibit cell fusion induced by herpes simplex virus type 1. Virology 129: 218–224

Nogueira N, Bianco C, Cohen Z (1975) Studies on the selective lysis and purification of *Trypanosoma cruzi*. J Exp Med 142: 224–229

Norrild B (1985) Humoral response to herpes simplex virus infections. In: Roizman B, Lopez C (eds) Immunobiology and prophylaxis of human herpes infections. The herpes viruses, vol 4. Plenum, New York, pp 69–86

Nussenzweig V, Nussenzweig RS (1985) Circumsporozoite proteins of malaria parasites. Cell 42: 401–403

Olofsson S, Sjöblom I, Lundström M, Jeansson S, Lycke E (1983) Glycoprotein C of herpes simplex virus type 1: characterization of O-linked oligosaccharides. J Gen Virol 64: 2735–2747

Ottesen EA, Stanley AM, Gelfand JA, Gadek JE, Frank MM, Nash TE, Cheever AW (1977) Immunoglobulin and complement receptors on human eosinophils and their role in cell adherence to schistosomules. Am J Trop Med Hyg 26: 134–141

Ouaissi MA, Santoro F, Capron A (1980a) *Schistosoma mansoni*: ultrastructural damage due to complement on schistosomula in vitro. Exp Parasitol 50: 74–82

Ouaissi MA, Santoro F, Capron A (1980b) Interaction between *Schistosoma mansoni* and the complement system. Receptors for C3b on cercariae and schistosomula. Immunol Lett 1: 197–210

Ouaissi MA, Auriault C, Santoro F, Capron A (1981) Interaction between *Schistosoma mansoni* and the complement system: role of the IgG Fc peptides in the activation of the classical pathway by schistosomula. J Immunol 127(4): 1556–1559

Ozaki LS, Svec P, Nussenzweig RS, Nussenzweig V, Godson GN (1983) Structure of the *Plasmodium knowlesi* gene coding for the circumsporozoite protein. Cell 34: 815–822

Pereira L, Dondero DV, Gallo D, Devlin V, Woodie JD (1982) Serologic analysis of herpes simplex virus types 1 and 2 with monoclonal antibodies. Infect Immun 35: 363–367

Ramalho-Pinto FJ, McLaren DJ, Smithers SR (1978) Complement-mediated killing of schistosomula of *Schistosoma mansoni* by rat eosinophils in vitro. J Exp Med 147: 147–156

Reid KBM, Bentley DR, Campbell RD, Chung LP, Sim RB, Kristensen T, Tack BF (1986) Complement system proteins which interact with C3b or C4b. Immunol Today 7: 230–234

Richman DD, Buckmaster EA, Bell SE, Hodgman C, Minson AC (1986) Identification of a new glycoprotein of herpes simplex virus type 1 and genetic mapping of the gene that codes for it. J Virol 57: 647–655

Rimoldi MT, Sher A, Hieny S, Lituchy A, Hammer CH, Joiner K (1988) Developmentally regulated expression by *Trypanosoma cruzi* of molecules that accelerate the decay of complement C3 convertases. Proc Natl Acad Sci USA 85: 193–197

Rubio M (1956) Actividad litica de sueros normales sobre formas de cultivo y sanguineas de Trypanosoma cruzi. Bol Chil Parasitol 9: 62–69

Russell DG, Wright SD (1988) Complement receptor type 3 (CR3) binds to an Arg-Gly-Asp-containing region of the major surface glycoprotein, gp63, of *Leishmania* promastigote. J Exp Med 168: 279–292

Samuelson JC, Caulfield JP (1985) The cercarial glycocalyx of *Schistosoma mansoni*. J Cell Biol 100: 1423–1434

Samuelson JC, Caulfield JP (1986) Cercarial glycocalyx of *Schistosoma mansoni* activates human complement. Infect Immun 51/1: 181–186

Santoro F, Lachmann PJ, Capron A, Capron M (1979) Activation of complement by *Schistosoma mansoni* schistosomula: killing of parasites by the alternative pathway and requirement of IgG for classical pathway activation. J Immunol 123(4):1551–1557

Santoro F, Ouaissi MA, Pestel J, Capron A (1980) Interaction between *Schistosoma mansoni* and the complement system: binding of C1q to schistosomula. J Immunol 124(6): 2886–2891

Schulz TF, Dierich MP, Yefenof E, Klein G (1980) C3-activating proteases on human lymphoblastoid cells superinfected with Epstein-Barr virus. Cell Immunol 51: 168–172

Seidel-Dugan C, Ponce de Leon M, Friedman HM, Fries LF, Frank MM, Cohen GH, Eisenberg RJ (1988) C3b receptor activity on transfected cells expressing glycoprotein C of herpes simplex virus types 1 and 2. J Virol 62: 4027–4036

Sher A, Hieny S, Joiner K (1986) Evasion of the alternative complement pathway by metacyclic trypomastigotes of *Trypanosoma cruzi*: dependence on the developmentally regulated synthesis of surface protein and N-linked carbohydrate. J Immunol 137: 2961–2967

Sixbey JW, Davis DS, Young LS, Hutt-Fletcher L, Tedder TF, Rickinson AB (1987) Human epithelial cell expression of an Epstein-Barr virus receptor. J Gen Virol 63: 805–811

Sjöblom I, Lundström M, Sjögren-Jansson E, Glorioso JC, Jeansson S, Olofsson S (1987) Demonstration of highly carbohydrate dependent epitopes in the herpes simplex virus type 1-specified glycoprotein C. J Gen Virol 68: 545–554

Smiley ML, Friedman HM (1985) Binding of complement component C3b to glycoprotein C is modulated by sialic acid on herpes simplex virus type 1-infected cells. J Virol 55: 857–861

Smiley ML, Hoxie JA, Friedman HM (1985) Herpes simplex virus type 1 infection of endothelial, epithelial and fibroblast cells induces a receptor for C3b. J Immunol 134: 2673–2678

Spear PG (1984) Glycoproteins specified by herpes simplex virus. In: Roizman B (ed) The herpesviruses, vol 3. Plenum, New York, pp 315–356

Spear PG (1985) Antigenic structure of herpes simplex viruses. In: Van Regenmortel MHV, Neurath AR (eds) Immunochemistry of viruses; the basis for serodiagnosis and vaccines. Elsevier Science, Amsterdam, pp 425–446

Stanley HA, Mayes JT, Cooper NR, Reese RT (1984) Complement activation by the surface of *Plasmodium falciparum* infected erythrocytes. Mol Immunol 21: 145–150

Swain MA, Peet RW, Galloway DA (1985) Characterization of the gene encoding herpes simplex virus type 2 glycoprotein C and comparison with the type 1 counterpart. J Virol 53: 561–569

Tarleton RL, Kemp WM (1981) Demonstration of IgG-Fc and C3 receptors on adult *Schistosoma mansoni*. J Immunol 126(1): 379–384

Tavares CAP, Gazzinelli G, Mota-Santos TA, Dias da Silva W (1978) *Schistosoma mansoni*: complement mediated cytotoxic activity in vitro and effect of decomplementation on acquired immunity in mice. Exp Parasitol 46: 145–151

Torpier G, Capron A, Ouaissi MA (1979) Receptor for IgG(Fc) and human β2-microglobulin on *S. mansoni* schistosomula. Nature 278: 447–449

Venkatesan S, Gershowitz A, Moss B (1982) Complete nucleotide sequences of two adjacent early vaccinia virus genes located within inverted terminal repetitions. J Virol 44: 637–646

Wenske EA, Bratton MW, Courtney RJ (1982) Endo-β-*N*-acetylglucosaminidase H sensitivity of precursors to herpes simplex virus type 1 glycoproteins gB and gC. J Virol 44: 241–248

Wright SD, Weitz JI, Huang AJ, Levin SM, Silverstein SC, Loike JD (1988) Complement receptor type three (CD 11b/CD 18) of human polymorphonuclear recognizes fibrinogen. Proc Natl Acad Sci USA 85: 7734–7738

Zezulak KM, Spear PG (1984) Mapping of the structural gene for the herpes simplex virus type 2 counterpart of herpes simplex virus type 1 glycoprotein C and identification of a mutant which does not express this glycoprotein. J Virol 49: 741–747

Zweig M, Showalter SD, Bladen SV, Heilman CJ, Hampar B (1983) Herpes simplex virus type 2 glycoprotein gF and type 1 glycoprotein gC have related antigenic determinants. J Virol 47: 185–192

Structure and Function of C3a Anaphylatoxin*

T. E. HUGLI

1 Introduction 181
2 Chemical Structure of C3a 182
3 Synthetic Peptide Studies 184
4 Conformation of C3a Molecule 187
5 Fuctional Aspects of C3a 190
6 Receptors to C3a Anaphylatoxin 197
7 Assays and Clinical Studies 200
8 Concluding Remarks 203
References 203

1 Introduction

The C3a molecule is one of three activation fragments from the complement cascade, a family of factors that include C3a, C4a, and C5a. Early recognition that more than one bioactive fragment was generated during complement activation cascade must be credited to COCHRANE and MÜLLER–EBERHARD (1968) and to DIAS DA SILVA et al. (1967). These investigators were the first to use purified components of human complement to demonstrate, and then to characterize, the activation fragment from C3. In previous studies complement was activated in serum, and the bioassays that were used detected only the factor C5a (actually C5a$_{des\ Arg}$). DIAS DA SILVA et al. realized, even using purified components of the complement C1 esterase, C4, C2, and C3, that the anaphylatoxin released from C3 was not stablized unless the digest was acidified to pH 2–5, according to earlier observations of STEGEMANN et al. (1964). Later work by BOKISCH and MÜLLER-EBERHARD (1970) would explain why the acid treatment was successful. Acidification presumably prevented residual carboxypeptidase (serum carboxypeptide N) in the isolates from removing an essential C-terminal arginine and inactivating the newly formed C3a anaphylatoxin. It was this lability of C3a bioactivity in serum that had prevented its discovery prior to isolation and activation of the purified components. These investigators went on to separate an active principal from the larger protein components in the reconstituted C3 – C3

Department of Immunology, Research Institute of Scripps Clinic, La Jolla, CA 92037, USA
* This research was supported by PHS Grants AI 17354, HL 16411, and HL 25658. Publication number 5851

convertase system by gel filtration. They demonstrated that the lower molecular weight factor: (a) induced smooth-muscle contraction and was tachyphylactic to itself but not cross-tachyphylactic to the C5-derived anaphylatoxin; (b) enhanced vascular permeability when injected into skin; (c) degranulated guinea pig ileal mast cells; and (d) promoted histamine release from rat peritoneal mast cells. DIAS DA SILVA's group originally termed the active fragment F(a)C'3, i.e., activation (a) fragment from the third component of complement. COCHRANE and MÜLLER-EBERHARD identified the fragment as the C3 anaphylatoxin, assuming it to be an analog to the "classical" C5a anaphylatoxin. We will refer to the fragment here primarily as C3a because the biologic action of the factor is not anaphylactoid in nature, and anaphylatoxin is functionally an inaccurate designation. However, the recognized nomenclature from common usage still refers to fragments C3a, C4a, and C5a collectively as anaphylatoxins.

It is important to note then that the historic beginnings of C3a research derived essentially from two studies in 1967–1968 in which convincing evidence was first given that both C3 and C5 could serve as anaphylatoxinogens. Recognition that a single serum enzyme was responsible for inactivating the small fragment released from C3 led to development of successful methods for isolating C3a anaphylatoxin directly from complement – activated serum (VALLOTA and MÜLLER-EBERHARD 1973). Availability of milligram quantities of C3a as a stable, intact, and fully active molecule permitted extensive structural and functional studies to be accomplished during the following decade.

Virtually all that is known about the C3a molecule has been learned over the past 20 years. The primary structures of C3a molecules obtained from five different species have now been determined completely, including that of human C3a. A crystallographic and detailed nuclear magnetic resonance (NMR) solution analysis of human C3a exists to provide additional conformational data. Synthetic peptide analogs of C3a, now numbering in the hundreds, have been produced to mimic activity and to elucidate subtle structure-function relationships. Finally, evidence is accumulating that will help to identify and characterize C3a receptors on cells and to understand cellular activation mechanisms that are induced by this humoral mediator. It should be noted here that the less active factor C4a interacts with the same receptor population as does C3a according to its biologic attributes (GORSKI et al. 1981; HUGLI et al. 1983; CUI et al. 1983). The present chapter is designed to outline these and other aspects of C3a and to provide a status report on the nature of this bioactive fragment. C3a is released from component C3 as a potentially important physiologic product of the complement cascade. C3a is one of the few active factors in blood that can be generated in micromolar concentrations, and yet it is under such rigorous control by serum carboxypeptidase that clinical consequences of extensive C3 activation remain difficult to assign to this anaphylatoxin.

2 Chemical Structure of C3a

Detailed analyses of the chemical structures of several C3a molecules have been compiled through the combined efforts of numerous laboratories over the past 15 years. Chemical analysis of the human factor began with the isolation of milligram quantities of intact C3a from complement-activated serum (HUGLI et al. 1975b). The procedure required that an efficient inhibitor of serum carboxypeptidase N be present during activation to protect the C3a activity. The competitive inhibitor 6-aminohexanoic acid was used based on the work of VALLOTA et al. (1973). Both porcine and human C3a were purified from complement-activated serum using a multistep gel filtration and ion exchange isolation scheme. Soon thereafter the primary structure for human C3a was determined and reported (HUGLI 1975). Elucidation of the amino acid sequence of human C3a was followed by reports of porcine C3a (CORBIN and HUGLI 1976), rat C3a (JACOBS et al. 1978), and guinea pig C3a (STIMLER et al. 1989). All determined by conventional protein sequencing techniques. A cDNA-derived sequence of mouse C3a was reported by WETSEL et al. (1984). A summary of these five sequences is given in Fig. 1.

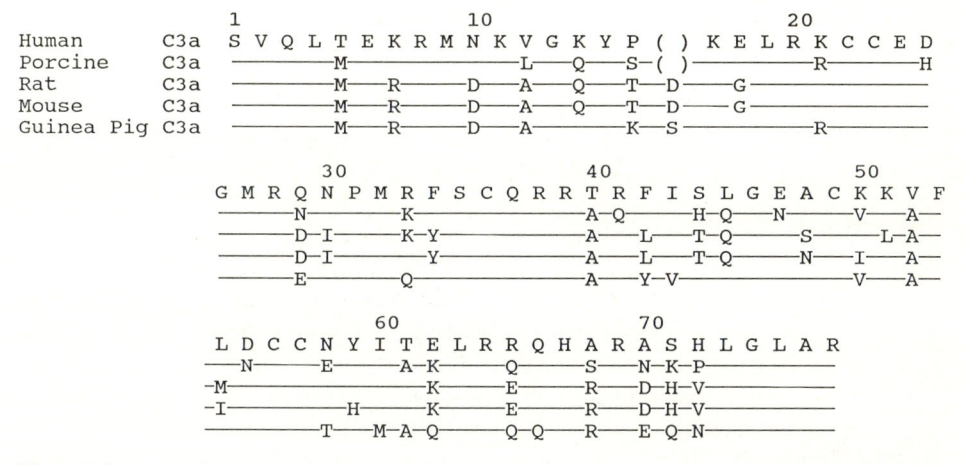

```
                    1                   10                        20
Human       C3a   S V Q L T E K R M N K V G K Y P ( ) K E L R K C C E D
Porcine     C3a   ─────────M─────────────L──Q───S─( )─────────R───────H
Rat         C3a   ─────────M──R──────D───A──Q───T──D────G─────────────
Mouse       C3a   ─────────M──R──────D───A──Q───T──D────G─────────────
Guinea Pig  C3a   ─────────M──R──────D───A──────K──S──────────R───────

                      30                    40                    50
              G M R Q N P M R F S C Q R R T R F I S L G E A C K K V F
              ────N───────────K───────────A─Q────H─Q────N─────V───A─
              ────D─I─────────K─Y─────────A────L──T─Q────S────L─A─
              ────D─I─────────Y───────────A────L──T─Q────N────I───A─
              ────E───────────Q───────────A────Y─V────────────V───A─

                          60                70
              L D C C N Y I T E L R R Q H A R A S H L G L A R
              ────N─────E────A─K───────Q──────S────N─K─P─────
              ─M──────────────K────────E──────R────D─H─V─────
              ─I──────────H───K────────E──────R────D─H─V─────
              ────────────T───M─A─Q────Q─Q────R────E─Q─N─────
```

Fig. 1. Primary amino acid sequences of five C3a molecules. The sequence of human C3a was used for reference in this comparison. *Open bracket* between positions 16 and 17 in the human and porcine C3a sequences indicates a site at which a single residue insertion occurs in the C3a molecules from rat, mouse, and guinea pig. Note that the binding site C-terminal pentapeptide sequence of LGLAR has been rigorously conserved

Although the sequence of C3a from each species is distinct, there are several common features that appear to be important for maintaining function. The C-terminal pentapeptide sequence LGLAR has been conserved in C3a from all species analyzed, and the active binding site of the molecule is known to reside at the C-terminal portion of the molecule. Six cysteinyl residue positions have been conserved, and they participate in the formation of three intrachain disulfide bonds, largely responsible for the folding pattern of this relatively compact molecule. These covalent disulfide bonds must contribute to the unusual stability of the native molecule. One residue deletion (insertion) is observed when human/porcine C3a is compared with the other C3a molecules. This difference results in human and porcine C3a containing 77 residues while rat, mouse, and guinea pig C3a are each 78 residues in length. Functional consequences of the many primary structural differences observed between C3a from various species appear to be minor since rat, porcine, and human C3a vary not more than two- or three fold in expression of activity.

In general, there are fewer substitutions occurring adjacent to the cysteinyl residues, and at either the N- or C-terminal ends, than elsewhere in the C3a molecule (see Fig. 1). The conserved patterns observed from C3a sequence comparisons reflect regions where intrachain contacts are important for polypeptide folding and indicate surfaces that are intimately involved in receptor-ligand interactions. Many of the residue replacements are conservative of side-chain size, polarity, or shape, and when considered in this manner they tend to minimize sequence differences between the linear structures of these five characterized C3a molecules. Therefore, the extent of structural variation between molecules of C3a from different species appears to be relatively small. Unlike rat C4a (Cui et al. 1988) and human C5a (Fernandez and Hugli 1978), both molecules which contain oligosaccharide units, none of the C3a molecules examined to date have contained carbohydrate.

Basic information obtained from primary structural analysis of C3a is essential for completing the conformational analysis of the molecule and yet is very limiting in providing insights of the folded form.

3 Synthetic Peptide Studies

A most interesting and unexpected advance in C3a chemistry was provided by the discovery that synthetic fragments of the intact molecule could express a biologic activity that was qualitatively identical to that of the natural factor. The first report that synthetic peptide analogs of C3a were biologically active also established that a synthetic octapeptide, based on the C-terminal sequence of human C3a (ASHLGLAR), was 1%–2% as potent as natural C3a for inducing a spasmogenic response (Hugli and Erickson 1977). Later studies examined activity as a function of peptide length (Caporale et al. 1980) and demonstrated that a minimum length for activity was four or five residues (0.005%–0.2% active), and that activity increased as the peptide length increased. At 13 residues (C3a 65–77, RQHARASHLGLAR) the peptide analog exhibits approximately 5% as much activity as does intact C3a on a molar basis (Table 1, group I). It was these data that implied that both sequence and conformation might be important parameters for activity since the folded structure of C3a was 20 times more active than the 13-residue fragment of irregular secondary structure.

A study by Unson et al. (1984) focused on the structural requirements of the model C3a pentatpeptide LGLAR. These studies involved systematic replacement of lipophilic residues at each of the four positions adjacent to the arginine to ascertain their influence on activity. In summary, it was found that arginine is an essential residue and cannot be substituted by Lys (Caporale et al. 1980) or by D-Arg or Arg-NH$_2$ (Erickson et al. 1981), indicating the importance of chirality, side-chain shape, and the negatively charged carboxylate group (Table 1, group II). Substitutions at the other four residue positions are tolerated based on proximity to the essential

Table 1. Biologically active synthetic C3a peptides

Peptide		Relative Activity (%)	
		Ileal assay[a]	Platelet assay[b]
Human C3a		100	100
Human C3a$_{des\ Arg}$		<1	
C3a peptides	74 77		
Group I	G–L–A–R	0.005	
	73		
	L–G–L–A–R	0.2	0.01
	70		
	A–S–H–L–G–L–A–R	2–3	0.4
	65		
	R–Q–H–A–R–A–S–H–L–G–L–A–R	4–8	
	57		
	C–N–Y–I–T–E–L–R–R–Q–H–A–R–A–S–H–L–G–L–A–R	50–100	
C3a peptide analogs			
Group II	A–S–H–L–G–L–A–R–G	0.03	0.03
	A–S–H–L–G–L–A	<0.01	
	Ac–A–L–G–L–A–R*	<0.0025	
	Ac–A–L–G–L–A–K	<0.0045	
	Ac–A–L–G–L–A–R–NH$_2$	<0.001	
Group III	<u>A</u>–G–L–A–R	0.04	
	L–<u>A</u>–L–A–R	0.5	
	L–<u>G</u>–A–A–R	<0.002	
	L–G–<u>L</u>–<u>G</u>–R	<0.012	
	<u>A</u>–L–G–L–<u>A</u>–R		0.06
Group IV	Ahx–A–L–G–L–A–R		0.2
	Fmoc–Ahx–A–L–G–L–A–R		13.2
	Nap–Ahx–A–L–G–L–A–R		23.0
	Y–R–R–G–R–A–A–A–L–G–L–A–R		8.3
	Fmoc–Y–R–R–G–R–A–A–A–L–G–L–A–R		56.4
	Fmoc–Ahx–Y–R–R–G–R–A–A–A–L–G–L–A–R		130.0
	Nap–Ahx–Y–R–R–G–R–A–A–A–L–G–L–A–R		603.0

[a] Activity was measured using the guinea pig ileal assay procedure.
[b] Activity was measured by the ATP-release assay using guinea pig platelets (Gerardy-Schahn et al. 1988). Group I represents peptides based on the sequence of human C3a. Group II represents a series of C3a peptides modified at the C-terminal arginyl position. Group III represents peptide analogs of human C3a 73–77 containing single site replacements. Note that only replacement of an alanine for glycine (C3a position 74) resulted in activity enhancement. Group IV represents a series of C3a analog peptides to which a lipophilic organic group has been added to the N-terminus. Underlined residues indicate replacements or additions in the natural C3a sequence. Symbols used: *Ahx*, α-aminohexyl; *Fmoc*, fluorenylmethoxycarboxyl; *Nap*, 2-nitro-4-azidophenyl; *R**, D-arginine; *R–NH$_2$*, arginine-amide; and *Ac*, acetyl.

C:\WP\C3a. Peptide

C-terminal arginine. For example, substitution of the penultimate alanine by a somewhat more bulky serine (LGLSR) led to 99% reduction in the activity. All replacements at this alanine position, including glycine, led to dramatic loss in activity. While replacement of glycine by serine (LSLAR) reduced activity of the resultant C3a pentapeptide by only 50%, and replacement by alanine actually enhanced it nearly threefold. Substitution of the middle leucine by methionine, valine, alanine, or phenylanine virtually inactivated the analog peptide. Therefore, it was concluded that each residue in the pentapeptide LGLAR was important for eliciting a biologic response (contraction of guinea pig ileal strips); however, residues nearest to the essential arginine were most important. Mediation of signal events at putative cellular C3a receptors depends on this unique sequence of LGLAR at the C-terminus of C3a for optimal activity. The activity studies with synthetic analogs can be extended to other assay systems such as ^{51}Cr release and noncytolytic serotonin release from guinea pig platelets (BITTER-SUERMANN et al. 1980), where the relative activities observed between the synthetic factors and natural C3a correspond to those obtained using the ileal assay.

Studies describing side-chain requirements for the essential binding site region in C3a led investigators to focus attention on the conformational requirements of the molecule. Using the powerful tool of synthetic peptide analogs again to probe the question of conformational involvement, a 21-residue synthetic C3a peptide was prepared and found to exhibit nearly full activity (50%–100%) on a molar basis (LU et al. 1984). This peptide C3a 57–77 represents the linear sequence of human C3a from the last cysteinyl residue in the molecule at position 57 to the C terminus (CNYITELRRQHARASHLGLAR). The model was chosen for three reasons: (a) the C-terminal fragment was sufficiently long to form stabilized regular structure such as α-helix in free solution; (b) the sequence contained a tyrosine to facilitate ^{125}I labeling for tracer purposes; and (c) the amino-terminal residue contained a free SH group for coupling various reporter compounds. The 21-residue fragment of C3a (C3a 57–77) was predicted to contain α-helical conformation when in the intact C3a molecule, according to the analysis of CHOU and FASMAN (1978). It was determined that the peptide failed to assume a helical conformation in aqueous solution, but it could be induced into a helical form in trifluoroethanol and water, according to the circular dichroism (CD) spectra. This result seemingly supported the hypothesis that fully active C3a depended on both the binding site sequence at the C-terminus and a helical conformation adjacent to the binding site residues for proper orientation of the interactive side chains. This concept appeared to be supported by the work of HOEPRICH and HUGLI (1986) that showed incorporation of helix-enhancing residues such as aminobutyric acid into C3a 57–77 maintained a high level of activity while insertion of prolyl residues to disrupt helix formation significantly reduced biologic activity of the 21-residue analog. Unfortunately the peptides designed specifically to enhance helix were not helical in aqueous solution without the addition of trifluoethanol.

In summary, the various studies with synthetic peptides have tended to support the hypothesis that C3a contains a linear C-terminal binding site activity of perhaps 5 residues that is essential for activity. The region adjacent to the pentapeptide needs either to be helical or must easily assume a helical conformation. Perhaps the synthetic peptides assume a helical form at the receptor surface, and this interaction could explain why the highly active synthetic peptide C3a 57–77 is not a helical peptide in aqueous solution. Unfortunately the contribtuion of conformation to activity of C3a remains unsolved.

A recent report by GERARDY-SCHAHN et al. (1988) describes the attachment of hydrophobic groups, such as fluorenylmethoxycarbonyl, 2-nitro-4-azidophenyl, fluoresceinyl, and rhodaminyl, to the N-terminus of hexa- and octapeptide analogs of C3a with dramatic enhancement in activity. Addition of 2-nitro-4-azidophenyl to a C3a analog (YRRGRAAALGLAR) through an aminohexyl linkage group enhanced activity by nearly 100-fold and resulted in a synthetic analog of C3a with six times the activity of the natural factor (Table, 1, group IV). This analog is not expected to assume a helical conformation, yet it has extraordinary activity. There are no binding date to accompany the functional data, therefore it remains unclear whether the mechanism of activity is the same as that for other sequence analogs of C3a. Perhaps the hydrophobic groups being added to these C3a peptides have non specific membrane interactions that increase the effective concentration of the analogs at the cell surface without a corresponding increase in binding affinities for the C3a receptor. In such a case, the apparent increase in activity would be a local concentration effect and not a true enhancement in ligand binding, as the activity data might imply. Nevertheless, these analogs provide valuable new tools for exploring the interactions between the ligand C3a and the receptor-bearing cells. It is clear that applications of synthetic peptide technology have contributed greatly to our understanding of complex structure-function relationships in the C3a molecule.

4 Conformation of C3a Molecule

Once primary structures for C3a were determined, it immediately became possible to examine secondary and tertiary folding patterns of the molecule. Secondarily, it was important to establish that conformation was a significant parameter for functional expression of C3a, and to what extent folding governed activity. Initial studies estimated the content of regular structure in the native C3a molecule using CD as a procedure to discern relative contributions of α, β, and irregular structure. The CD spectra for both human and porcine C3a gave nearly identical profiles indicating rather classical double-negative adsorption bands of elipticity at 208 and 222 nm, which signifies α-helix in model structures (HUGLI et al. 1975a). The calculations of helical content in human and porcine C3a suggested 40%–45% of the residues participate in regular helical structure while very low levels of α and β structure were indicated in C3. Considering the placement and restraints from internal disulfide bonds it was speculated that the amino- and carboxyterminal ends of the C3a molecule were the most likely regions to be involved in helix.

The CD measurements were used to correlate structural perturbations with the biologic behavior of the molecule. It was shown that exposure of C3a to extremes of pH (pH 1–9.6) had only minor effects on the CD spectra, and much like exposure to reducing agent (0.02 M mercaptoethanol) and/or guanidinium hydrochloride, proved fully reversible in terms of biologic activity. Other treatments such as heating for several hours at 100 °C or reduction and alkylation led to permanent unfolding of C3a based on the CD spectra and also inactivated the molecule to greater than 90%. These results were interpreted as evidence that a conformational component of C3a was required for full activity. Correspondingly, removal of the essential C-terminal arginine of human C3a by carboxypeptidase B treatment, to form C3a$_{des\ Arg\ 77}$, showed no difference in the CD spectra when compared with that of intact C3a. The functional role of the C-terminal arginine does not appear to be in stabilizing native structure but rather in direct interaction with the C3a receptor.

Confirmation that a high helical content exists in C3a was provided by the crystallographic analysis of human C3a by HUBER et al. (1980). This elegant work showed for the first time the architecture of a folded anaphylatoxin molecule. The presumed similar placement of disulfide linkages in C3a and C5a permitted GREER (1985) to use the backbone structure of C3a as a model for developing the folded structural portrait of C5a.

Crystals of C3a$_{des\ Arg\ 77}$ were grown in 3 M phosphate at pH 4.5 for diffraction studies (PÀQUES et al. 1980), and later intact C3a was co-crystallized with the des Arg derivative indicating that the two forms were isomorphous. It should be noted that the C3a crystals were grown at pH 4.5 because of difficulty in handling crystals grown at or adjusted to pH 7.0. This would imply that at or near physiologic pH conditions the molecular form of C3a may be significantly different from that analyzed at pH 4.5. Nevertheless, the X-ray analysis data obtained for C3a at pH 4.5 represented a reference folded structure for later comparisons. The C3a molecule crystalizes as a dimer under these conditions in a P 4, 2, 2 space group with lattice constants of a = b = 44.8 Å and C = 107.8 Å and a molecule dimension of $42 \times 22 \times 16$ Å. The structure of C3a was determined at a 3.2 Å resolution, and a molecular model was constructed from these X-ray coordinates. The crystal structure analysis established that the disulfide bond arangement was Cys-22 to Cys-49, Cys-23 to Cys-56, and Cys-36 to Cys-57, which defines the intrachain disulfide "knot". There is every reason to believe that these particular covalent side-chain linkages exist in the same arrangement in all of the anaphylatoxins (e.g., C3a, C4a, and C5a) from every species assuming they contain the six cysteines. The major α-helical segment were identified in the molecule as Try-15 to Met-27 and Gly-46 to Ser-71, representing the amino- and carboxyterminal ends of the molecule, as suggested earlier. It is interesting to note that the helical content as estimated from the crystalline structure is 56%, cosiderably higher than that obtained from CD measurements (40%–45%). This could suggest that either the molecule is somewhat more "relaxed" in solution, or that contributions of the three disulfide bonds in the CD spectra influence the measured elipticity contributed by the helical structures. In any case, the folded C3a molecule (Fig. 2), as determined in crystalline form, is portrayed by an ill-defined flexible N-terminal portion of 14 residues followed by a 13-residue α-helix (15–27), a loop and pseudoregular region folded antiparallel to the first helix ending in another loop contains 18 residues, and ending in a long 26-residue α-helix with an irregular C-terminal hexapeptide region containing the essential binding site residues LGLAR. It is this dagger-shaped molecule that represents the folded form of the natural product C3a. The crystallographic model of C3a is missing an N-terminal region that was not defined because of segmental motion: orientation of this segment of the structure and its contacts with the well-defined conformation of the main C3a 14–77 portion are apparently not extractable from the X-ray data.

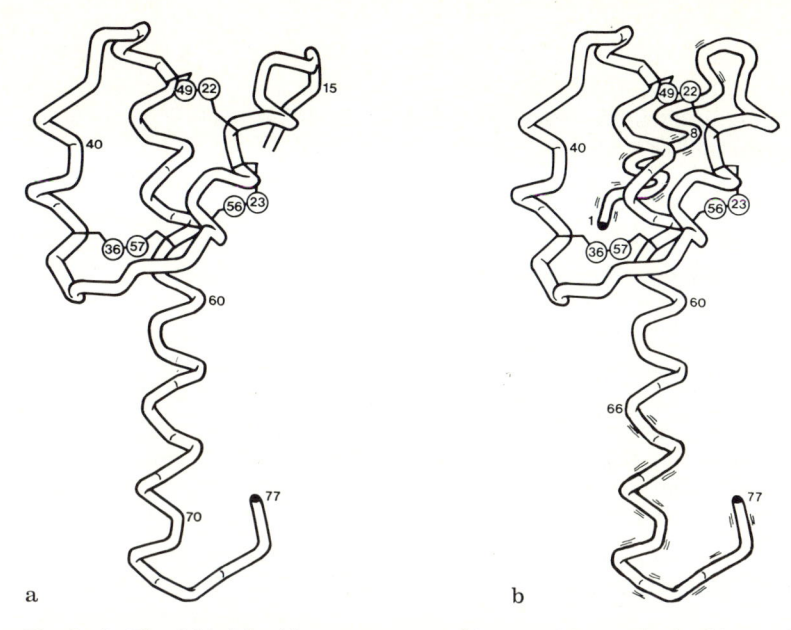

Fig. 2a, b. The folded backbone structure of human C3a. **a** The backbone three-dimensional structure of C3a on the crystalline molecule as determined by X-ray technique (HUBER et al. 1980). This analysis led to assignments of the three disulfide bonds and identified regions of α-helical structure. The N-terminus of C3a was ill defined in these crystals grown under acidic pH conditions (pH 4.5). The long helix containing residues 47–71 causes the C-terminal binding site to extend away from the globular "core" of the molecule. **b** The global fold structure of C3a deduced from the NMR contacts identified in solution. The backbone structure of C3a residues 15–77 in solution was quite similar to that in the crystalline structure except for the C terminus which displayed greater flexibility of residues 66–71 than observed in crystals. Also, the N-terminal portion of C3a assumes a helix from residues 8 to 15 with flexibility of residues 1–8. The segment defined by residues 1–15 appears to fold back onto the "core" of the molecule resulting in a compact conformation

Overall, the coordinate structure of C3a based on X-ray analysis data was a major contribution to the field. For the first time a concrete model existed that could be used to test hypotheses concerning structural requirements for activity of the anaphylatoxin.

Proton NMR studies of C3a appeared in the literature as early as 1985 (MUTO et al. 1985). These preliminary reports relied an the X-ray data of HUBER et al. (1980) to design specific experiments for testing the conformation at the C-terminal portion of C3a. The NMR data provided spectral assignments for residues Ala-48, His-67, His-72, Gly-74, Leu-75 and Ala-76 on the basis of limited carboxypeptidase digestions. These NMR data were interpreted as indicative that a regular conformation exists N-terminal to His residue 72 in C3a, consistent with the helical structure in the X-ray model, and that the residues C-terminal to residue 72 were exposed to solvent and assume a highly flexible (irregular) behavior. This data was extended by recent reports of detailed NMR analyses of human C3a (NETTESHEIM et al. 1988; CHAZIN et al. 1988). The two-dimensional NMR analysis of C3a in solution by NETTESHEIM et al. (1988) locates three definite helical regions in C3a as segments 17–28, 36–43, and

47–66. These segments identified by NMR compare favorably with those identified in crystals of C3a, except that the C-terminal helix in crystalline C3a extended five additional residues and contained residues 47–71. In addition, a well-defined helix was observed for residues 8–15 in solution, the segment that was invisible in the crystalline form of C3a.

The total involvement of residues in helical conformation for C3a in solution is estimated at nearly 60% by NMR, which is considerably higher than that estimated by CD. The work of CHAZIN et al. (1988) differs somewhat from the NETTERSHEIM studies in that the assignments of true helix were inclusive of residues 9–15 (5–8 transitional), 19–28, (40–43 transitional), 47–66 (67–70 transitional). In this case, the residues defined as definitely helical account for 37 of 77 residues, or 48% helix. It is clear that both X-ray and NMR analysis of C3a agreed on the general segments of structure that assume α-helical conformation and differ only in the assigned length of each helix. A further observation by the Chazin study (CHAZIN et al. 1988) indicates placement of the N-terminal portion of the molecule (see Fig. 2) against the main body of the molecule and between the two main helices. The N-terminal segment containing residues 1–15, including a helix from 8 to 15, remains docked to the stabilized body of C3a molecule for only part of the time and spends a significant portion of the time in dynamic motion. Futher evidence that helix I (8–15) folds back against the core of the molecule is provided by the independent observation that Tyr-59 iodinates approximately five times more efficiently than does Tyr-15 (HUGLI et al. 1975b), suggesting that the side chain of Tyr-15 is not exposed or is as readily accessible to the surface of the molecule as would be the case if this segment protruded out into solution.

In summary, the native conformation of the C3a molecule is nearly solved, and a highly developed model has evolved from the combined crystal and solution studies. We can say that the folded structure contains defined helical segments with known disulfide cross-links, and both the N- and C-terminal regions are helical, ending in each case in 6–8 residues that appear highly flexible. It is not clear yet that docking of the N-terminal segments influence activity, but this compact form does exist in solution and may assist in stabilizing the C-terminal region which appears to make the major contribution to activity.

5 Functional Aspects of C3a

Several reviews have outlined the various activities that are generally attributed to the anaphylatoxins (HUGLI and MÜLLER-EBERHARD 1978; HUGLI 1981, 1984, 1986; WEIGLE et al. 1983). However, none of these overviews have concentrated specifically on the factor C3a. The spectrum of C3a effects measured in vivo are now recognized to represent quite a diverse collection of observations. For example, purified C3a has been shown to induce neutrophilia and thrombocytopenia at levels of 30 µg/kg (HOFFMANN et al. 1988) and was shown to be lethal in guinea pigs under defined conditions at the mg/kg level (HUEY et al. 1983). When 2–3 mg C3a was injected into adult guinea pigs, death occurred within minutes, but only if a carboxypeptidase inhibitor had been previously introduced to prevent C3a inactivation from enzymatic removal of the essential C-terminal arginine. Since guinea pig platelets are activated by C3a (BECKER et al. 1978), much of the in vivo response in this animal model could be a result of species-specific mediator release from platelets. Consequently, the lethal response observed in guinea pigs may not occur in other anaimals unless they have the platelet C3a receptor. The guinea pig airway was demonstrated to be hyperresponsive to C3a in animals pretreated with either histamine of capsaicin, a substance P releasing factor. These results suggest that substantial synergy may exit in vivo between certain mediators and C3a (WATSON et al. 1988). Relatively dramatic responses can also be demonstrated in man, a visible wheal and flare occurs almost immediately when submicromolar levels of C3a are injected into human skin (LEPOW et al. 1970; WUEPPER et al. 1972). This is one of the few in vivo responses to C3a that does not require a carboxypeptidase inhibitor for expression; however, many of the potential actions of C3a are muted by rapid inactivation following generation of introduction in the circulation.

Both lethal and nonlethal effects of C3a are presumed to be induced by the recruitment of numerous cell-derived mediators such as vasoamines, prostaglandins, and leukotrienes (JOSÉ et al. 1981; WUEPPER et al. 1972; STIMLER et al. 1983). Since the guinea pig is particularly sensitive to vasoamines and has lungs that react as shock organs when challenged by histamine, the mechanism of C3a-induced death in these animals appears to be asphyxia or respiratory failure. The skin reaction in man is partially mediated by histamine but may have other mediator involvement. Other animals such as rats, which are relatively insensitive to histamine, fail to respond markedly to C3a either during exposure in vivo or to tissue isolates. This rather anecdotal observation tends to support the hypothesis that C3a molecules from various species differ only slightly in their biologic activity; however different animals or animal tissues may exhibit a wide range of responsiveness to the factor regardless of its source. This diversity in action requires one to define with great care the animal being studied and not presume that observed effects in an animal model will relate to responses in man.

One in vivo effect of C3a occurs in the microvasculature where topical (extravascular) applications result in pronounced alterations in local hemodynamics. MAHLER et al. (1975) examined C3a-induced events in rabbits using an omentum preparation and observed reduced vascular micropressure and decreased diameters of terminal or feeding arterioles lasting only seconds, with a rebound effect on arteriolar diameter indicating a dilation. The α-adrenergic blocking agent phentolamine eliminated the microvascular effect of C3a, suggesting a neurologic mechanism. The hamster cheek pouch is another microvascular preparation that responds to topical C3a application producing vasoconstriction, transient platelet aggregation and increased leakage of the postcapillary venules (BJÖRK et al. 1985). These effects were observed using either 2–10 nM C3a or 500 nM synthetic C3a octapeptide. Antihistamine (mepyramine) reduced the permeability component of the C3a effect by approximately half without affecting other responses. It was noted that human C3a, being active at 10–20 nM range, was nearly as effective as human C5a on this particular preparation. Topical application may explain the relatively high activity of C3a since it tends to cause more pronounced effects when administered extravascularly rather than by an intravascular mode, even when the carboxypeptidase inhibitor is present. These observations may be important insights into the true function of C3a, which may be to exert its primary effects outside of the vasculature.

Actions of C3a on isolated vessels from the guinea pig were examined by MARCEAU and HUGLI (1984) who found a hierachy of responses for the vessels tested. The portal vein was most responsive to C3a followed by the pulmonary artery and the thoracic aorta. Pharmacologic analysis of C3a-induced vasoconstriction effects indicated that prostaglandins and leukotrienes were the major mediators of the response. A detailed review of the many effects of anaphylatoxins of circulation was recently published (MARCEAU et al. 1987).

One of the most provocative actions recently attributed to C3a was a possible effect on brain tissue. Psychopharmacologic activity was observed in rats when C3a was injected into the perifornical hypothalamic region (SCHUPF et al. 1983). These studies conclude that when sated rats are drug-induced to eat or to drink by treatment with norepinephrine and carbamylcholine, respectively, behavior was augmented significantly by an injection of 1 µg C3a into the hypothalamic region. The factor C3a did not cause direct enhancement of the eating or drinking behavior, but it was able to enhance the effect on drug-stimulated consumption. Since this is a site-directed injection procedure, it may explain why this in vivo effect of C3a is relatively insensitive to the serum carboxypeptidase. In these preliminary studies the administration of haloperidol, a dopamine receptor antagonist, blocked C3a-induced increases in norepinephrine-stimulated food intake and phentolamine, an α-adrenengic antagonist, reversed C3a-induced carbamylcholine- stimulated drinking by the animals. These actions are consistent with mechanisms involving catecholamine receptor systems and relate to the α-adrenergic effects of C3a on microcirculation mentioned above (see MAHLER et al. 1975).

Monitoring in vivo effects of C3a, except those in skin, has been impeded by the presence of serum carboxypeptidase N that inactivates this molecule in seconds (HUGLI 1978). Therefore most of the biologic characterizations of C3a in the literature have been accomplished using isolated tissues. The three most commonly studied tissues are intestinal, cardiopulmonary, and vascular preparations (see Table 2). The original assay for C3a was developed based on the ability of this factor to induce contraction of guinea pig ileum. The C3a contractile response in the ileum is largely a histamine-driven mechanism (COCHRANE and MÜLLER-EBERHARD 1968), and isolated C3a is active in the 10^{-6}–10^{-7} M concentration range. Specificity of the response was demonstrated both by the inability of C3a$_{des\ Arg}$ to contract tissue or to block C3a-induced contractions. Further evidence for the specific nature of this effect was provided by synthetic C3a analogs (HUGLI and ERICKSON 1977) that were true agonists and proved to be tachyphylatic for the natural factor.

Contraction of parenchymal lung tissues by C3a was characterized by STIMLER et al. (1981) as a predominantly nonhistaminic response. The mechanism was investigated further (STIMLER et al. 1983) and found to depend on prostaglandin release. Specificity of the C3a response on guinea pig lung was validated using synthetic C3a oligopeptide (HUEY et al. 1984a).

Data for C3a effects on human lung preparations are scarce; however, one report does indicate an activity for mucosal glycoprotein release (MAROM et al. 1985). Human C3a and not C3a$_{des\ Arg}$ stimulated mucous glycoprotein synthesis by human lung explants in a dose- (5–20 µg/ml) and time-dependent- (0–1 h) manner. The effect

Table 2. Biologic actions of the C3a anaphylatoxin

Response	Animals	Comments
I. *In vivo*		
Lethal	Guinea pig	SCPN-INH[a] required, lungs were congestive, intraalveolar hemorrhage, edema, bronchospasms, death results within 1–4 min, high level response (2–3 mg/kg).
Skin wheal and flare	Man Rabbit Guinea pig	Activity between 10^{-6} – 10^{-7} M C3a and C3a$_{des\,Arg}$ is less active, BUT WEAKLY INACTIVE. Visible response maximal at 5–10 min and resolves in 30 min, diminished by antihistamines and local anesthetics.
Microvascular effects	Hamster Rabbit	Vasoconstriction, platelet aggregation, increased post-capillary venular permeability.
Bronchoconstriction	Guinea pig	Pretreatment of histamine or substance P markedly enhances effect. Synergy for mediator action proposed. Low level response (10–200 µg/kg).
Neutropenia, Neutrophilia, Thrombocytopenia	Guinea pig	SCPN-INH augments the response, synthetic C3a peptides also effective, may be specific, low level response (30 µg/kg).
Central nervous system activity	Rat	Potentiates drug-induced eating and drinking behavior. Effects inhibited by catecholamine receptor antagonists.
II. *In vitro*		
LUNG		
parenchymal contraction	Guinea pig	Prostaglandin and leukotriene release.
Hyperresponsiveness	Guinea pig	Synergy with histamine or substance P pre-treatment.
Mucous glycoprotein secretion contraction	Human	Non-histamine, non-leukotriene, non-macrophage-dependent mucus secretagogue pathway.
HEART		
dysfunctions	Guinea pig	Tachycardia may be histamine dependent, coronary vasoconstriction may depend on prostanoids and/or leukotrienes.
atrial contraction	Guinea pig	Non-histamine dependent, partially inhibited by prostaglandin and leukotriene antagonists.
IMMUNOREGULATION	Human Mouse Rat	Macrophage dependent, suppression of polyclonal antibody response, inhibits LIF[b] generation, inhibit NK cell cytotoxicity, induces IL-1 release, enhances prostaglandin release.
CELLS		
platelet activation	guinea pig human (?)	Serotonin release, aggregation, may be species specific, receptors demonstrated only for guinea pig platelets.
mast cell release	Rat Human	Release phenomena that may be non-specific
basophil/eosinophil release	Human	Basophilic leukkocytes release histamine, both cell types bind the ligand.

[a] SCPN-INH, serum carboxypeptidase inhibitor; [b] LIF, leukocyte inhibition factor

was eliminated by exposure of the C3a to an anti-C3a immunoadsorbant. Attempts to elucidate a mechanism of action were unsuccessful since antagonists and inhibitors of various pathways failed to identify the means by which C3a induces mucous secretion. It was speculated that C3a may act directly on the mucous glands or goblet cells; however, cell-derived mediator pathways were not eliminated as potentially responsible.

The effects of C3a anaphylatoxin on guinea pig atria were described by HUEY et al. (1984b). The spontaneous contraction was increased by more than 70% at optimal levels of C3a (10^{-7} M), and inhibitors such as cimetidine, indomethacin, and FPL 55712 were only partially effective when used separately; however, in combination their effectiveness was significantly enhanced, an effect which is also seen in other tissues. Similar studies by DEL BALZO et al. (1985) concluded that C3a stimulates cardiac tissue to exhibit tachycardia while reducing atrioventricular conduction and decreasing ventricular contractile force, coronary vasoconstriction, and histamine release as a result of leukotriene, prostaglandin, and vasoamine release. Interrelationships between these different mediator cascades is illustrated by a report that atria from antigen-sensitized guinea pigs exhibit enhanced rates of contraction, that histamine is released, and that C3a is generated when challenged by a specific antigen (DEL BALZO et al. 1988). The histamine release was enhanced when serum carboxypeptidase inhibitor was present, as was the chronotropic (rate) and inotropic (magnitude) response. Although the ability of C3a to induce greater release of mediators or to enhance responsiveness of sensitized atrial tissue was not examined, the implications are that generation of C3a may be potentiated in immediate hypersensitivity reactions, and that influences of the generated C3a on the tissue may also be enhanced.

Initial reports of a potential role for C3a in immunomodulation identified inhibition of lymphocyte blastogenesis (NEEDLEMAN et al. 1981) and suppression of antigen-specific and polyclonal antibody responses (MORGAN et al. 1982) as C3a-dependent effects. Suppression of the humoral immune response by synthetic analogs of C3a verified the specificity of this phenomenon (MORGAN et al. 1983). A macrophage-dependent inhibition of natural killer cell cytotoxicity was reported as a function mediated by C3a in vitro (CHARRIANT et al. 1982). Human mononuclear leukocytes generated reduced levels of leukocyte inhibition factor in the presence of C3a (10^{-8} M) or synthetic C3a (70–77) at 10^{-7}–10^{-8} M. Leukocyte inhibition factor inhibits neutrophil migration and is produced by mitogen stimulation of mononuclear leukocytes (PAYAN et al. 1982). The target cell for this C3a effect has not been identified, but it could easily originate at the macrophage. Other cell-derived mediators are known to be released by C3a from adherent mononuclear cells (macrophages) including interleukin 1 (IL-1; HAEFFNER-CAVAILLON et al. 1987) and thromboxane A_2 (HARTUNG et al. 1983).

In an overview considering the regulation of immune responses by complement factors, WEIGLE et al. (1983) summarized the status of C3a involvement at the time and noted that suppression of the murine immune response was T-cell dependent and results from Lyt-1^-2^+ suppressor cells. A later study gave details of that mechanism as the generation of suppressor T cells (nonspecific Lyt-2^+) at an early phase in the C3a response requiring interaction between T cells and macrophages (MORGAN et al. 1985a). Suppression of the human in vitro polyclonal antibody response occurs through the generation of nonspecific OKT8⁺ suppressor T cells (MORGAN et al. 1985b). Although IL-2 or IL-2 producing cells can override the suppression induced by C3a, the effect of C3a is not to block or reduce IL-2 synthesis. Recent studies suggest that the influence of macrophages on the C3a response may be the release of endogenous prostaglandins, such as PGE_2, which are known to inhibit polyclonal antibody secretion in vitro (MORGAN 1987). This hypothesis is supported by the effectiveness of indomethacin in reversing C3a suppression of the immune response. These data fail to show convincingly that only prostaglandins are involved in the suppressive effect. The evidence does favor some involvement of prostaglandins along with other factors released from C3-stimulated macrophages in suppressing the immune response. The role of C3a in suppression and that of C5a in immune enhancement (HUGLI and MORGAN 1984) are perhaps consistent with the relative modes for generation of these factors during complement activation. Classical pathway activation as a manifestation of autoimmunity or excessive antibody production leads mainly to C3a production without efficient C5 conversion (WAGNER and HUGLI 1984). It may be beneficial to promote down-regulation (e.g., suppression of the humoral immune response) under these conditions of activation. STRUNK and WEBSTER (1985) have shown that C3a or $C3a_{des\ Arg}$ can inhibit C3 cleavage by the classical pathway but not by the alternative pathway, suggesting a selective feedback control mechanism for C3a generation. Conversely, activation of the alternative pathway generates significant levels of C5a and occurs in response to the invasion of microbial agents whose presence dictates a need for enhaced immunoresponsiveness. Since C5a enhancement overrides the C3a suppressive effects, at least under in vitro conditions, the role of each anaphylatoxin may be important according to the paricular condition of complement recruitment.

Biologic evidence for platelet activation by C3a was reported for guinea pig platelets (BECKER et al. 1978a; MEUER et al. 1981) and for human platelets (POLLEY and NACHMAN 1983). The guinea pig platelet response has been characterized both as a release reaction (e.g., serotonin release) and as a potent stimulator of cellular aggregation (ZANKER et al. 1982). The action of C3a on guinea pig platelets is specific for intact C3a; $C3a_{des\ Arg}$ is ineffective as in other cellular systems such as the macrophage. Furthermore, the C3a response occurs independently of the C5a effect. Guinea pig C3a was used to elicit the response with guinea pig platelets, and levels above $10^{-8}\ M$ C3a were active. It is important to note that C3a from other species (e.g., human and porcine) are equally effective in activating the guinea pig platelet. The studies carried out in BITTER-SUERMANN's laboratory failed to find a corresponding activity with human platelets.

Work by POLLY and NACHMAN (1983) demonstrated both C3a-induced aggregation and serotonin release from human platelets. These studies had an unusual requirement that the responsive platelets needed to be isolated by gel filtration with minimal washing, and both C3a and $C3a_{des\ Arg}$ were effective. The possibility exists that the C3a-induced behavior of human platelets described by POLLY and NACHMAN represents a nonspecific response. Platelets express heparin on their surface, and washing could remove this polyanionic substance to which the polycationic C3a (or $C3a_{des\ Arg}$) may bind. If binding of C3a to surface heparin promotes aggregation and release phenomena, then this might explain the action of C3a ($C3a_{des\ Arg}$) on these cells. It has been difficult to demonstrate these same C3a activation events with platelets in plasma, perhaps as a result of interference with the heparin sites by serum proteins. Another parameter of these studies that seems questionable is the extremely high sensitivity of the human cells to C3a; both aggregation and release occur at concentrations 10^{-2}–10^{-3} times lower than those observed with guinea pig platelets. The key questions arising from the observed human platelet response to C3a is whether it represents a specific receptor-mediated action, and, if not, is it a valid in vivo response.

The interaction of C3a with mast cells is perhaps analogous to the human platelet situation in that the binding and activation may be a nonspecific response. Studies using isolated rat mast cells have shown histamine release of up to 30% by C3a in the 10^{-6}–10^{-5} M concentration range (JOHNSON et al. 1975). Similar concentrations of $C3a_{des\ Arg}$ were unable to induce histamine release; however, higher levels of the des Arg derivative were not examined, and uptake of C3a and $C3a_{des\ Arg}$ by the mast cell were nearly identical. Later studies by GERVASONI et al. (1986) suggested that the mechanism for C3a uptake by the rat mast cell was nonspecific. The rat mast cell binds high levels of C3a, and the bound ligand is then rapidly degraded by the enzyme chymase on the cell surface. Since chymase is conjugated to heparin proteoglycan in an insoluble complex at the cell surface, it is speculated that highly cationic C3a binds to the heparin which is proximal to the chymase and is thereby susceptible to degradation. This conversion of C3a may be an important control mechanism in preventing elevated levels of C3a from accumulating and activating the mast cell by nonspecific means.

Cross-linking experiments have failed to demonstrate specific receptors for C3a on the mast cell (FUKUOKA and HUGLI 1986). Therefore the mechanisms of activation of C3a on the rat mast cell remains a mystery, unless the effect is simply that of a polycation which resembles the action of protamines, compound 48/80, and other polyamine-like substances. The wider question is whether tissue mast cells in man, such as the pulmonary and cutaneous mast cells, differ from rat cells in respect to specific C3a receptors. It is known that C3a causes cutaneous mast cells to degranulate., and that the skin reaction in man is histamine dependent. It is not known whether mast cells are the immediate target cells for C3a in these responses. There does exist an enzyme called tryptase in human skin mast cells that is capable of both generating C3a from C3 and degrading the C3a once it is formed (SCHWARTZ et al. 1983). Tryptase is complexed to heparin proteoglycan at the surface of the human connective tissue mast cell in the same manner as chymase is conjugated on the surface of the rat mast cell and mucosal mast cells. It is interesting to note that the two cell types that appear to respond nonspecifically to C3a (i.e., human platelets and rat mast cells) each have heparin on their surfaces. Also, the skin response to intradermal injection of C3a is somewhat greater than that to $C3a_{des\ Arg}$, but the fact that $C3a_{des\ Arg}$ is active at all suggests two components to the response, one specific and one nonspecific. The specific response may or may not be mast cell targeted while the nonspecific response effect could be a mast cell event.

Although it has been speculated that C3a might interact in a selective manner with granulocytes, endothelial cells, muscle cells, and other tissue cell types, there is presently no firm evidence for direct activation of any of these cells by the C3a anaphylatoxin. Therefore the list of actual cellular targets of C3a is rather short and indicates an area in need of further investigation.

6 Receptors to C3a Anaphylatoxin

Current evidence of cell types that actually have receptors for C3a is quite limited. Functional evidence for receptor-mediated activation of macrophages (WEIGLE et al. 1983; HARTUNG et al. 1983; HAEFFNER-CAVAILLON et al. 1987) and platelets (BECKER et al. 1978b) appears to be convincing since isolated and purified cells were used in these studies. Exclusive of these two cell types, no other isolated cell type has been shown to possess a specific binding component (i.e., receptor) on its surface for the C3a molecule. In fact, the evidence for receptors on monocytes and macrophages is based largely on functional data which demonstrate specificity using either active synthetic C3a analogs of the inactive des Arg derivative of C3a, as was the case for guinea pig peritoneal macrophages (HARTUNG et al. 1983). Cultured human monocytes (adherent cells) were shown to induce both synthesis and release of IL-1 when stimulated by a mixture of C3a and $C3a_{des\ Arg}$; however, the levels of release were widely variant (HAEFFNER-CAVAILLON et al. 1987), suggesting that other factors may be involved. Endotoxin of lipopolysaccharide is a strong activator of these cells, and in combination with C3a a significant enhancement in IL-1 release occurs, perhaps as a consequence of up-regulation of receptor expression. This phenomena may be used to explain the wide variations observed as due to the uncontrolled preinduction status of the cultured cells. Unfortunately, these cells were only 85% monocytes, and therefore evidence that human monocytes bear C3a receptors remains inconclusive. Evidence for a C3a receptor on peripheral blood mononuclear cells based on chemical cross-linkage experiments, similar to those used in characterizing C5a receptors on neutrophils (HUEY et al. 1986) have failed to identify a binding component on the cell surface. Similar studies would be required to estabilsh whether lymphocytes have C3a receptors, as suggested by various studies (PAYAN et al. 1982). The indications that specific receptors for C3a exist on lymphocytes are based on biologic data that are perhaps clouded by the presence of monocytes.

Earlier data suggesting C3a activation of neutrophils appears to have been confused by either C5a contamination of the C3a or impure cellular preparations. Attempts to cross-link C3a to surface binding components on the human neutrophil have failed (FUKUOKA and HUGLI 1988) using the same techniques that were successful with the C5a ligand (HUEY and HUGLI 1987). This techique is vulnerable both to low binding affinity and to low levels of the receptor; therefore a negative result fails to prove that receptors are indeed absent. Other granulocytes such as eosinophils and basophils may possess C3a receptors since functional responses seem to indicate a specific action by this ligand (GLOVSKY et al. 1979). Recent evidence that human eosinophils have high-affinity C5a receptors makes this cell type a likely candidate for C3a receptors (GERARD et al. 1989).

One cell type from which convincing evidence for C3a receptors was obtained is the guinea pig platelet. Initial indications that C3a receptors exist on these platelets were based solely on functional results. Isolated guinea pig platelets at high purity exhibit serotonin release stimulated by C3a but not by $C3a_{des\ Arg}$ (BECKER et al. 1978a). This specificity is a major criterion for concluding that receptors are present, and that the activation process is receptor mediated. Conclusions that "high-and low-affinity" receptor types exist may be questioned in light of the potential for interactions between C3a and heparin on the platelet surface. Certainly the high-affinity interaction appears to be specific and receptor mediated. Strains of guinea pigs with platelets deficient in the C3a receptor have been described (BITTER-SUERMANN and BURGER 1986).

Direct evidence for a C3a receptor on guinea pig platelets was provided by chemical cross-linkage of the radiolabeled ligand to a binding component on the cell surface (FUKUOKA and HUGLI 1988). The estimated K_d for C3a (human) binding to the receptor was 8×10^{-10} M, with as few as 1200 sites per cell. These estimates of binding sites and affinity are suspect since the nonspecific "low-affinity" binding is predominant and obscures an accurate Scatchard-type analysis. When radiolabeled C3a was cross-linked to guinea pig platelets using a homobifunctional reagent, bands were detected in the acrylamide gels. The ligand-receptor complex appears diffuse in sodium dodecyl sulfate polycrylamide gel electrophoresis, indicating two main binding components of M_r 95000 and 105000, exclusive of the ligand (see Fig. 3). These high molecular weight platelet components are the putative C3a receptor.

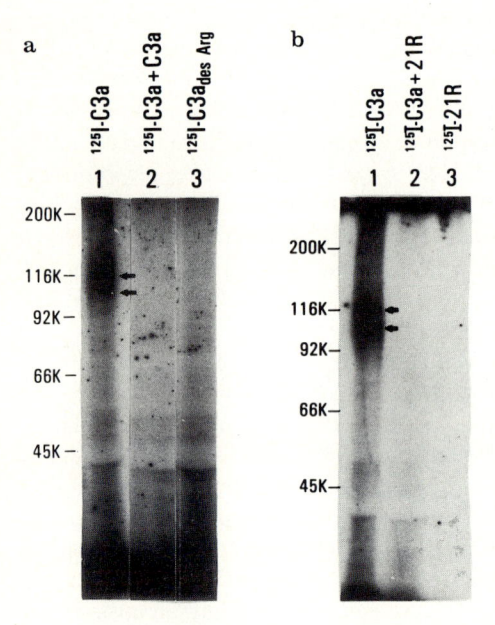

Fig. 3a, b. Cross-linking experiments with human C3a and guinea pig platelets. a Lane 1, Autoradiography of 8% SDS-PAGE gels containing SDS extracts from guinea pig platelets (2×10^8 cells) incubated with ^{125}I-labeled C3a (^{125}I-C3a) and cross-linked with bis(sulfosuccinimidyl) suberate (BS$_3$). Note the double bands at 115000 and 105000 mn (arrows). Lane 2, material from platelets exposed tom a 500-fold molar excess of C3a prior to addition of the ^{125}I-labeled C3a and BS$_3$. Lane 3, that cross-linking failed to occur when ^{125}I-labeled C3a$_{des Arg}$ was substituted for ^{125}I-labeled C3a in the cross-linking procedure. b Lane 1, a platelet extract from cells after cross-linkage of ^{125}I-labeled C3a with BS$_3$. Lane 2, an extract from cells exposed to a 1000-fold molar excess of synthetic C3a 57–77 (21R) before ^{125}I-labeled C3a and BS$_3$ were added. Lane 3, unlike ^{125}I-labeled C3a, the synthetic ^{125}I-labeled C3a 57–77 peptide did not cross-link to the C3a receptor

The cross-linking pattern was seen only under conditions of intact C3a. Neither was inactive product C3a$_{des Arg}$ cross-linked, nor could it compete with the cross-linking of C3a to the putative receptor. This is as it should be if the binding component identified on the platelet is truly the C3a receptor that relates to specific functional responses observed with these cells. Further evidence was given for C3a specificity based on the fact that the 21-residue synthetic C3a peptide (C3a 57–77) competed effectively with natural C3a preventing cross-linkage from occurring with the radiolabeled C3a ligand (Fig. 3). Both rat and pig C3a cross-link to the guinea pig platelet receptor (Y. FUKUOKA, Personal communication), indicating that C3a from various species bind interchangeably to the platelet receptor.

Similar cross–linking experiments using human platelets have failed to indicate the presence of C3a receptors (Fig. 4). Consequently, the existence of specific C3a receptors on human platelets is in doubt, and so then is the role and importance of C3a on human platelets in general. It is paradoxical that the cell type for which the best evidence of C3a receptors now exists may have little biologic significance in man.

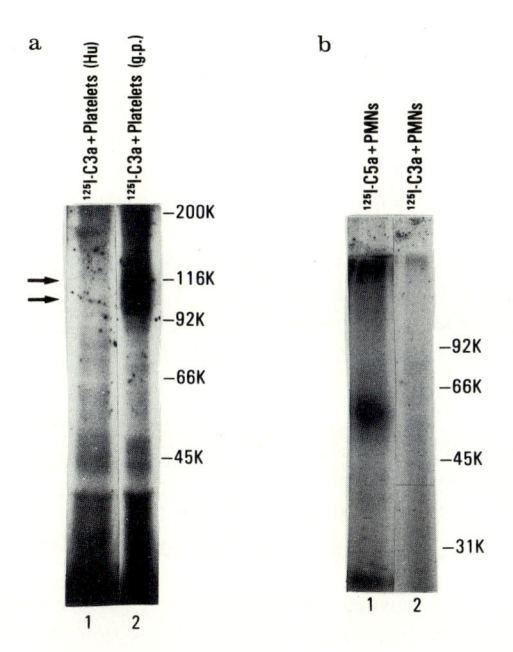

Fig. 4a, b. Attempts to cross-link human C3a with human platelets and neutrophils. a Autoradiographs of 8% SDS-PAGE gels containing SDS extracts from human platelets (2×10^8 cells; *lane 1*) and in guinea pig platelets (2×10^8 cells; *lane 2*) after incubation with ^{125}I-labeled C3a and BS$_3$. A cross-linked C3a-binding component was seen only in the guinea pig platelets. b Autoradiography of extracts from human polymorphonuclear leukocytes (*PMNs*) after incubation with ^{125}I-lebeled C5 (^{125}I-*C5a*; *lane 1*) or ^{125}I-labeled C3a (^{125}I-*C3a*; *lane 2*) and the cross-linker ethylene glycol bis(succinimidylsuccinate) indicate that leukocyte receptors only to C5a can be demonstrated

Mast cells are perhaps the most likely target for the C3a molecule in terms of the known spasmogenic and vascular permeability effects of anaphylatoxins. Several in vivo and in vitro actions appear to involve vasoamines and are assumed to require mast cell recruitment. The issue of whether mast cells are recruited by C3a in a direct or an indirect mechanism remains unsolved. An indirect mechanism could involve C3a stimulation of cells that are proximal to the mast cells in tissues, and that release factors capable of activating these mast cell. Even if the action of C3a on mast cells is direct, it need not be specific or receptor-mediated. In fact, the action of C3a or rat mast cells appears to be a direct nonspecific type of activation since C3a$_{des\ Arg}$ is nearly as effective as C3a (HERRSCHER et al. 1986) in stimulating noncytotoxic release of preformed mediators. Also, attempts to cross-link C3a to surface components on rat mast cells resulted only in the linkage of C3a to the enzyme chymase but not to candidates for receptor molecules (Y. FUKUOKA, Unpublished Observations). This preliminary result certainly does not exclude mast cells, particularly human mast cells, from the list of C3a receptor bearing cells. It does however point up a need for further studies to elucidate the exact mechanisms for C3a effects on isolated and resident tissue mast cells.

7 Assays and Clinical Studies

The first nonbiologic assay for C3a was designed as a radioimmunoassay (RIA) and was adapted for use with plasma and other biologic fluids (HUGLI and CHENOWETH 1981). Parameters were established such as antibody specificity for C3a, optimal sample handling procedures (e.g., disodium ethylenediaminetetraacetate addition and rapid removal of the cells) and acidification to remove the precursor C3, so that C3a could be measured in any body fluid. This RIA technique affords a reliabke assay for plasma C3a levels with baseline concentrations of a factor less than 200 ng/ml, or 0.3% of the total C3a in fully activated plasma. The protocol for a C3a RIA was further outlined by CHENOWETH and HUGLI (1982) and a discussion of possible clinical applications utilizing the RIA procedure was included. It was noted that a combination of C3a and C4a measurements could provide evidence of not only the extent of activation by also an indication of the pathways involved. Detection of only C3a indicates alternative pathway activation while the presence of both C3a and C4a confirms that the classical pathway was also involved. Time course studies of complement activation were monitored by the RIA procedure in serum activated by zymosan, cobra venom factor, or heat-aggregated IgG (WAGNER and HUGLI 1984). These were the first time-course studies in which generation of all the anaphylatoxins were monitored simultaneously in a single sample. Based on these assay procedures, a commercial RIA kit was developed by Upjohn Diagnostics (SATOH et al. 1983) and is currently available through Amersham Corp. (Arlington Heights, IL, USA; RPA-518).

Most of the clinical C3a data currently published have been collected using conventional RIA procedures. Several modifications have been made in the RIA procedure including the use of Norit charcoal to bind the free C3a, thus avoiding the need for either a second antibody or *Staphylococcus aureus* in the assay procedure. This method permits direct measurement of C3a levels based on dilution of the radioactivity introduced to the samples in the form of tracer C3a (LAMCHE et al. 1988).

Enzyme-linked immunosorbent assay (ELISA) systems have recently been developed for C3a utilizing monoclonal antibodies raised to denatured C3 (NILSSON et al. 1988), to intact C3 (BURGER et al. 1987), to C3a (KLOS et al. 1988), and to the synthetic C3a octapeptide (C3a 69–76; BURGER et al. 1988). The monoclonal antibody raised to denatured C3 detects C3a but not native C3, permitting direct analysis of the anaphylatoxin in biologic samples without the prior removal of precursor C3. This method is effective only if the C3 remains undenatured. The monoclonal antibodies raised to C3a (868) or to native C3 (H13) are specific for C3a and were used to develop ELISA systems, but in these case the C3 was removed by precipitation prior to measurement. Therefore, the advantages of an ELISA methodology were largely lost. A more promising system was described by BURGER et al. (1988) using synthetic C3a analogs to produce monoclonal reagent that apparently reacts only with C3a without cross-reacting with C3. If this reagent proves to be a true neoantigen, then an ELISA system based on this monoclonal antibody may well be the system of choice. Prior C3 removal from the biologic samples would not be required for analysis of the factor C3a$_{des\,Arg}$ using the neoantigen ELISA.

Numerous clinical samples have been examined based on a variety of C3a assay procedures. The earliest observations were those of CHENOWETH et al. (1981) who observed significant elevations of C3a levels ($p < 0.0001$) in the plasma of adults undergoing cardiopulmonary bypass. Elevation of C3a was dependent on the duration of bypass (i.e., surgical procedure), and the activation was due largely to surface contact with the nylon-mesh liner of the bubble oxygenation unit and to a lesser extent on the gaseous oxygenation process. It had been known that bypass procedures and renal dialysis led to neutropenia and pulmonary vascular leucostasis, effects largely attributed to the complement activation product C5a (CRADDOCK et al. 1977). What had not been accomplished was a method to accurately monitor the activation products in plasma or an explanation for the mechanism of activation. Since plasma C5a levels do not accurately reflect the activation status of complement due to a rapid uptake of C5a by neutrophils and macrophages, C3a has proven to be a much better indicator of the extent of in vitro activation. Likewise, the relatively large quantities of C3a that can be generated makes this anaphylatoxin the factor of choice for monitoring low-level activation in vitro as well as in vivo.

Complement activation in renal dialysis patients, as in those with cardiobypass, was a result of surface activation, in this case the dialyzer membrane was found responsible (CHENOWETH et al. 1983). The clinical effects of complement activation such as post-pump syndrome in cardiobypass patients refers to phenomena observed upon restoration of cardiopulmonary circulation. The term first-use syndrome characterizes a response which consists of shortness of breath, angina or chest tightness, back pain, nausea, vomiting, and hypotension often associated with new cuprophane membrane filters being used in hemodialysis. These syndromes have largely been eliminated from the clinical scene since the underlying mechanisms were elucidated. Informed changes in materials (IVANOVICH et al. 1985) and practices (SUZUKI et al. 1987) can now be made to improve these treatments, based largely on RIA data for the anaphylatoxins. It is widely recognized that blood contact with foreign surfaces can lead to pronounced complement activation (SPENCER et al. 1986). A careful analysis of the complement activation data related to extracorporeal circulation was provided by GARDINALI et al. (1986).

The predictive value of plasma C3a (and C5a) levels in monitoring various clinical conditions has received considerable attention. It was initially suggested that complement activation may signal the onset of adult respiratory distress syndrome (ARDS), an often fatal consequence of acute trauma or sepsis. This syndrome has a hypotensive component in part caused by enhanced vascular permeability. Since the anaphylatoxins exhibit known vasoactive functions, it was hypothesized that they may be directly involved in promoting the syndrome in man. Studies by MAUNDER et al. (1984) and KETAI and GRUM (1986) failed to support this hypothesis of anaphylatoxins as direct indicators of adult respiratory distress syndrome. The indicator value of C3a may better monitor development of multisystem organ failure as a consequence of severe injury or infection (HEIDEMAN and HUGLI 1984) and could prove particularly useful in evaluation of therapies for trauma patients. An analogous conclusion was reached by investigators (BELMONT et al. 1986) examining C3a levels in the plasma of patients with systemic lupus erythematosus (SLE). The C3a (and C5a) levels were significantly elevated in acutely ill SLE patients and in moderately active SLE patients, but not in patients with inactive SLE. Whether or not the anaphylatoxin levels relate to microvascular activity and participation in the disease process, it seems clear that C3a levels are useful in predicting severity of the disease. Elevated C3a has been observed in the plasma (KAPP et al. 1984) and synovial fluid (HERMANN et al. 1988) of patients with chronic polyarthritis.

Another clinical area in which the C3a anaphylatoxin may prove useful in monitoring the diasease state is dermatology. Several reports have demonstrated that anaphylatoxins are present in psoriatic scales and sera of patients with psoriasis (TAKEMATSU et al. 1986; OHKOKCHI et al. 1985) as well as in atopic dermatitis (KAPP et al. 1985).

Measurements of C3a in the cerebral spinal fluid from patients suffering from acute monophasic Guillain-Barré syndrome indicated elevated C3a levels when compared with controls. Since plasma levels were normal in these patients, it was speculated that activation of C3 occurred locally (extravascularly) and was a consequence of enhanced vascular permeability. The pathway of activation could be either direct alternative pathway involvement with myelin or classical pathway activation by immune complexes formed between myelin and autoimmune IgM. These mechanisms could perhaps be distringuished by measuring both the C3a and C4a levels, unless tissue proteases are also involved.

Various agents have been examined for their ability to generate the C3a molecule in vitro. A number of radiologic contrasts agents were examined in vitro (DAWSON et al. 1983) because it has long been known that shock reactions to intravenous administration of contrast agents correlate with complement activation (WESTABY et al. 1985). The in vitro data indicate that all contrast agents examined caused activation of C3. The effects of iodinated contrast media on RIA measurements of C3a were discussed by EATON et al. (1987). Other materials that are introduced into man as therapeutic agents such as protein A from *S. aureus*, an agent used in cancer therapy, also proved to be efficient activators of complement (LANGONE et al. 1984). Allergens show a wide diversity of complement activation in vitro and may play a separate role from that of IgE in the allergy-like symptoms induced by inhalation of various common agents (NAGATA and GLOVSKY 1987). Even cigarette smoke has been shown to enhance plasma levels of C3a in animal models (KIHIRA et al. 1987).

The fact that nearly all foreign surfaces or agents are likely to induce complement activation in vivo may limit the usefulness of anaphylatoxin measurements; however the sensitive signal that is provided by monitoring C3a release in vitro offers a distinct advantage in screening therapeutic agents for undesirable side effects. The assay also has promise for testing biocompatability of materials before they are routinely introduced to the patient.

8 Concluding Remarks

The C3a fragment released from C3 during complement activation has been characterized in detail, and three-dimensional models of its structure in crystal and in solution now exist. The unique structural motif of the folded C3a molecule appears to be shared only by the anaphylatoxins and defines a group of serum proteins that function as humoral hormones. The family of humoral factors including C3a, C4a, and C5a are ligands having specific receptors on selected cell types. Interactions between the C3a/C4a molecules and cell surface receptors stimulate the various biologic actions that are attributable to this factor. Cell types bearing receptors to C3a have been identified, although a complete list is yet to be compiled. Numerous tissue responses to C3a have been documented, and in some cases the mechanisms are understood. It seems apparent that most of the activities associated with C3a are elicited by cell-derived mediators released from the target cells. Effects of C3a in vivo are less well defined in terms of their physiologic or pathophysiologic importance. Clinical conditions that result from extensive complement activation have complex sequelae that are not easily assigned as developing as a consequence of exposure to C3 cleavage products. The possible role or roles for C3a in vivo remain an enigma. What is becoming clear is that the presence of high levels of C3a in bodily fluids signals abnormal immunologic events. The capability of monitoring C3a levels in patients with acute disease may provide a further means of assessing therapeutic efficacy and will eventually aid in defining the true biologic role of this most abundant of the anaphylatoxins.

References

Becker S, Meuer S, Hadding U, Bitter-Suermann D (1978a) Platelet activation: a new biological activity of guinea-pig C3a anaphylatoxin. Scand J Immunol 7: 173–180

Becker S, Hadding U, Schorlemmer HU, Bitter-Suermann D (1978b) Demonstration of high-affinity binding sites for C3a anaphylatoxin on guinea-pig platelets. Scand J Immunol 8; 551–555

Belmont HM, Hopkins P, Edelson HS, Kaplan HB, Ludewig R, Weissmann G, Abramson S (1966) Complement activation during systemic lupus erythematous. C3a and C5a anaphylatoins circulate during exacerbations of disease. Arthritis Rheum 29: 1085–1089

Bitter-Suermann D, Burger R (1986) Guinea pigs deficient in C2, C4, C3 or the C3a receptor. Prog. Allergy 39: 134–158

Bitter-Suermann D, Becker S, Meuer S, Schorlemmer HU, Hadding U, Andreatta R (1980) Comparative study on biological effects of the guinea pig complement-peptide C3a-related synthetic oligopeptides. Mol Immunol 17: 1257–1261

Björk J, Hugli TE, Smedegård G (1985) Microvascular effects of anaphylatoxins C3a and C5a. J Immunol 134: 1115–1119

Bokisch VA, Müller-Eberhard HJ (1970) Anaphylatoxin inactivator of human plasma: Its isolation and characterization as a carboxypeptidase. J Clin Invest 49: 2427–2436

Burger R, Bader A, Kirschfink M, Rother U, Schrod L, Wörner I, Zilow G (1987) Functional analysis and quantification of the complement C3 derived anaphylatoxin C3a with a monoclonal antibody. Clin Exp Immunol 68: 703–711

Burger R, Zilow G, Bader A, Friedlein A, Naser W (1988) The C terminus of the anaphylatoxin C3a generated upon complement activation represents a neoantigenic determinant with diagnostic potential. J Immunol 141: 553–558

Caporale LH, Tippett PS, Ericson BW, Hugli TE (1980) The active site of C3a anaphylatoxin. J Biol Chem 255: 10758–10763

Charriaut C, Senik A, Kolb JP, Barel M, Frade R (1982) Inhibition of in vitro natural killer activity by the third component of complement: role for the C3a fragment. Proc Natl Acad Sci USA 79: 6003–6007

Chazin WJ, Hugli TE, Wright PE (1988) ^{1}H NMR studies of human C3a anaphylatoxin in solution: sequential resonance assignments, secondary structure, and global fold. Biochemistry 27: 9139–9148

Chenoweth DE, Hugli TE (1982) Assays for chemotactic factors and anaphylatoxins. In: Immunologic analysis. Masson, Paris, pp 227–237 (Recent progress in diagnostic laboratory immunology, chap 22)

Chenoweth DE, Stewart RW, Cooper SW, Blackstone EH, Kirklin JW, Hugli TE (1981) Complement activation during cardiopulmonary bypass: evidence for generation of C3a and C5a anaphylatoxins. N Engl J Med 304: 497–503

Chenoweth DE, Cheung AK, Henderson LW (1983) Anaphylatoxin formation during hemodialysis: effects of different dialyzer membranes. Kidney Int 24: 764–769

Chou PY, Fasman GD (1978) Empirical predictions of protein conformation. Annu Rev Biochem 47: 251–276

Cochrane CG, Müller-Eberhard HJ (1968) The derivation of two distinct anaphylatoxin activities from the third and fifth components of human complement. J Exp Med 127: 371–386

Corbin NC, Hugli TE (1976) The primary structure of porcine C3a anaphylatoxin. J Immunol 117: 990–995

Craddock PR, Hammerschmidt D, White JG, Dalmasso AP, Jacob HS (1977) Complement (C5a)-indicated granulocyte aggregation in vitro. A possible mechanism of complement-mediated leukostasis and leukopenia. J Clin Invest 60: 260–264

Cui L, Ferreri K, Hugli TE (1988) Structural characterization of the C4a anaphylatoxin from rat. Mol Immunol 25: 663–671

Dawson P, Turner MW, Bradshaw A, Westaby S (1983) Complement activation and generation of C3a anaphylatoxin by radiological contrast agents. Br J Radiol 56: 447–448

Del Balzo UH, Levi R, Polley MJ (1985) Cardiac dysfunction caused by purified human C3a anaphylatoxin. Proc Natl Acad Sci USA 82: 886–890

Del Balzo UH, Polley MJ, Levi R (1988) Activation of the third complement component (C3) and C3a generation in cardiac anaphylaxis: histamine release and associated inotropic and chronotropic effects. J Pharmacol Exp Ther 246: 911–916

Dias da Silva W, Eisele JW, Lepow IH (1967) Complement as a mediator of inflammation. III. Purification of the activity with anaphylatoxin properties generated by interaction of the first four components of complement and its identification as a cleavage product of C'd. J Exp Med 126: 1027–1048

Eaton S, Tsay HM, Yost F, Tweedle M (1987) Effect of iodinated contrast media on a radioimmunoassay of C3a (Abstract). Clin Chem 33: 1470

Erickson BW, Fok K-F, Khan SA, Lukas TJ, Molinar RR, Munoz H, Prystowsky MB, Unson G, Volk-Weiss J, Hugli TE (1981) Synthetic studies of serum complement. In: Rich DH, Gross E (eds) Peptides: synthesis-structure-function. Pierce Chemical, Rockford, IL, pp 525–534

Fernandez HN, Hugli TE (1978) Primary structural analysis of the polypeptide portion of human C5a anaphylatoxin. I. Polypeptide sequence determination and assignment of the oligosaccharide attachment site in C5a. J Biol Chem 253: 6955–6964

Fukuoka Y, Hugli TE (1986) Degradation of C3a anaphylatoxins by rat mast cells. Fed Proc 45: 381 (Abstract)

Fukuoka Y, Hugli TE (1988) Demonstration of a specific C3a receptor on guinea pig platelets. J Immunol 140: 3496–3501

Gardinali M, Cicardi M, Agostini A, Hugli TE (1987) Complement activation in extracorporeal circulation. Physiological and pathological implications. Pathol Immunopathol Res 5: 352–370

Gerard NP, Hodges MK, Drazen JM, Weller PF, Gerard C (1989a) Characterization of a receptor for C5a anaphylatoxin on human eosinophils. J Biol Chem 264: 1760–1766

Gerard NP, Lively M, Gerard C (1989b) The structure of guinea pig C3a. Mol Immunol 27 (in press)

Gerardy-Schahn R, Ambrosius D, Casaretto M, Grötzinger J, Saunders D, Wollmer A, Brandenburg DM, Bitter-Suermann D (1988) Design and biological activity of a new generation of synthetic C3a analogues by combination of peptidic and non-peptidic elements. Biochem J 255: 209–216

Gervasoni JR Jr, Conrad DH, Hugli TE, Schwartz LB, Ruddy S (1986) Degradation of human anaphylatoxin C3a by rat peritoneal mast cells: a role for the secretory granule enzyme chymase and heparin proteoglycan. J Immunol 136: 285–292

Glovsky MM, Hugli TE, Ishizaka T, Lichtenstein LM, Erickson BW (1979) Anaphylatoxin-induced histamine release with human leucocytes. J Clin Invest 64: 804–811

Gorski JP, Hugli TE, Müller-Eberhard HJ (1979) C4a: The third anaphylatoxin of the human complement system. Proc Natl Acad Sci USA 76: 5299-5302

Greer J (1985) Model structure for the inflammatory protein C5a. Science 228: 1055–1060

Haeffner-Cavaillon N, Cavaillon JM, Laude M, Kazatchkine MD (1987) C3a (C3a des Arg) induces production and release of interleukin 1 by cultured human monocytes. J Immunol 139: 794–799

Hartung H-P, Bitter-Suermann D, Hadding U (1983) Induction of thromboxane release from macrophages by anaphylatoxic peptide C3a of complement and synthetic hexapeptide C3a 72–77. J Immunol 130: 1345–1349

Hartung HP, Schwenke C, Bitter-Suermann D, Toyka KV (1987) Guillain-Barré syndrome: activated complement components C3a and C5a in CSF. Neurology 37: 1006–1009

Heidemann M, Hugli TE (1984) Anaphylatoxin generation in multisystem organ failure. J Trauma 24: 1038–1043

Hermann E, Vogt P, Hagmann W, Dunky A, Muller W (1988) Synoviaspiegel von Interleucin-1 und C3a bei chronischer Polyarthritis, Psoriasisarthritis und aktivierter Arthrose. Z Rheumatol 47: 20–25

Herrscher R, Hugli TE, Sullivan TJ (1986) Anaphylatoxin C3a induced mediator relese from mast cells. Fed Proc 45: 243 (Abstract)

Hoeprich PD, Hugli TE (1986) Helical conformation at the carboxyl-terminal portion of human C3a is required for full activity. Biochemistry 25: 1945–1950

Hoffmann T, Böttger C, Baum HP, Messner M, Hadding U, Bitter-Suermann D (1988) In vivo effects of C3a on neutrophils and its contribution to inflammatory lung processes in a guinea-pig model. Clin Exp Immunol 71: 486–492

Huber R, Scholze H, Páques EP, Deisenhofer J (1980) Crystal structure analysis and molecular model of human C3a anaphylatoxin. Hoppe-Seylers Z Physiol Chem 361: 1389–1399

Huey R, Hugli TE (1987) Characterization of the chemotactic C5a receptor on human neutrophils. In: Disabato G (ed) Methods in enzymology, vol 150. Academic, New York, pp 615–627

Huey R, Bloor CM, Kawahara MC, Hugli TE (1983) Potentiation of the anaphylatoxins in vivo using an inhibitor of serum carboxypeptidase N (SCPN). I. Lethality and effects of pulmonary tissue. Am J Pathol 112: 48–60

Huey R, Erickson BW, Bloor CM, Hugli TE (1984a) Concentration of guinea pig lung by synthetic oligopeptide related to human C3a. Immunopharmacology 8: 37–45

Huey R, Bloor CM, Hugli TE (1984b) Effects of human anaphylatoxins on guinea pig atria. Immunopharmacology 8: 147–154

Huey R, Fukuoka Y, Hoeprich PD, Hugli TE (1986) Cellular receptors to the anaphylatoxins C3a and C5a. Biochem Soc Symp 51: 69–81

Hugli TE (1975) Human anaphylatoxin (C3a) from the third component of complement: primary structure. J Biol Chem 250: 8293–8301

Hugli TE (1978) Chemical aspects of the serum anaphylatoxins. In: Reisfeld RA, Inman FP (eds) Contemporary topics in molecular immunology, vol 7. Plenum, New York, pp 181–214

Hugli TE (1981) The structural basis for anaphylatoxin in chemotactic functions of C3a and C5a. CRC Crit Rev Immunol 1 (4): 3231–3366

Hugli TE (1984) Structure and function of the anaphylatoxins. Springer Semin Immunopathol 7: 193–219

Hugli TE (1986) Biochemistry and biology of anaphylatoxins. Complement 3: 111–127

Hugli TE, Chenoweth DE (1981) Biologically active peptides of complement: techniques and significance of C3a and C5a measurements. In: Nakamura RM, Dito WR, Tucker ES III (eds), Immunoassays: clinical laboratory techniques for the 1980s, vol 4. Liss, New York, pp 443–460

Hugli TE, Erickson BW (1977) Synthetic peptides with the biological activities and specificity of human C3a anaphylatoxin. Proc Natl Acad Sci USA 74: 1826–1830

Hugli TE, Müller-Eberhard HJ (1978) Anaphylatoxins: C3a and C5a. In: Kunkel HG, Dixon FG (eds) Advances in immunology, vol 26. Academic, New York, pp 1–53

Hugli TE, Morgan WT, Müller-Eberhard HJ (1975a) Circular dichroism of C3a anaphylatoxin: effects of pH, heat, guanidinum chloride and mercaptoethanol on conformation and function. J Biol Chem 250: 1479–1483

Hugli TE, Vallota EH, Müller-Eberhard HJ (1975b) Purification and partial characterization of human and porcine C3a anaphylatoxin. J Biol Chem 250: 1472–1478

Hugli TE, Kawahara MS, Unson CG, Molinor RL, Erickson BW (1983) The active site of human C4a anaphylatoxins. Mol Immunol 20: 637–645

Ivanovich P, Hammerschmidt DE, Chenoweth D, Klinkmann H, Vidovic D (1985) Blood-membrane interaction: C3a, an indicator of biocompatibility. Life Support Syst 3: 394–403

Jacobs JW, Rubin JS, Hugli TE, Bogardt RA, Mariz IK, Daniels JS, Daughaday WH, Bradshaw RA (1978) Purification, characterization, and amino acid sequence of rat anaphylatoxin (C3a). Biochemistry 17: 5031–5038

Johnson AR, Hugli TE, Müller-Eberhard HJ (1975) Release of histamine from rat mast cells by the complement peptides C3a and C5a. Immunology 28: 1067–1080

José PJ, Forrest MJ, William TJ (1981) Human C5a$_{des\,Arg}$ increases vascular permeability. J Immunol 127: 2376–2380

Kapp A, Meske-Brand S, Maly FE, Muller W (1984) Complement activation in patients with chronic polyarthritis measured by the level of complement fraction C3a in plasma. Z Rheumatol 43: 103–105

Kapp A, Wokalek H, Schopf E (1985) Involvement of complement in psoriasis and atopic dermatitis — measurement of C3a and C5a, C3, C4 and C1 inactivator. Arch Dermatol Res 277: 359–361

Ketai LH, Grum CM (1986) C3a and adult respiratory distress syndrome after massive transfusion. Crit Care Med 14: 1001–1003

Kihira Y, Kuratomi Y, Matsuoka R, Kitamura S (1987) Cigarette smoking and lung metabolism — effects of complement C3a and C5a in rabbits. J Thorac Dis (Japan) 25: 643–648

Klos A, Ihring V, Messer M, Grabbe J, Bitter-Suermann D (1988) Detection of native human complement compoennt C3 and C5a and their primary activation peptide C3a and C5a anaphylatoxic peptides by ELISA with monclonal antibodies. J Immunol Methods 111: 241–252

Lamche HR, Paul E, Schlag G, Redl H, Hammerschmidt DE (1988) Development of a simple radioimmunoassay for human C3a. Inflammation 12: 265–276

Langone JJ, Das C, Bennett D, Terman DS (1984) Generation of human C3a, C4a and C5a anaphylatoxins by protein A of *Staphylococcus aureus* and immobilized protein A reagents used in serotherapy of cancer. J Immunol 133: 1057–1063

Lepow IH, Kretschmer KW, Patrick RA, Rosen FA (1970) Gross and ultrastructural observations on lesions produced by intradermal injection of human C3a in man. Am J Pathol 61: 13–24

Lu ZX, Fok KF, Erickson BW, Hugli TE (1984) Conformation analysis of COOH-terminal fragments of human C3a: evidence of ordered conformation in an active monosapeptide. J Biol Chem 259: 7367–7370

Mahler F, Intaglietta M, Hugli TE, Johnson AR (1975) Influences of C3a anaphylatoxin compared to other vasoactive agents on the microcirculation of the rabbit omentum. Microvasc Res 9: 345–356

Maunder RJ, Harlan JM, Talucci RC, Stager MA, Reed RL, PEP, PE, Hudson LD (1984) Measurement of C3a and C5a in high-risk patients does not predict ARDS. Am Rev Respir Dis 129: A104

Marceau FM, Hugli TE (1984) Effect of C3a and C5a anaphylatoxins on guinea pig isolated blood vessels. J Pharmacol Exp Ther 230: 749–754

Marceau F, Lundberg C, Hugli TE (1987) Effect of the anaphylatoxins on circulation. (Short review). Immunopharmacology 14: 67–84

Marom Z, Shelhamer J, Berger M, Frank M, Kaliner M (1985) Anaphylatoxin C3a enhances mucous glycoprotein release from human airways in vitro. J Exp Med 161: 657–668

Meuer S, Ecker U, Hadding U, Bitter-Suermann D (1981) Platelet-serotonin release by C3a and C5a: two independent pathways of activation. J Immunol 126: 1506–1509

Morgan EL (1987) The role of prostaglandins in C3a-mediated suppression of human in vitro polyclonal antibody response. Clin Immunol Immunopathol 44: 1–11

Morgan EL, Weigle WO, Hugli TE (1982) Anaphylatoxin-mediated regulation of the immune response. I. C3a-mediated suppression of human and murine humoral immune responses. J Exp Med 155: 1412–1426

Morgan EL, Weigle WO, Erickson BW, Fok K-F, Hugli TE (1983) Suppression of humoral immune responses by synthetic C3a peptides. J Immunol 131: 2258–2261

Morgan EL, Thoman ML, Weigle WO, Hugli TE (1985a) Human C3a-mediated suppression of the immune response. I. Suppression of murine in vitro antibody responses occurs through the generation of nonspecific Lyt-2 suppressor T cell. J Immunol 134: 51–57

Morgan EL, Thoman ML, Hobbs MV, Weigle WO, Hugli TE (1985b) Human C3a-mediated suppression of the immune response. II. Suppression of human in vitro polyclonal antibody responses occurs through the generation of nonspecific OKT8[+] suppressor T cells. Clin Immunol Immunopathol 37: 114–123

Muto Y, Fukumoto Y, Arata Y (1985) Proton nuclear magnetic resonance study of the third component of complement: solution conformation of the carboxyl-terminal segment of C3a fragment. Biochemistry 24: 6659–6665

Nagata S, Glovsky MM (1987) Activation of human serum complement with allergens. J Allergy Clin Immunol 80: 24–32

Needleman BW, Weiler JM, Feldbush TL (1981) The third component of complement inhibits human lymphocyte blastogenesis. J Immunol 125: 1586–1589

Nehesheim DG, Edalji RP, Mollison KW, Greer J, Zuiderweg RP (1988) Secondary structure of complement component C3a anaphylatoxins in solution as determined by NMR spectroscopy: differences between crystal and solution conformations. Proc Natl Acad Sci USA 85: 5036–5040

Nilsson B, Svensson KE, Inganäs M, Nilsson UR (1988) A simplified assay for the detection of C3a in human plasma employing a monoclonal antibody raised against denatured C3. J Immunol Methods 107: 281–287

Ohkohchi K, Takematsu H, Tagami H (1985) Increased anaphylatoxins (C3a and C4a) in psoriatic sera. Br J Dermatol 113: 189–196

Páques E, Scholze H, Huber R (1980) Purification and crystallization of human anaphylatoxin C3a. Hoppe-Seylers Z Physiol Chem 361: 977–980

Payan DG, Trenthan DE, Goetzl EG (1982) Modulation of human lymphocyte function by C3a and C3a (70–77). J Exp Med 156: 756–765

Polley MJ, Nachman RL (1983) Human platelet activation by C3a and C3a des arg. J Exp Med 158: 603–615

Satoh PS, Yonker TC, Kane DP, Yeagley BW (1983) Measurement of anaphylatoxins: an index for activation of complement cascades (1983) Biotechniques 1: 90–95

Schupf N, Williams CA, Cox J, Hugli TE, (1983) Psycho-pharmacological activity of anaphylatoxin C3a in rat hypothalamus. J Neuroimmunol 5: 305–316

Schwartz LB, Kawahara MS, Hugli TE, Fearon DT, Austen KF (1983) Generation of C3a anaphylatoxin from human C3 by human mast cell trypstase. J Immunol 130: 1891–1895

Spencer PC, Schmidt B, Gurland HJ (1986) Determination of plasma C3a des Arg levels after blood contact with foreign surfaces. Artif Organs 10: 61–63

Stegemann H, Vogt W, Friedberg KD (1964) Über die Natur des Anaphylatoxins. Hoppe-Seylers Z Physiol Chem 337: 269–276

Stimler NP, Hugli TE, Bloor CM (1980) Pulmonary injury induced by C3a and C5a anaphylatoxins. Am J Pathol 100: 327–340

Stimler NP, Brockelhurst WE, Hugli TE, Bloor CM (1981) Anaphylatoxin mediated contraction of guinea pig lung strips: a non-histamine tissue response. J Immunol 126: 2258–2261

Stimler NP, Bloor CM, Hugli TE (1983) C3a-induced contraction of guinea pig lung parenchymal: role of cyclooxygenase metabolites. Immunopharmacology 5: 251–257

Strunk RC, Webster RO (1985) Inhibition of cleavage of the third component of human complement (C3) by its small cleavage fragment C3a: inhibition occurs with the classical-pathway, but not the alternative-pathway, C3 convertase. Mol Immunol 22: 37–43

Suzuki Y, Uchida J, Tsuji H, Kuzuhara K, Hara S, Nihei H, Ogura Y, Otsubo O, Mimura, N. (1987) Acute changes in C3a in an anaphylactoid reaction in hemodialysis patients. Tohoku J Exp Med 152 (1): 35–45

Takematsu H, Ohkohchi K, Tagami H (1986) Demonstration of anaphylatoxins C3a, C4a and C5a in the scales of psoriasis and inflammatory pustular dermatoses. Br J Dermatol 114: 1–6

Unson CG, Erickson BW, Hugli TE (1984) Active site of C3a anaphylatoxin: contributions of the lipophilic and orienting residues. Biochemistry 23: 585–589

Vallota EH, Müller-Eberhard HJ (1973) Isolation and characterization of a new and highly active form of C5a anaphylatoxin from epsilon-aminocapric acid-containing porcine serum. J Exp Med 137: 1109–1123

Wagner JL, Hugli TE (1984) Radioimmunoassay for anaphylatoxins in biological fluids: a sensitive method for determining complement activation. Anal Biochem 136: 75–88

Watson JW, Drazen JM, Stimler-Gerard NP (1988) Synergism between inflammatory mediators in vivo. Am Rev Respir Dis 137: 636–640

Weigle WO, Goodman MG, Morgan EL, Hugli TE (1983) Regulation of immune response by fragments of the C3 components of complemnt. Springer Semin Immunopathol 6: 173–194

Westaby S, Dawson P, Turner MW, Pridie RB (1985) Angiography and complement activation. Evidence for generation of C3a anaphylatoxins by intravascular contrast agents. Cardiovas Res 19: 85–88

Wetsel RA, Lundwall A, Davidson F, Gibson T, Tack BF, Fey GH (1984) Structure of murine complement component C3. J Biol Chem 259: 13857–13862

Wuepper RB, Bokisch VA, Müller-Eberhard HJ, Stoughton RB (1972) Cutaneous responses to human C3 anaphylatoxin. Clin Exp Immunol 11: 13–20

Zanker B, Rasokat H, Hadding U, Bitter-Suermann D (1982) C3a induced activation and stimulus specific reversible densitization of guinea pig platelets. Agents Action [Suppl] 11: 147–157

Zuiderweg ERP, Mollison KW, Henkin J, Carter GW (1988) Sequence-specific assignments in the ^{1}H NMR spectrum of the human inflammatory protein C5a. Biochemistry 27: 3568–3580

Molecular Modelling of C3 and its Ligands

R. B. Sim[1] and S. J. Perkins[2]

1 Structures of Complement Proteins: Prospects for Tertiary Structure Determination 209
2 Domain Structures Present in Proteins that Interact with C3 212
2.1 Complement Control Protein Repeat 212
2.2 Thrombospondin Repeat 214
2.3 Von Willebrand Factor Repeat 214
2.4 Low-Density Lipoprotein Receptor Domain Type A 215
2.5 Serine Protease Domain 215
2.6 Factor I/C7 Repeat 216
2.7 Metal-Binding and Other Domains 216
2.8 Summary 216
3 Ultrastructure of Proteins that Interact with C3 216
3.1 CR1, CR2, MCP, DAF, and Factor H 216
3.2 Factors B and I 217
3.3 Conglutinin 217
3.4 Properdin 218
3.5 CR3, CR4, and Other C3 Receptors 218
4 The Structures of C3, C4, and C5 218
References 220

1 Structures of Complement Proteins: Prospects for Tertiary Structure Determination

Progress in determining the primary structures of complement components has been very rapid during the 1980s, and, with a few minor gaps, the amino acid sequences of all of the recognised soluble components of the system have been determined from protein or DNA sequencing. The primary structures of many cell-surface components, including complement receptor types 1–4 (CR1, CR2, CR3, CR4), membrane cofactor protein (MCP) and decay-accelerating factor (DAF), all of which interact with fragments of C3, have also been derived from DNA sequences (for review see R. D. Campbell et al. 1988; Reid and Day 1989; Day 1989).

---

[1] MRC Immunochemistry Unit, Department of Biochemistry, Oxford University, South Parks Road, Oxford OX1 3QU, UK
[2] Department of Biochemistry and Chemistry, Royal Free Hospital School of Medicine, University of London, Rowland Hill Street, London NW3 2PF, UK

Despite efforts in many laboratories, however, progress in determining tertiary structures at high resolution (e.g. by X-ray crystallography) has been disappointing. Only the structure of the small C3a fragment has been established in this way (HUBER et al. 1980). The relatively large size of many complement components, together with moderately high glycosylation, conformational instability (for example, in C3 and C4) or susceptibility to random proteolysis, makes these proteins unattractive candidates for crystallization. Information on tertiary structures of complement proteins is essential if we are to reach an understanding of the extensive protein-protein interactions, probably involving large areas of the protein surfaces, which are a dominant feature of the system.

In the absence of crystallography data, useful information at lower resolution can be obtained from simple hydrodynamic studies, and, for at least the larger proteins, from electron microscopy. Data from these techniques can be refined and considerably extended by low-angle X-ray and neutron-scattering techniques. Scattering methods have the advantage that they are performed with soluble samples, often in conditions close to physiological, and are not influenced by artefacts created during sample preparation, a frequent problem in electron microscopy. Scattering techniques can also indicate features of molecular flexibility and, in appropriate circumstances, conformational changes in solution. Such techniques can be used to examine complex formation when two or more proteins are mixed and may provide invaluable information on the formation of complexes between, for example, C3b and factor H or factor B (PERKINS 1988).

A further opportunity to pursue the determination of tertiary structure has arisen from the striking mosaic and multi-domain characteristics of most complement proteins, features which can be deduced from the amino acid sequences (R. D. CAMPBELL et al. 1988; DAY 1989). With the unfortunate exceptions of C3, C4 and C5, most complement proteins for which amino acid sequence is available are made up of small units of structure, which often correspond to independently folding domains and are referred to as modules (PATTHY 1985), structural motifs, repeats or domains. A summary of features of structural motifs in complement proteins is shown in Table 1. Further details of those occurring in proteins which interact with C3 are presented in the next section. Many of the small structural units shown in Table 1, and particularly the EGF (epidermal growth factor), CCP (complement control protein repeat or short consensus repeat), LDLR (low density lipoprotein receptor repeat, type A) and the TP (thrombospondin repeat) structures are of suitable size for attempts at high-resolution tertiary-structure determination by two-dimensional nuclear magnetic resonance (2D-NMR) techniques (WUTHRICH 1986; CLORE and GRONENBORN 1987; I. D. CAMPBELL et al. 1989). With available information on cDNA sequences of complement proteins, coding sequences for individual domains can be tailored into expression vectors, and the domains synthesised in simple eukaryotic systems. Provided that the expressed domains undergo correct assembly of disulphide bridges, fold homogeneously, and do not require complex post-translational modifications, their structures can be determined by 2D-NMR. Alternatively, quantities of some domains, sufficient for 2D-NMR analysis, may be obtainable by proteolysis of the protein in which they are contained. The tertiary structure of epidermal growth factor has been obtained in this way (for review see I. D. CAMPBELL et al. 1989) and serves as a basis on which to model the structure of EGF domains in C1r, C1s and

Table 1. Structural units recognised at the amino acid sequence level in complement proteins

Unit	Abbreviation	Length and conserved features	Protein in which unit is present, and number of units per polypeptide
Complement control protein repeat (short consensus repeat)	CCP (SCR)	60–75 residues; four Cys, one Trp	CR1 (30), H (20), CR2 (15/16), C4bp (8 + 3)[a], DAF, MCP (4 each), B, C2 (3 each), C1r, C1s (2 each), C6, C7 (2 each)
Epidermal growth factor domain	EGF	35–40 residues; six Cys	C1r, C1s, C6, C7, C8α, C8β, C9 (1 each)
Low-density lipoprotein receptor domain, type A	LDLR	40 residues; six Cys	I (2), C6, C7, C8α, C8β, C9 (1 each)
Thrombospondin repeat	TP	60 residues; six Cys, three Trp	P (6), C6, C7, C8α, C8β (2 each), C9
Von Willebrand factor repeat	VWF	170–220 residues	CR3, CR4, C2, B (1 each)
Serine protease domain	SP	210–240 residues	D, C1r, C1s, C2, BI (1 each)
C1r/C1s-specific repeat	C1r/s	106–114 residues; two Cys	C1r, C1s (2 each)
Factor I/C7 repeat	I/C7	60–70 residues; eight Cys, one Trp	I, C6, C7 (1 each)
Metal (divalent cation) binding domains	MB	60 residues	CR3, CR4 α-chains (7 each)[b]

Data for C6 are taken from Chakravarti et al. (1989).
[a] C4bp has 8 CCPs on each of 7 identical polypeptides, plus 3 or more CCPs in an additional subunit (Reid and Day 1989);
[b] Each chain contains 7 repetitive structures, of which only 3 are likely to bind divalent cations

C6–C9. Substantial progress has been made in expressing and analysing one CCP structure from human factor H (BARON et al. 1989), and resolution of a small number of these domains will provide a basis for modelling any of the CCPs from CR1, CR2, H, C4bp, etc. Further information on the way in which domains fit together to form the whole protein can be obtained from 2D-NMR and from X ray and neutron scattering. In this way, an approach can be made to predicting, with high accuracy, the complete tertiary structure of some of the complement proteins. Similar methods are being applied successfully to proteins of the coagulation system (for review of domain structures of coagulation proteins, see PATTHY 1985; FURIE and FURIE 1988).

The proportion of the primary structure of complement proteins which is occupied by recognizable domains or repetitive structures is shown in Table 2. Those components which are made up largely of such structures are good candidates for the domain-by-domain approach to determining tertiary structure.

Table 2. Proportion of primary structure occupied by recognised domains or repetitive structures

Protein	Proportion	Domain type
C3, C4, C5	5%	C3a, C4a, C5a
CR3, CR4 α-chains	30%	VWF, MB
C9	30%	LDLR, EGF, TP
C1q	30%	Collagen
C8 α- and β-chains	45%	LDLR, EGF, TP
C6, C7	55%	CCP, TP, EGF, LDLR, I/C7
I	70%	LDLR, SP, I/C7
DAF, MCP	70%	CCP
P	80%	TP
B, C2	90%	CCP, VWF, SP
H, CR1, CR2, C4bp	>90%	CCP
C1r, C1s	>90%	CCP, EGF, SP, C1r/s
D	100%	SP

Estimates for C6–C9 do not take account of a region of homology with perforin

2 Domain Structures Present in Proteins that Interact with C3

Of the structures described in Tables 1 and 2, those occuring within proteins which interact with C3 include CCP, TP, VWF, LDLR, serine protease (SP), factor I/C7, and metal-binding (MB) domains. The characteristics of these structures have been reviewed in detail by DAY (1989).

2.1 Complement Control Protein Repeat

The commonest domain type in complement proteins is the CCP (DAY 1989; Table 1). It also occurs in many non-complement proteins, including β2 glycoprotein I, interleukin 2 receptor, thyroid peroxidase, cartilage proteoglycan core protein, coagulation factor XIII β-chain, a 35-kDa protein from vaccinia virus, factor C protease of horseshoe crab, and haptoglobin (DAY 1989). This unit is rather variable in length, of about 56–75 amino acids (average 60 or 61), with four conserved cysteines and conserved acidic, proline, tryptophan and Tyr/Phe residues. The consensus sequence is shown in Fig. 1. Disulphide bridges are formed between cysteines 1 and 3, 2 and 4 within the CCP, indicating strongly that each CCP folds independently of others (DAY 1989). With few exceptions, each CCP is encoded as a single exon at the gene level. Proteins which contain many CCPs (such as factor H, β2I and C4bp) can be shown by simple hydrodynamic methods to have very elongated structures, and this is confirmed by electron microscopy and scattering techniques (for summary see SIM et al. 1986; PERKINS et al. 1986). This has led to the general view that CCP units are assembled in proteins like small beads on a string, each unit having dimensions of about $45 \times 20 \times 20$ Å, where 45 Å is the length (PERKINS et al. 1986).

```
CCP     xCxxPPxIxNGxIxxxxxxxYxxGExVxYxCxxGYxxxxGxxxIxCxxxGxWxxxxPxCxx

TP      DxGWxxWSxWxxCSxTCGxGVxxxRxRxCN---5to10---CxGxxxExxxCxxQxxC

LDLR    xxCxxxxxxFxCxxxGxCIxxxxxCNGDxDCxDxSxDExxxxC

I/C7    xxCxxWxxxxxxxxxCVCxxxxxCxxxxxxxCxxxxxxxxxxxxxCxxxxxxCxxxxxxxxxxxxCxxxxx
```

Fig. 1. Consensus sequences for the CCP, TP, LDLR and I/C7 structures. Conserved amino acids are shown in one-letter code. *Dashed lines* with numbers in the TP sequence represent gaps of variable length; X, non-conserved positions

Prediction of secondary structure from primary structure is not expected to have an accuracy of greater than 50%–60% where single sequences are involved. However since the sequences of well over 100 CCPs are now available, secondary structure predictions of much greater statistical validity can be made. Secondary structure prediction methods, in which predictions for 101 CCP sequences were averaged, in combination with Fourier transform infrared spectroscopy (FTIR) of factor H (which contains 20 CCPs) has resulted in a satisfactory model for the secondary structure of the CCP. FTIR spectra are consistent with extensive anti-parallel β-strand secondary structure. Structure prediction indicates clearly that residues 21–51 of the 61-residue (average) CCP form four strands of β structure and four β turns (PERKINS et al. 1988). Residues 7–20 and 52–61 are also reasonably compatible with the β-sheet structure. The model accounts for the pattern of conserved alternating hydrophobic and hydrophilic residues within the CCP, the hydrophobic residues being at the interface between two β sheets. The semi-conservation of glycine residues at turn positions is also consistent with the model, as are the positions of some insertions in the sequences, at predicted external loops. Further high-resolution structural determination by 2D-NMR (BARON et al. 1989) is expected to refine this relatively simple model.

Although CCPs occur in many proteins which bind C3 or C4 fragments, it is clear that not all proteins containing CCPs bind such fragments. It is also evident that within C3 or C4 fragment-binding proteins, such as CR1, CR2 or factor H, only a few of the many CCPs present are likely to be involved in binding these ligands. In factor H, the major C3b binding site is likely to be located in one or more of CCPs 4,5 or 6 (ALSENZ et al. 1985) although a subsidiary C3b binding site may occur more towards the C-terminus. In CR1 two C3b and one C4b binding regions have been detected, involving in total about six of the CCPs in the molecule (KLICKSTEIN et al. 1988). Evidently CCPs in DAF, MCP, CR2 and C4bp are involved in binding C3 fragments or C4b, and CCPs at the amino-termini of C2 and factor B are also involved in C4b and C3b binding. It can be suggested that CCPs in C1r and C1s may interact with C4b, although there is no experimental evidence to support this hypothesis. Similary, CCPs in C6 and C7 could be involved in binding C5, a homologous of C3 and C4. As with the immunoglobulin domain, however, it is to be expected that structures mediating many different functions may be superimposed on the basic β-sheet framework structure of the CCP. Further functions for CCPs remain to be investigated. Further very similar structures which contain six rather than four cysteine residues have recently been reported in the platelet and endothelial cell proteins ELAM1 and GMP140, and in the lymph node homing receptor MEL-14 (for summary see REID and DAY 1989).

2.2 Thrombospondin Repeat

The thrombospondin repeat first observed as a repetitive feature in thrombospondin, a 420-kDa adhesion protein, is present in properdin (GOUNDIS and REID 1988) and C6–C9 (Table 1). A consensus sequence for this repeat is shown in Fig. 1. Like the CCP, the TP repeat is about 60 amino acids long but contains six highly conserved (but not invaried) Cys residues three Trp residues and conserved Pro, Ser, Gly and Arg residues. The position of disulphide bridges is unknown. A combination of averaged secondary structure prediction backed up by FTIR of properdin (which contains six TPs), has been performed (PERKINS et al. 1989a) in a similar way to that reported above for CCPs. The results indicate a structure with 19%–38% β sheet, and an unusually high percentage of β turn (57%–66%). The high amount of turn structure is consistent with Gly, Pro, Cys and Ser being the four most abundant amino acids in properdin.

Estimates from electron microscopy of properdin and consideration of the secondary structure indicate that the TP is similar in size (3.3–4.3 nm) to a CCP, and, as with the CCP, that contiguous TPs give rise to a very elongated protein structure. The TP structure is of appropriate length for tertiary structure determination by 2D-NMR and is likely to fold as an independent domain.

Again, it has been suggested that since TPs in properdin are likely to mediate binding to C3b, similar structures in two malaria parasite proteins may possess this function (GOUNDIS and REID 1988). Similarly it is possible that TPs in C6–C9 may have a role in C5 binding, but there is as yet no experimental evidence for these suggestions.

2.3 Von Willebrand Factor Repeat

The VWF domain is a region of about 220 amino acids which occurs as a triple repeat in Von Willebrand factor. Single regions homologous to this are present in each of the α-chains of CR3, CR4 (p150,95) and LAF-1 but not in other integrins described to date. A single such unit is also found in the N-terminal region of Bb and, with relatively poor sequence identity, in the corresponding region of C2. This structure contains no highly conserved cysteines, and there is insufficient information available to draw conclusions as to whether it folds homogeneously as one independent domain.

Because of this, and because of its relatively large size, it is not a good candidate for structural determination by 2D-NMR. Since this structure is present in B, CR3, CR4 and in a modified form in C2, it is possible that it mediates magnesium ion dependent binding to C3b, iC3b or C4b. Potential magnesium ion binding sites within VWF domains have been suggested (CORBI et al. 1988).

2.4 Low-Density Lipoprotein Receptor Domain Type A

Low density lipoprotein receptor contains seven contiguous repetitive units about 40 amino acids long, each with six cysteine residues (for summary see STANLEY et al. 1986). The LDLR repeat occurs twice in factor I but not in other proteins which interact with C3. It is also present in C6–C9. A consensus sequence is shown in Fig. 1. There is only fragmentary information on the disulphide-bridging pattern, intron-exon arrangement or function of this structure.

2.5 Serine Protease Domain

The SP domain is a well-characterised large structural unit which contains numerous conserved residues, including the Asp-His-Ser charge relay catalytic system, and residues determining the specificity of the serine proteases. The tertiary structures of small serine proteases such as trypsin and chymotrypsin have been determined to high resolution by X-ray crystallography (for review see YOUNG et al. 1978) and provide a basis for modelling the structures of the SP domains of all the complement proteases. Among the proteases which act on C3, the SP domain of factor I contains an Asp residue at position 501, the region associated with binding of the side chain of the substrate amino acid residue on the amino-terminal side of the bond cleaved. This is consistent with the specificity of factor I for cleaving arginyl bonds. The factor I SP domain contains, in comparison with other SPs, an additional disulphide bridge which is also present in urokinase and tissue plasminogen activator (CATTERALL et al. 1987).

The SP domain of factor B, of which a model has been constructed based on trypsin and chymotrypsin coordinates (CAPORALE et al. 1984), and that of C2 are in the C-terminal half of Bb and C2a and are unusual in that activation of these proenzymes does not involve proteolytic cleavage at the amino-terminus of the SP domain. A general mechanism ocurring after activation of serine proteases, involving conformational changes partly due to salt bridging of the newly created amino-terminal residue of the SP domain (usually Ile or Val) cannot occur in C2 or factor B, as the site of proteolytic activation of these molecules is distant in the primary sequence from the SP domain. In both factor B and C2, the SP domains contain a number of small insertions in the primary sequence not seen in other serine proteases, and one or more of these may have a role in interaction with substrate (BENTLEY and CAMPBELL 1986).

2.6 Factor I/C7 Repeat

A region in factor I (residues 29–93) originally interpreted as being an EGF domain (CATTERALL et al. 1987) is now recognised as a longer type of repetitive sequence, of which two examples are found at the C-terminus of C7 (residues 678–751 and 752–821 and in C6 (DAY 1989; R. G. DISCIPIO, personal communication). Each unit of this type contains eight conserved Cys residues and one conserved Trp residue. The disulphide-bridging pattern and the function of this structure are unknown. The structure is of suitable size for analysis by 2D-NMR if it does fold as an independent domain.

2.7 Metal-Binding and Other Domains

CR3 and CR4 contain within their α-chains seven repeating elements 47–64 residues in length. These lack conserved cysteine residues. The fifth, sixth and seventh such repeats contain within them divalent cation binding consensus sequences similar to the EF hand. The three-dimensional structure of the putative calcium ion and magnesium ion binding regions of these domains could be modelled on the known tertiary structure of other calcium and magnesium ion binding proteins such as parvalbumin and calmodulin (PYTELA 1988; CORBI et al. 1988).

Other structures of a repetitive or predictable nature within the proteins which interact with C3 are the cysteine-rich segments of the β-chain of CR3 or CR4 (LAW et al. 1987) and the collagenous region of conglutinin (STRANG et al. 1986; THIEL and REID 1989).

2.8 Summary

Of the structural motifs discussed above, the SP domains and portions of the MB regions can be modelled on existing crystallographic data. The CCP, TP, LDLR and I/C7 structures are suitable in size for attempts at high-resolution tertiary structure determination by 2D-NMR. The VWF homology is too large for this technique to be applied, and it is not yet possible to predict whether this type of sequence folds as an independent domain. As noted in Table 2, however, C3, C4 and C5 are not made up of numerous small domains, and so crystallisation of the whole molecules or of the large domains corresponding to C3c and C3d still remains a favourable approach.

3 Ultrastructure of Proteins that Interact with C3

3.1 CR1, CR2, MCP, DAF, and Factor H

Each of the proteins CR1, CR2, MCP, DAF and factor H is made up entirely or predominantly of CCP units (Tables 1, 2). Hydrodynamic measurements on factor H (SIM and DISCIPIO 1982) indicate a very elongated structure (frictional ratio 2.1), and recent electron microscopy shows a thin thread-like structure with a slight kink or bend near the middle, giving a predominant flattened V shape. The overall length is in the range of 60–80 nm (A. J. DAY, personal communication). Electron microscopy of CR1 has also been reported (BARTOW et al. 1989), again indicating a filamentous structure, 80–90 nm long and about 3 nm thick, with a small globular end. In this case a "string of beads" appearance was noted, each bead (possibly corresponding to one CCP) being about 3 nm in diameter. The overall length of CR1, which has 30 CCPs, suggests a rather smaller length for each CCP (3 nm) than the estimates of about 4.5 nm based on C4bp structure (PERKINS et al. 1986). An earlier brief report of electron microscopy of factor H suggested a length of only 28 nm, with a large globular portion (SMITH et al. 1984a). This may have arisen from a sample preparation artefact.

CR2, consisting of 15 or 16 CCPs, is likely to be similar in shape to H and CR1. MCP and DAF, with 4 CCPs each, might be expected to have an elongated N-terminal region, but the effect on the overall shape of the extensive O-glycosylation of the C-terminal halves of these molecules is difficult to predict.

3.2 Factors B and I

The two proteases which act on C3 and its fragments are of similar size, but in terms of domain structures the only common feature is the SP domain. The SP domain of B has been computer-modelled (CAPORALE et al. 1984), and the same could be done for the factor I SP domain. Electron microscopy of factor B, Bb and C3bBb complexes (SMITH et al. 1984b: VOGEL et al. 1984) indicates that factor B is globular, about 8–9 nm diameter, and appears as a three-lobed structure. Each lobe or domain is 4 nm in diameter. It is likely that one lobe corresponds to the three CCPs in the Ba region, since the Bb fragment has only two lobes, of similar size. Possibly, therefore, the three CCPs in Ba do not adopt an elongated shape, as in factor H, but rather interact non-covalently with each other to form a more compact structure. The two lobes in Bb are connected by a short linking strand, and only one lobe makes contact with C3b. These two lobes may correspond to the SP domain (C-terminal) and the VWF repeat-containing region (N-terminal)

3.3 Conglutinin

The ultrastructure of bovine conglutinin has been investigated (STRANG et al. 1986) and has been described as a tetramer of four "lollipop" structures emanating from a central hub. It is suggested that the conglutinin monomer (40-kDa) polypeptides associate in threes, and the 180-residue-long collagenous regions near the N-termini intertwine to form a collagen triple helix, while the C-terminal regions of the three polypeptides combine to form a globular head. This represents one "lollipop". Four of these structures associate to form the native molecules (480 kDa). The recently described human conglutinin is likely to have a similar shape (for discussion see THIEL and REID 1989).

3.4 Properdin

The properdin monomer consists largely of six TP units and, like the CCP-containing proteins, is very elongated. Electron microscopy (SMITH et al. 1984c) indicates that the monomer polypeptide (56 kDa) is a flexible rod about 24 nm long, with a sharp bend in the middle. These associate to form cyclic dimers, trimers, tetramers and higher oligomers. Trimers appear to be the commonest form. The head-to-tail intersubunit contacts form a small loop structure.

3.5 CR3, CR4, and Other C3 Receptors

CR3 and CR4 are both members of the integrin family, and although there are no ultrastructural data for these two receptors, the structurally related fibronectin receptor has been examined by electron microscopy. At high detergent concentration (where, presumably, oligomerisation via interaction of membrane-spanning segments was avoided), it was seen as having a U, or staple shape. The portion at the bend of the U corresponded to a thick globular head, approximately 8×12 nm, with the straight parts of the U formed from two "tails" about 2 nm thick and 20 nm long (Nermut et al. 1988). It appears that one tail is made up of β-chain sequence, the other of α-chain, and the globular portion contains segments of both chains.

The identification of a fifth type of C3 receptor (CR5) at the molecular level is still not fully established. JOHNSON et al. (1989) have hypothesised that intercellular adhesion molecule-1 (ICAM-1), a member of the immunoglobulin superfamily, may be a C3 fragment-binding protein. The hypothesis was made on the basis of two observations: (a) an antibody to ICAM-1 has been reported to inhibit C3b binding to cells; (b) it was suggested that ICAM-1 contains a CCP unit, like CR1, factor H, etc. The latter suggestion is incorrect, as there is no statistically valid homology between ICAM-1 and CCP sequences, but the studies with antibodies merit further investigation.

4 The Structures of C3, C4, and C5

C3, C4 and C5 are homologues and also interact with each other, since it is now apparent that in forming the C5 convertases activated C3 binds covalently to a surface-bound C4b or C3b molecule, and that the C4b–C3b or C3b–C3b complexes form a binding site for C5 (TAKATA et al. 1987). Electron microscopy of C3, C3 fragments and homologues (VOGEL et al. 1984, and as summarised by PERKINS and SIM 1986) have given variable results. Extensive solution scattering studies of C3 and C4 and their fragments and of C5 (PERKINS and SIM 1986: PERKINS et al. 1989b, c) show that C3, C4 and C5 can all be represented by a similar two-domain shape, consisting of a flat ellipsoid about 18 nm long, 2 nm thick and 8–10 nm wide, with a smaller flat domain of $2 \times 4 \times 9$ nm (Fig. 2). The larger domain represents the C3c or C4c region of the molecule, and the smaller is the C3d or C4d region. No movement of the domains relative to each other was detected on conversion of C3 or C4 to the form in which the thiolester is hydrolysed (C3u or C3b-like C3 and the equivalent in C4), but on proteolytic activation and removal of C3a or C4a there is a large conformational change, with the two domains moving closer together. Two interpretations of this movement, both compatible with the experimentally obtained scattering curves, are shown in Fig. 2. These models are similar in shape to some of the electron microscope images seen by VOGEL et al. (1984). The single major difference between C3 and C4 observed in these studies was that C4c, unexpectedly, exists in solution as a non-covalent dimer, while C3, C3b, C3c, C3d, C4, C4d and C5 were all monomeric.

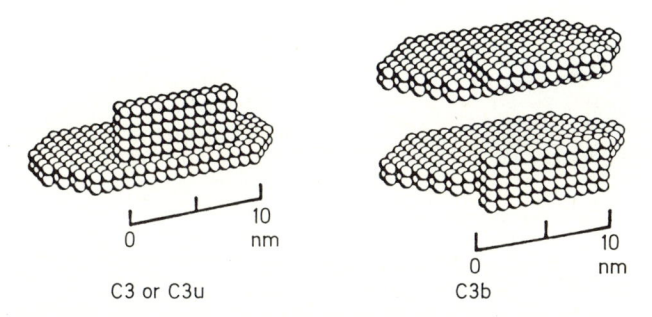

Fig. 2. Models of C3 derived from scattering studies. As noted in the text, C3 and C3 in which the thiolester has been cleaved have the same conformation at this resolution. In C3b the two domains move closer to each other; two possible conformations are shown. The equivalent forms of C4 and C5 have shapes essentially identical to those shown here

C3, C4 and C5, unlike most complement components, do not have a recognisable multi-domain structure at the amino acid sequence level. At the ultrastructural level they appear to consist of only two large domains. They are therefore, as noted above, poor candidates for determination of tertiary structure with a segment-by-segment approach. The three-dimensional structure of one small region of C3 and C4 is, however, of exceptional interest and can be modelled. This region is the thiolester site, which consists only of the highly conserved pentapeptide sequence Gly-Cys-Gly-Glu-Gln, with the thiolester formed between the SH of Cys and the side-chain carbonyl of Gln. C5 lacks the Gln residue and does not form a thiolester. The thiolester is extremely reactive once C3 or C4 has been activated by proteolysis and has a half-life of only about 0.1 ms (SIM et al. 1981). It is in contrast relatively unreactive in the native proteins. Since in the native proteins the thiolester can be attacked only by low molecular weight nucleophiles (e.g. ammonia, water), it is clear that the thiolester is shielded from the environment by the folding of the polypeptide chain. Once C3 or C4 is activated, the thiolester appears to become exposed; however, in this condition it is several orders of magnitude more reactive than a simple thiolester. Thus a mechanism of charge transfer or polarisation must exist to increase the electrophilic properties of the carbonyl group of the thiolester. This mechanism may come into existence only when C3 is activated or may pre-exist in the native molecule. One simple explanation for the activation of the carbonyl group is the formation of a hydrogen bond between the OH group of the Glu side chain and the oxygen of the thiolester carbonyl (Fig. 3; DAVIES and SIM 1981). Models of the pentapeptide structure,

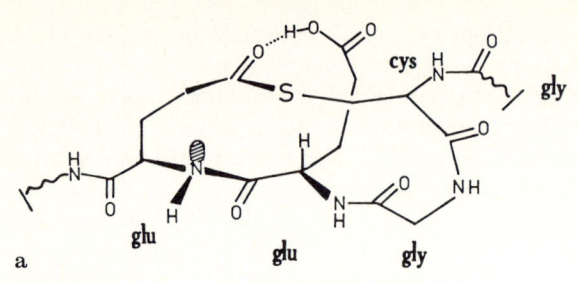

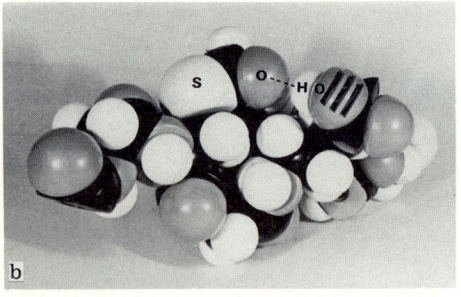

Fig. 3a, b. The conformation of the thiolester in C3. **a** Conformational drawing. *Dotted line*, hydrogen bond. **b** A space-filling model showing the position of the thiolester and the hydrogen bond. Details are given in the text

with the hydrogen bond appropriately placed (Fig. 3) have two attractive features, in addition to providing the required enhancement in reactivity. These are (a) the planar thiolester is completely shielded from solvent on one side by the pentapeptide itself, thus requiring only a single protein segment to shield the other side; (b) the normally planar peptide bond between the Glu and Gln residues is distorted, an essential feature in explaining the unusual autolytic cleavage reaction which occurs when C3 or C4 is denatured (SIM and SIM 1983). In this reaction, nucleophilic attack by the distorted peptide bond NH on the thiolester carbonyl results in peptide bond cleavage and formation of a new cyclic N-terminus. This small segment of C3 can therefore be modelled satisfactorily.

References

Alsenz J, Schulz TF, Lambris JD, Sim RB, Dierich MP (1985) Structural and functional analysis of complement factor H with the use of different enzymes and monoclonal antibodies. Biochem J 232: 841–850
Baron M, Willis AC, Bazar K, Campbell ID, Sim RB, Day AJ (1989) NMR studies of the structure of the complement control protein repeat. Complement Inflammation 6: 311 (abstr)
Bartow T, Klickstein LB, Wong W, Roux K, Fearon DT (1989) Analysis of the tertiary structure of CR1 by electron microscopy. FASEB J 3: A501 (abstr)
Bentley DR, Campbell RD (1986) C2 and factor B: structure and genetics. Biochem Soc Symp 51: 7–18
Campbell ID, Cooke RM, Baron M, Harvey TS, Tappin J (1989) The solution structures of EGF and TGF alpha. Prog Growth Factor Res 1: 13–22

Campbell RD, Law SKA, Reid KBM, Sim RB (1988) Structure regulation and organisation of the complement genes. Annu Rev Immunol 6: 161–195

Caporale LH, Woods D, Gagnon J, Christie DL, Bing DH (1984) A computer-generated model of the serine proteinase domain of human complement factor B. Immunobiology 164: 213 (abstr)

Catterall CF, Lyons A, Sim RB, Day AJ, Harris TJR (1987) Characterisation of the primary sequence of human complement control protein factor I from an analysis of cDNA clones. Biochem J 242: 849–856

Chakravarti DN, Chakravarti B, Parra CA, Müller-Eberhard HJ (1989) Structural homology of complement protein C6 with other channel-forming proteins of complement. Proc. Natl. Acad. Sci USA 86: 2799–2803

Clore GM, Gronenborn AM (1987) Determination of three-dimensional structures of proteins in solution by NMR. Protein Engineering 1: 275–288

Corbi AL, Kishimoto TK, Miller LJ, Springer TA (1988) The human leukocyte adhesion glycoprotein Mac-1 (CR3, CD11b) alpha subunit. J Biol Chem 263: 12403–12411

Davies SG, Sim RB (1981) Intramolecular general acid catalysis in the binding reaction of alpha-2-macroglobulin and complement components C3 and C4. Biosci Rep 1: 461–468

Day AJ (1989) Structure and evolution of the complement system proteins. In: Sim RB (ed) Biochemistry and molecular biology of the complement system. Kluwer, Lancaster (in press)

Furier B, Furie BC (1988) Molecular basis of blood coagulation. Cell 53: 505–518

Goundis D, Reid KBM (1988) Properdin, the terminal complement components, thrombospondin and the circumsporozoite protein of malaria parasites contain similar sequence motifs. Nature 335: 82–85

Huber R, Scholze H, Paques EP, Deisenhofer J (1980) Crystallographic structure analysis and molecular model of human C3a anaphylotoxin. Hoppe-Seyler's Z Physiol Chem 361: 1389–1399

Johnson JP, Stade BG, Holzmann B, Schwable W, Riethmuller G (1989) De novo expression of ICAM-1 in melanoma correlates with increased risk of metastasis. Proc Natl Acad Sci USA 86: 641–644

Klickstein LB, Bartow T, Miletic V, Rabson LD, Smith JA, Fearon DT (1988) Identification of distinct C3b and C4b binding sites in the human C3b/C4b receptor by deletion mutagenesis. J Exp Med 168: 1699–1717

Law SKA, Gagnon J, Hildreth JEK, Wells CE, Willis AC, Wong AJ (1987) The primary structure of the beta subunits of the cell-surface adhesion proteins LFA-1, CR3 and p150,95 and its relationship to the fibronectin receptor. EMBO J 6: 915–919

Nermut MV, Green NM, Eason P, Yamada SS, Yamada KM (1988) Electron microscopy and structural model of fibronectin receptor. EMBO J 7: 4093–4099

Patthy L (1985) Evolution of the proteases of blood coagulation and fibrinolysis by assembly from modules. Cell 41: 657–663

Perkins SJ (1988) Structural studies of proteins by high flux X-ray and neutron solution scattering. Biochem J 254: 313–327

Perkins SJ, Sim RB (1986) Molecular modelling of human complement component C3 and its fragments by solution scattering. Eur J Biochem 157: 155–168

Perkins SJ, Chung LP, Reid KBM (1986) Unusual ultrastructure of C4bp by synchroton X-ray scattering and hydrodynamic analysis. Biochem J 233: 799–807

Perkins SJ, Haris PI, Sim RB, Chapman D (1988) A study of human complement factor H by FTIR and secondary structure averaging methods. Biochemistry 27: 4004–4012

Perkins SJ, Nealis AS, Haris PI, Chapman D, Goundis D, Reid KBM (1989a) The secondary structure in properdin and related proteins; a study by FTIR. Biochemistry 28: 7176–7182

Perkins SJ, Nealis AS, Sim RB (1989b) Molecular modelling of human complement component C4 and its fragments by X-ray and neutron solution scattering. Biochemistry (in press)

Perkins SJ, Smith KF, Nealis AS, Lachmann PJ, Harrison RA, Sim RB (1989c) Domain structures of the complement components C3, C4 and C5 by X-ray and neutron scattering. Complement Inflammation 6: 384 (abstr)

Pytela R (1988) Amino acid sequence of the murine Mac-1 alpha chain reveals homology with the integrin family and an additional domain related to von Willebrand factor. EMBO J 7: 1371–1378

Reid KBM, Day AJ (1989) Structure-function relationships of the complement components. Immunol Today 10: 177–180

Sim RB, DiScipio RG (1982) Purification and structural studies on the complement system control protein factor H. Biochem J 205: 285–293

Sim RB, Sim E (1983) Autolytic fragmentation of complement components C3 and C4 and its relationship to covalent binding activity. Ann N Y Acad Sci 421: 259–276

Sim RB, Twose TM, Paterson DS, Sim E (1981) The covalent binding reaction of complement component C3. Biochem J 193: 115–127

Sim RB, Malhotra V, Ripoche J, Day AJ, Micklem KJ, Sim E (1986) Complement receptors and related complement control proteins. Biochem Soc Symp 51: 83–96

Smith CA, Pangburn MK, Vogel C-W, Müller-Eberhard HJ (1984a) Structural investigations of properdin and factor H of human complement. Immunobiology 164: 298 (abstr)

Smith CA, Vogel C-W, Müller-Eberhard HJ (1984b) An electron microscopy study of the C3 convertases of human complement. J Exp Med 159: 324–329

Smith CA, Pangburn MK, Vogel C-W, Müller-Eberhard HJ (1984c) Molecular architecture of human properdin. J Biol Chem 259: 4582–4588

Stanley KK, Page M, Campbell AK, Luzio JP (1986) A mechanism for the insertion of C9 into target membranes. Mol Immunol 23: 451–458

Strang CJ, Slayter HS, Lachmann PJ, Davis AE (1986) Ultrastructure and composition of bovine conglutinin. Biochem J 234: 381–389

Takata Y, Kinoshita T, Kozono H, Takeda T, Tanaka E, Hong K, Inoue K (1987) Covalent association of C3b with C4b within C5 convertase of the classical pathway. J Exp Med 165: 1494–1507

Thiel S, Reid KBM (1989) Structures and functions associated with the group of mammalian lectins containing collagen-like sequences. FEBS Lett (in press)

Vogel C-W, Smith CA, Müller-Eberhard HJ (1984) Cobra venom factor: structural homology with the third component of complement. J Immunol

Wuthrich K (1986) NMR of proteins and nucleic acids. Wiley, New York

Young CL, Barker WC, Tomaselli CM, Dayhoff MO (1978) The serine proteases. In: Dayhoff MO (ed) Atlas of protein sequence and structure. NBRF, Washington, pp 73–94

C3 Deficiencies

D. Bitter-Suermann[1] and R. Burger[2]

1 Inherited Deficiencies of Human C3   224

2 C3 Deficiency in Guinea Pigs   226
2.1 Molecular and Genetic Basis of the C3 Deficiency   226
2.2 Biological Consequences of the C3 Deficiency   227
2.3 Historical C3-Deficient Guinea Pigs   228

3 C3 Deficiency in Dogs   229
3.1 Inheritance of C3 Deficiency   229
3.2 Antigenic and Functional Levels of C3   229
3.3 Disease Association in C3-Deficient Dogs   230
3.4 Defective Antibody Production in C3-Deficient Dogs   230

4 Hereditary C3 Hypocomplementemia in Rabbits   230

5 Acquired C3 Deficiencies   231

References   232

The third component of complement is central to both pathways of the complement cascade and in mediating or initiating the bulk of physiological and often pathological effects of this system; it is also dominant due to its abundant plasma concentration and the number of different epitopes and binding sites for a large number of cooperating and regulatory, soluble and membrane-bound proteins. Consequently, deficiencies in C3 would be expected to exhibit a life-threatening clinical picture. In the early years of discovering complement deficiency states it was thought that C3 deficiency in man does not exist because of the presumed inherent lethality. When the first human C3 deficiences were described, followed by reports of genetic defects of C3 in guinea pigs and dogs, a more realistic and differentiated picture as to the biological role of C3 in vivo emerged. This view was supported by in vivo data from transient C3 deficiencies secondary to massive activation of C3. Our knowledge about the central function of this component has been broadened through experimental in vivo models of C3 depletion and disease-associated C3 consumption or hypercatabolism, due to disturbances in the regulatory network which physiologically guarantees a delicate balance between activation and inactivation of C3.

---

[1] Institute of Medical Microbiology, Medizinische Hochschule Hannover, Konstanty-Gutschow-Str. 8, 3000 Hannover 61, FRG
[2] Department of Immunology, Robert Koch-Institute, Bundesgesundheitsamt, Nordufer 20, 1000 Berlin (West) 65

In this chapter we discuss primary human C3 deficiencies and genetic C3 deficiencies in animals and give an overview of secondary or acquired deficiencies in man and of experimental depletion of C3 in animals.

1 Inherited Deficiencies of Human C3

The first report on the probable existence of a genetically controlled human C3 deficiency was from ALPER et al. (1969) who presented a family with seven members exhibiting a heterozygous state, i.e., only half of normal C3 concentration in plasma due to one normal and one silent allele inherited in an autosomal codominant fashion. All affected members of this family were quite healthy. Three years later the same group of authors described the first case of total C3 deficiency (ALPER et al. 1972). This patient, a 15-year-old white South African girl was homozygous for a blank or null C3 allele and showed less than 0.1% plasma C3 in protein and hemolytic terms. The impressive family tree of this patient over three generations undoubtedly revealed the autosomal codominant inheritance of a blank C3 gene (C3 Q0) allelic to the structural genes $C3^F$ and $C3^S$. The patient's history showed numerous episodes of recurrent, life-threatening pyogenic infections, such as pneumonia, meningitis, otitis, osteomyelitis, peritonitis, and bacteremia caused by encapsulated bacteria, mostly pneumococci, *Haemophilus influenzae*, and meningococci.

A number of later C3-deficient patients showed impaired chemotactic responses and a sluggish neutrophilic response to infectious agents. Interestingly, some of the C3-deficient patients, in contrast to the original description, had about 1% C3-like hemolytic activity without antigenic C3 (DAVIS et al. 1977; OSOFSKY et al. 1977; KITAMURA et al. 1984b). In addition to the susceptibility to bacterial infections, which is especially prominent early in life (median age of the first episode is far younger than in other complement deficiencies), immune complex disease, vasculitis, systemic lupus erythematosus, and membranoproliferative glomerulonephritis are diagnosed with a significant frequency.

In 1984 ROSS and DENSEN summarized all genetic complement deficiencies reported so far in the literature; those of C3 were 14 in number. They noted a remarkable disequilibrium in the male/female ratio (1:6) in these patients. Since that time two more cases have been published (STRATE et al. 1987; BORZY et al. 1988), with the same overall clinical picture. None of the affected individuals was free of infectious or immune complex mediated sequelae from the C3 deficiency. This is in sharp contrast to most other complement deficiencies, in which a substantial number of persons do not suffer from their complement abnormality. In two C3-deficient patients the ability of their monocytes to synthesize C3 was studied, and it was shown that they produced C3 at a rate 20%–30% of that in normal controls (EINSTEIN et al. 1977).

What is the reason for this clinical syndrome involving such different entities as pyogenic infections and immune complex diseases? Obviously C3 cannot be bypassed. Therefore absence of C3 profoundly influences classical pathway functions and totally abrogates the alternative pathway of complement activation. No other complement can substitute the central humoral preimmune defense mediated by the C3-dependent alternative pathway amplification loop at the bacterial surface, which leads to opsonization, chemotaxis, and subsequent triggering of the lytic membrane attack mechanisms of the terminal complement sequence. This explains the severe pyogenic infections early in life in children at a time when protective antibodies and anamnestic responses are missing. The infectious agent cannot be eliminated until antibodies a few days after onset of infection are produced. It is at this time that the IgM-mediated activation of the classical pathway should operate. However, the lack of C3 also blocks the efficiency of this pathway. IgM is dependent on C3b-mediated opsonization of bacteria for further processing by phagocytic cells. Therefore, the IgM-mediated activation of C1, C4, and C2 stops just before the essential C3 step, and C4b can substitute for C3b only to a limited extent (because of a weak affinity for C3 receptors). Consequently the elimination of invasive encapsulated bacteria such as pneumococci and meningococci is further postponed to the IgG phase of the specific immune response. Here, IgG-Fc-dependent opsonization takes place although IgG-dependent classical pathway activation of C3 would increase phagocytic efficiency 10- to 100-fold.

Taken together, the blockade of C3b- and C3bi-mediated opsonization, the inability to initiate the generation of the lytic membrane attack complex, the lack of generation of the inflammatory and chemotactic anaphylatoxic peptides C3a (because of missing precursor molecule C3) and C5a (because of missing activation), and the absence of C3e and probably other C3 fragments to induce leukocytosis and to recruit polymorphonuclear neutrophilic leukocytes from the storage pool offer a plausible explanation for the infectious problems and history in these C3-deficient patients.

What effect does the lack of C3 have in predisposing to immune complex diseases? The crucial position of C3 in both pathways of complement activation, and thus its role in antibacterial defense, is further substantiated and generalized by its role in immune complex handling. In addition to opsonic requirements for immune complex processing, C3b is essential in preventing the generation and deposition of large immune complexes. SCHIFFERLI and YIU (1988) have extensively studied and described the mechanism by which complement, especially C3b, prevents local accumulation of immune complexes in tissues, enhances their clearance, inhibits immune precipitation, and thereby generates removal of soluble immune complexes by diffusion and by interaction (immune adherence) with CR1 on cells the erythrocytes having a transport function for immune complexes.

Finally, C3 plays a role in regulation of the humoral immune response, and the lack of C3 may thus contribute to the pathogenesis of autoimmune diseases. This issue is discussed below together with C3 deficiencies in guinea pigs and dogs, the only instance in which the influence of C3 deficiency on the immune homeostasis can be tested experimentally. Whereas C3-deficient patients were previously reported to show a normal humoral immune response (ALPER et al. 1976), C3-deficient patients were later shown to exhibit some abnormalities in qualitative (no isotype switch from IgM to IgG) rather than quantitative terms (OCHS et al. 1986). More detailed analysis has recently revealed that all deficiencies which prevent activation of C3 like C2 and C4 deficiencies (via the classical pathway) or C3 deficiency itself exhibit an imbalance in the production of IgG isotypes, especially a severe depression of IgG4, a subclass associated with secondary responses to T-dependent antigens (BIRD and LACHMANN 1988).

In addition to C3 deficiency itself, genetic deficiencies of two regulatory proteins involved in the control of C3b mimic most of the clinical picture characteristic for the absence of C3. The few patients with genetic deficiencies in factor H and factor I of the complement system lack C3 in the circulation (for review see ROSS and DENSEN 1984; DAY 1986; LACHMANN 1988). This is because of uncontrolled and permanent C3b-mediated (C3bBb) hyperactivation of C3 with subsequent consumption. Synthesis of C3 is normal, but in essence C3 is not available when needed.

2 C3 Deficiency in Guinea Pigs

Identification of a C3 deficiency in a guinea pig strain was the first description of a C3 deficiency in a small laboratory animal (BURGER et al. 1986; BITTER-SUERMANN and BURGER 1986). The C3-deficient guinea pigs were identified among a colony of inbred strain 2 guinea pigs. Strain 2 animals represent one of the few common inbred guinea pig strains available. The C3 deficiency was initially recognized by the impaired capacity of serum from these animals to serve as a source of complement in the standard assay for complement-mediated lysis. When leukemia cells were treated with appropriate antibody, followed by incubation with fresh serum from these animals, the cell lysis was considerably reduced compared to those receiving serum obtained from normal guinea pigs. The C3-deficient animals might originate from a single pair of guinea pigs. The animals were propagated by brother-sister mating. Functional titration revealed a reduced total hemolytic activity corresponding to about 34% of activity of components of the classical and alternative pathways and of the regulatory protein H, with the exception of C3. The hemolytic titer of C3 was reduced on average to about 5.67% of normal. Measuring C3 protein via its antigenic activity in an enzyme-linked immunosorbent assay (ELISA) gave a similar result (5%–10% of normal). The reduction in both antigenic and functional activity of C3 in the serum of these animals indicates that C3 is absent and argues against the presence of a major proportion of a functionally inactive C3 protein. There was no indication of C3 cleavage or any other abnormality of the C3 present in the C3-deficient serum on Western blot analysis or in an indirect sandwich ELISA using combinations of monoclonal antibodies to C3a and C3b determinants.

2.1 Molecular and Genetic Basis of the C3 Deficiency

At the molecular level the nature of the C3 deficiency is still unclear, but several possible explanations have been excluded (BURGER et al. 1986). No obvious defect in synthesis or secretion of C3 was observed in these animals. In cultured supernatants of peritioneal macrophages and of hepatocytes from C3-deficient animals comparable levels of C3 protein were found as in control cultures from normal animals. C3 was measured in these experiments by ELISA. Therefore, no conclusion regarding the functional activity of secreted C3 can be made because the monoclonal antibodies used in this ELISA system might also detect functionally inactive protein. An obvious alternative explanation for the reduced level of C3 is a genetic deficiency in one of the regulatory proteins controlling C3 turnover in vivo, as previously described in the human system. This mechanism was similarly excluded by measuring the catabolism of radiolabeled C3 in C3-deficient guinea pigs in comparison to normal guinea pigs. No difference in the plasma clearance of the radiolabeled C3 was found, arguing strongly against a defect in regulatory proteins. It should be emphasized, however, that in these experiments purified normal C3, i.e., that of serum from normal animals, was used. Therefore, no prediction about the turnover of the C3 from the C3-deficient animals can be made. Immunization of the C3-deficient guinea pigs with normal serum failed to induce any antibody reactive with components of normal serum. Thus, there seems to be no complete absence of a serum protein.

The C3-deficient animals possessed the expected pattern of genetically controlled cell surface markers present in inbred strain 2 animals, including proteins controlled by the major histocompatibility complex (MHC) as well as non-MHC coded proteins. The inheritance of the C3 deficiency and its genetic linkage to other immunologically relevant markers was analyzed by breeding studies (BURGER et al. 1986). The C3 deficiency is inherited in a codominant autosomal fashion and is not linked to the MHC. In addition, the C3-deficient trait is not linked to the C3a receptor deficiency which we described in the guinea pig (ZANKER et al. 1983; BITTER-SUERMANN and BURGER 1986).

Analysis at the cDNA level is obviously required in order to identify or exclude any structural defect in the C3 gene of C3-deficient animals. This work is currently in progress. C3 clones were obtained from cDNA libraries of C3-sufficient and of C3-deficient animals, and nucleotide sequencing is being performed (AUERBACH et al., manuscript in preparation). The preliminary sequencing data of the C3 clones revealed no major nucleotide differences resulting in an abnormal amino acid sequence (AUERBACH, personal communication). Additional studies regarding the fate of the C3 protein of C3-deficient animals in vivo and its immunochemical properties are required, for example, with regard to carbohydrate substitution, proteolytic fragmentation, and catabolism.

2.2 Biological Consequences of the C3 Deficiency

Examining hereditary deficiencies in complement proteins greatly facilitated analysis of the various biological functions of given proteins. The rare cases of these helped considerably in elucidating the role of complement components in host defense and immune reaction (ROTHER and ROTHER 1986). The bactericidal activity of C3-deficient serum as a parameter for nonspecific host defense mechanisms was determined, and the capacity for producing a specific immune response to selected antigens was measured.

The bactericidal effect of C3-deficient serum on a serum-sensitive rough strain of *Escherichia coli* K12 was reduced several-fold compared to normal serum (BURGER et al. 1986). C3-deficient serum had to be used at 8- to 15-fold higher concentrations to achieve a comparable effect. Nevertheless, under the conventional conditions of animal facilities, in the C3-deficient colony a higher frequency of bacterial infection was not observed. Obviously, even the low amount of 5% C3 maintains a sufficient level of host defense mechanisms to cope with the pathogenic micro-organisms present in the more or less protected environment of an animal breeding facility.

The C3-deficient guinea pigs, further, represent a model system to study the role of C3 in the induction of the specific immune response. A genetically controlled deficiency provides information about the contribution of a given component in a direct manner. In contrast, the artificial complement depletion by treatment with cobra venom factor (CVF) might have undesired side effects leading to poor control of factors unrelated to the intended depletion per se. In addition, a number of studies on the modulation of humoral and cellular immune response by C3 led to quite controversial results (BÖTTGER and BITTER-SUERMANN 1987). Therefore, an in vivo analysis in the C3-deficient guinea pigs seems particularly promising.

The immune response of the C3-deficient animals to several model antigens was analyzed and compared to control groups of normal complement-sufficient strain 2 animals. This control via the appropriate inbred strain from which the deficiency was derived excludes the effects of unrelated genetically controlled phenomena such as the Ir gene. Upon immunization with the bacteriophage ϕX174 a marked impairment of the antibody response was observed (BÖTTGER et al. 1986b). IgM formation in the primary response was reduced. In the secondary response after a booster injection, the amplification of the antibody titer and the isotype switch from IgM to IgG did not occur. This impairment resembles the reduced antibody response in C4- or C2-deficient guinea pigs. In these two deficient strains the impairment is compensated by increasing the antigen dose (BÖTTGER et al. 1985). Overcoming the reduced capacity for antibody formation in the C3-deficient animals proved more difficult. Increasing the antigen concentration led only to a partial normalization in antibody formation. The same was true when the antigen ϕX174 was given with complete Freund's adjuvant, or when the animals were reconstituted to at least a certain extent with C3 during the primary immunization (KLEINDIENST et al. 1987). Thus, the requirement for C3 to effect an immune response seems to be quite stringent.

The antibody response to sheep red blood cells was similarly impaired in C3-deficient animals. In contrast, no reduction was found in antibody formation to the antigen ovalbumin (KÖHLER et al., manuscript in preparation). This may be due to the fact that the animals had a low level of antibodies to ovalbumin prior to the immunization, and that an amnestic response actually did occur that was less dependent on C3. The cellular site and the nature of this impairment in immune response remains to be analyzed. The critical role of C3 may be the development of B-cell memory, the generation of B-cell growth factors, the trapping and processing of antigen or immune complexes and localization in germinal centers, and finally modulation of macrophage function. The C3-deficient guinea pigs should provide a tool to address these questions and other biological phenomena in an experimental animal.

2.3 Historical C3-Deficient Guinea Pigs

In retrospect, a complement deficiency described in guinea pigs in the first quarter of this century may actually have been a complete C3 deficiency (MOORE 1919). These animals were identified by their lack of hemolytic complement activity and were classified as deficient in C3. However, at that time the term C3 was applied to a functionally defined moiety of the complement system which includes — as we know today — not only the C3 protein but also that of C5, C6, C7, C8, and C9. Mixing experiments showed that heat-inactivated but not zymosan or cobra venom treated serum did restore the hemolytic function of the deficient serum. This finding is compatible with a real C3 deficiency. Unfortunately, this guinea pig strain was lost in the mid 1930s, possibly due to an epidemic infection with group C hemolytic streptococci, a well-known pathogen in guinea pigs.

3 C3 Deficiency in Dogs

In 1981 WINKELSTEIN et al. reported on a colony of Brittany spaniels which were studied for hereditary canine spinal muscular atrophy; some of the animals had suffered during the course of inbreeding from recurrent local and systemic bacterial infections. The authors investigated the immune functions of these animals including the complement system and noted a markedly reduced hemolytic complement activity due to the absence of C3. Recently they summarized their current knowledge on the nature and inheritance of this C3 deficiency (WINKELSTEIN et al. 1986). Below we describe the essential features of this genetic disorder, which in contrast to the above-described partial C3 deficiency in guinea pigs leads to increased susceptibility for infections.

3.1 Inheritance of C3 Deficiency

In dogs C3 deficiency is inherited in an autosomal recessive pattern due to the homozygosity of a null gene allelic to the structural genes $C3^F$ or $C3^S$. As in man and guinea pigs, C3 is not linked to the MHC in dogs (JOHNSON et al. 1986). Heterozygotes have C3 antigen levels that range from 50% of the mean for normal dogs up to nearly normal levels. Segregation from the susceptibility for hereditary canine spinal muscular atrophy, inherited as an autosomal dominant trait, seems possible but has not yet been established. A molecular-genetic analysis to clarify the structural basis for this deficiency has thus far not not published.

3.2 Antigenic and Functional Levels of C3

Neither by immunoprecipitation methods nor by ELISA has antigenic canine C3 been detectable (<0.003% of normal serum C3), but no discrimination between lack of synthesis and lack of secretion can yet be made. All other complement components are normal or nearly normal. Increased turnover of C3 by inhibitory factors is not detectable, and substitution with purified normal dog C3 completely restores the functional C3 activity of these sera without increased consumption (WINKELSTEIN et al. 1982). Interestingly, the sera of C3-deficient dogs have some residual C3-like hemolytic activity (6%–10% of C3 titer in normal dog sera) which is not absorbable by anti-canine C3. This activity coincides in gel filtration with normal C3 activity. The activity is inhibited by incubation of erythrocyte antibody complexes and deficient dog serum in 0.01 M ethylenediaminetetraacetate (JOHNSON 1987). The C3-like activity, as with that of normal C3, is also inhibited by treatment of serum with 2 M methylamine. In considering the alternatives of (a) the existence of an aberrant C3 protein and (b) a completely C3-unrelated new protein for which appropriate detection conditions are not yet available, the authors favor the latter. They speculate that this may be in line with a residual (1%) C3-like hemolytic activity in some human C3 deficiencies (see above). This may be explained by a direct but inefficient activation of C5 by the classical pathway C3 convertase (KITAMURA et al. 1984a). Opsonic and chemotactic activities directly dependent on C3 or indirectly on its role in activation of C5 are markedly deficient.

3.3 Disease Association in C3-Deficient Dogs

There seems to be a remarkable similarity between C3-deficient dogs and humans with regard to susceptibility to pyogenic infections with encapsulated bacteria and to renal involvement such as in membranoproliferative glomerulonephritis (BLUM et al. 1985).

3.4 Defective Antibody Production in C3-Deficient Dogs

Total IgM levels in C3-deficient dogs were found to be significantly higher than those in normal or heterozygous dogs. In line with previously described abnormalities of humoral immune responses in C3-depleted animals (with CVF) and C3-deficient guinea pigs, O'NEIL et al. (1988) reported on markedly reduced and delayed primary antibody responses to T-dependent and T-independent antigens (bacteriophage ϕX174, SRBC, DNP-Ficoll) with a delayed and reduced switch from IgM to IgG. After secondary immunization C3-deficient dogs produced more IgM and less IgG than normal dogs in the control group. Neither intramuscular immunization nor the intravenous administration of increased antigen doses could correct the defect in the antibody response. Again, this is compatible with the results in C3-deficient guinea pigs (KLEINDIENST et al. 1987). Reconstitution experiments with purified C3 or serum as a source of C3 were not performed because of short supply.

4 Hereditary C3 Hypocomplementemia in Rabbits

Recently in a strain of C8-deficient New Zealand white rabbits animals were found with about 10% residual C3 activity and C3 antigen, not in linkage with this additional complement deficiency (KOMATSU et al. 1988). The C3 hypocomplementemia is inherited as a simple autosomal codominant trait. Purified rabbit C3 dose-dependently restored the antigenic and functional defect with no evidence of increased turnover and catabolism. The health status differed from that in rabbits without C3 hypocomplementemia in that during the first 3 months there was a significantly reduced survival rate. No reason (infectious agents?) or disease association was given. Similarities and differences to the C3-deficient guinea pigs must await further studies in both species.

5 Acquired C3 Deficiencies

A few clinical and experimental situations are known in which a significantly depressed plasma C3 titer (below 10% of normal) is the consequence of prolonged or transitory hyperactivation of this crucial complement component. All these situations have in common dysregulated C3 convertase of either the classical or the alternative pathway in vivo. This means that the complex enzyme C4b2a or C3bBb escapes the normal control of dissociation and cleavage by fluid-phase and cell-bound regulatory proteins (factors I and H, CR1). This imbalance can be initiated by autoantibodies of the IgG isotype against neodeterminants of both convertases leading to a stabilization of their C3-cleaving function instead of their inactivation. The autoantibodies against cell-bound or fluid-phase C3bBb are termed C3 nephritic factors (C3-NeF) and those against C4b2a termed C4NeF because of their association with a clinical syndrome dominated by a membranoproliferative, hypocomplementemic glomerulonephritis (DAHA 1988a, b). The half-life of cell-bound C3bBb is increased by C3-NeF from 4 min to 50 min.

The second, experimental instance of stabilized C3 convertase in vivo and in vitro is due to a substitution of C3b in the C3bBb enzyme by the snake analogue of mammalian C3, CVF (ALPER and BALAVITCH 1976). It seems reasonable, although it has yet to be proven, that phylogenetically conserved regions in the two proteins C3b and CVF interact equally well with factor B, whereas the fluid-phase regulatory proteins factors H and I obviously achieved more specialization during evolution and are restricted to the mammalian complement proteins. Therefore CVFBb convertase has a half-life in the circulation of more than 20 h compared to one in C3bBb of less than 5 min. This is the reason for the prolonged decomplementation of C3 in experimental animals for about 5–7 days after two to three consecutive doses of CVF intravenously or intraperitoneally within 12 h. After 1 week a strong humoral immune response against CVF leads to a neutralization and clearance of the CVFBb enzyme. Most experiments have been done with CVF purified from the crude venoms of either the Indian cobra *Naja naja* or the Egyptian cobra *Naja haje*. While in concert with factor B these act similarly in the cleavage of C3, they differ in their efficiency in the cleavage of C5 (for review see SCHULTZ 1986).

The in vivo depletion of C3 by CVF has a profound effect on primary antibody responses, mainly to T-dependent antigens, as was initially reported by PEPYS (1974, 1976), and mimics and resembles the impaired humoral immune responses in an inherited C3-deficient situation. By analogy, the CVF-induced disturbances may therefore be taken as evidence for the pivotal role of C3 in the induction of antibody responses (BÖTTGER et al. 1986a) of B-cell memory, antigen retention, and immunoregulatory effects in vivo at the level of B cells and follicular dendritic cells, both carrying CR1 and CR2; this was recently summarized by KLAUS (1988). Whereas the detrimental effects of complement activation in vivo are well known but often overestimated (such as in the pathological inflammatory sequelae of immune complex generation), the beneficial role of complement (especially C3) and of immune complexes in antigen removal and in augmenting the humoral immune response is often overlooked (NYDEGGER and KAZATCHKINE 1986). Therefore, examining acquired or induced C3 depletion contributes to our knowledge as to the key role that C3 plays in the induction, processes, and perpetuation of immune complex diseases associated with inherited complement deficiencies, including those of C3.

References

Alper CA, Badavitch D (1986) Cobra venom factor: evidence for its being altered cobra C3 (the first component of complement). Science 191: 1275–1276
Alper CA, Propp RP, Klemperer MR, Rosen FS (1969) Inherited deficiency of the third component of human complement (C'3). J Clin Invest 48: 553–557
Alper CA, Colten HR, Rosen FS, Rabson AR, Macnab GM, Gear JS (1972) Homozygous deficiency of C3 in a patient with repeated infections. Lancet 2: 1179–1181
Alper CA, Colten HR, Gear JSS, Rabson AR, Rosen FS (1976) Homozygous human C3 deficiency: the role of C3 in antibody production, C1s-induced vasopermeability, and cobra venom-induced passive hemolysis. J Clin Invest 57: 222–229
Bird P, Lachmann PJ (1988) The regulation of IgG subclass production in man: low serum IgG4 in inherited deficiencies of the classical pathway of C3 activation. Eur. J Immunol 18: 1217–1222
Bitter-Suermann D, Burger R (1986) Guinea pigs difecient in C2, C4, C3 or the C3a receptor. Prog Allergy 39: 134–158
Blum JR, Cork LC, Morris JM, Olson JL, Winkelstein JA (1985) The clinical manifestations of a genetically determined deficiency of the third component of complement in the dog. Clin Immunol Immunopathol 51: 204–315
Böttger EC, Bitter-Suermann D (1987) Complement and the regulation of humoral immune response. Immunol Today 8: 261–264
Böttger EC, Hoffmann T, Hadding U, Bitter-Suermann D (1985) Influence of genetically inherited complement deficiencies on humoral immune response in guinea pigs. J Immunol 135: 4100–4107
Böttger EC, Hoffmann T, Metzger S, Hadding U, Bitter-Suermann D (1986a) The role and mechanism of cobra venom factor-induced suppression of the humoral immune response in guinea pigs. J Immunol 137: 1280–1285
Böttger EC, Metzger S, Bitter-Suermann D, Stevenson G, Kleindienst S, Burger R (1986b) Impaired humoral immune response in complement C3-deficient guinea pigs: absence of secondary antibody response. Eur J Immunol 16: 1231–1235
Borzy MS, Gewurz A, Wolff L, Houghton D, Lovrien E (1988) Inherited C3 deficiency with recurrent infections and glomerulonephritis. Am J Dis Child 142: 79–83
Burger R, Gordon J, Stevenson G, Ramadori G, Zanker B, Hadding U, Bitter-Suermann D (1986) An inherited deficiency of the third component of complement, C3, in guinea pigs. Eur J Immunol 16: 7–11
Daha MR (1988a) The complement system in human and experimental glomerulonephritis. Baillieres Clin Immunol Allergy 2: 505–521
Daha MR (1988b) C3 nephritic factor: In: Rother K, Till Go (eds) The complement system. Springger, Berlin Heidelberg New York, pp 463–469
Davis AE, Davis JS, Rabson AR et al. (1977) Homozygous C3 deficiency: detection of C3 by radioimmunoassay. Clin Immunol Immunopathol 8: 543–550
Day NK (1986) Deficiencies of regulator proteins factor I and H. Prog Allergy 39: 335–338
Einstein P, Jansen PJ, Ballow M, Davis AE, Davis JS, Alper CA, Rosen FS, Colten HR (1977) Biosynthesis of the third component of complement (C3) in vitro by monocytes from both normal and homozygous C3-deficient humans. J Clin Invest 60: 963–969
Johnson JP (1987) C3-like activity in C3-deficient dog serum. Complement 4: 53–60
Johnson JP, McLean RH, Cork LC, Winkelstein JA (1986) Animal model: genetic analysis of an inherited deficiency of the third complement in Brittany spaniel dogs. Am J Med Genet 25: 557–562

Kitamura H, Matsumato M, Nagaki K (1984a) C3 independent immune hemolysis. Hemolysis of EAC1,4,oxy2 cells in C5–C9 without participation of C3. Immunology 53: 575–582

Kitamura H, Nishimukai Y, Sano, Nagaki K (1984b) Study on a C3-like factor in the serum of a C3 deficient subject. Immunology 51: 239–245

Klaus GBB (1988) Role of complement in the induction of antibody responses In: Rother K, Till Go (eds) The complement system. Springer, Berlin Heidelberg New York, pp 327–337

Kleindienst S, Böttger EC, Bitter-Suermann D, Schäfer R, Burger R (1987) Stringent requirement for the complement component C3 during the induction of the humoral immune response (Abstract). Immunology 175: 365

Komatsu M, Yamamoto K, Nakano Y, Nakazawa M, Ozawa A, Mikami H, Tomita M, Migita S (1988) Hereditary C3 hypocomplementemia in the rabbit. Immunology 64: 363–368

Lachmann PJ (1988) Deficiencies of factor I and factor H. Rother K, Till Go (eds) In: The complement system. Springer, Berlin Heidelberg, New York, pp 458–461

Moore HD (1919) Complementary and opsonic functions in their relations to immunity. A study of the serum of guinea pigs naturally deficient in complement. J Immunol 4: 425–441

Nydegger UE, Kazatchkin MD (1986) Modulation by complement of immune complex processing in health and disease in man. Prog Allergy 39: 361–392

Ochs HD, Wedgewood RJ, Heller SR, Beatty PG (1986) Complement, membrane glycoproteins and complement receptors. Their role in regulation of the immune response. Clin Immunol Immunopathol 40: 94–104

O'Neil KM, Ochs HD, Heller SR, Cork LC, Morris JM, Winkelstein JA (1988) Role of C3 in humoral immunity: defective antibody production in C3-deficient dogs. J Immunol 140: 1939–1945

Osofsky SG, Thompson BH, Lint TF, Gewurz H (1977) Hereditary deficiency of the third component in a child with fever, skin rash, and arthralgias: response to transfusion of whole blood. J Pediart 90: 180–186

Pepys MB (1974) Role of complement in the induction of antibody production in vivo. J Exp Med 140: 126–134

Pepys MB (1976) Role of complement in the induction of immunological responses. Transplantation 32: 93–120

Ross SC, Densen P (1984) Complement deficiency states and infection: epidemiology, pathogenesis and consequences of neisserial and other infections in an immune deficiency. Medicine 63: 243–273

Rother K, Rother U (1986) Hereditary and acquired complement deficiencies in animals and man. Prog Allergy 39: 1–7

Schifferli JA, Yin CNG (1988) The role of complement in the processing of immune complexes. Baillieres Clin Immunol Allergy 2: 319–334

Schultz DR (1986) Induced deficiences in animals and man. Prog Allergy 39: 101–122

Strate M, Olsen H, Teisner B (1987) Bacterial capacity against *Neisseria meningitidis* of normal human serum and sera with functional deficiencies of the third and eighth complement factor. Eur J Clin Invest 17: 226–230

Winkelstein JA, Collins-Cork L, Griffin DE, Griffin JW, Adams RJ, Price DL (1981) Genetically determined deficiency of the third component of complement in dog. Science 212: 1169–1170

Winkelstein JA, Johnson JP, Swift AJ, Ferry F, Yolken R, Cork LC (1982) Genetically determined deficiency of the third component of complement in the dog: in vitro studies on the complement system and complement-mediated serum activities. J Immunol 129: 2598–2602

Winkelstein JA, Johnson JP, O'Neil KM, Cork LC (1986) Dogs deficient in C3. Prog Allergy 39: 159–168

Zanker B, Engelberger W, Hadding U, Burger R, Bitter-Suermann D (1983) Combined genetic deficiency of anaphylatoxin C3a receptors and complement C4 in guinea pigs (Abstract). Immunobiology 165: 380

Structural and Functional Analysis of C3 Using Monoclonal Antibodies

J. Alsenz[1]*, J. D. Becherer[1], B. Nilsson[2], and J. D. Lambris[1]

1   Introduction   235
2   Generation of Anti-C3 Monoclonal Antibodies   235
3   Structural and Functional Analysis of C3 Using Anti-C3 Monoclonal Antibodies   236
3.1 Detection of Conformational Changes in C3 and its Fragments   236
3.2 Analysis of the C3 Functions   237
3.3 Mapping of the Epitopes Recognized by the Monoclonal Antibodies   239
4   Analysis of the C3 Complement Activation Products   243
5   Conclusions   244
References   245

1 Introduction

Monoclonal antibodies (MoAbs) have greatly facilitated the structural and functional analysis of proteins in general and of the third protein of complement (C3) in particular. Various aspects of the structure and functions of C3 have been addressed using MoAbs; these include: (a) the study of conformational changes occurring in the C3 molecule and its fragments during complement activation, (b) the analysis of the interactions of C3 with other complement components and receptors as well as with proteins of foreign origin, and (c) the detection of C3 activation products in biological fluids. The purpose of this review is to summarize the contribution that MoAbs have made in understanding the structure and functions of C3.

2 Generation of Anti-C3 Monoclonal Antibodies

Although the generation of MoAbs has become a routine procedure, a few interesting points have been observed concerning the nature of the immunizing antigen and the subsequent specificity of the anti-C3 MoAb. In the most commonly used immunization

---

[1] Basel Institute for Immunology, Grenzacherstr. 487, 4005 Basel, Switzerland
[2] Department of Clinical Immunology and Transfusion Medicine, University Hospital, Uppsala, Sweden
* J. A. has an EMBO long-term fellowship (ALTF 298-1987). The Basel Institute for Immunology was founded and is supported entirely by F. Hoffmann-LaRoche Ltd. Co., Basel, Switzerland

protocol, intact C3 (BURGER et al. 1982) or fluid-phase (TAMERIUS et al. 1982; HACK et al. 1988) or surface-bound (LACHMANN et al. 1980) C3 fragments were used as immunizing antigens. Although the antibodies so produced may have interesting features with respect to C3 functions, often the recognized epitopes are difficult to map due to their dependence on the tertiary structure of the molecule. Thus synthetic peptides (BURGER et al. 1988) or expressed segments of C3 (MA et al. 1985) have been used to generate MoAbs of predetermined specificity (LAMBRIS et al. 1989). The former have been especially useful in generating MoAbs which recognize exclusively C3a. Another interesting observation is that when sodium dodecyl sulfide denatured C3 was used as an immunizing antigen, the produced MoAbs preferentially recognized epitopes expressed by surface-bound C3 fragments (NILSSON et al. 1987; NILSSON et al. 1989b). An important consideration in generating MoAbs to C3 is the screening procedure used. This is the result of the different conformations assumed by C3 and depends on whether the molecule is in its native form or is degraded to its fragments; the latter possibility exists in two forms, surface-bound and fluid-phase (see below).

3 Structural and Functional Analysis of C3 Using Anti-C3 Monoclonal Antibodies

3.1 Detection of Conformational Changes in C3 and Its Fragments

During complement activation the C3 molecule undergoes gross conformational changes as it is processed by the enzymes of the complement cascade. This is evident from the ability of the generated fragments (C3a, C3b, iC3b, C3c, C3dg, C3d) to bind differentially to various complement components and receptors (LAMBRIS 1988; BECHERER et al. 1989a, 1989b). The changes in C3 conformation accompanying the proteolytic cleavage events have been detected using an assortment of chemical probes and spectral and solution scattering techniques (MOLENAAR et al. 1975; ISENMAN and COOPER 1981; ISENMAN 1983; PERKINS and SIM 1986; PANGBURN et al. 1981). Although these techniques allow the analysis of conformational changes of fluid-phase C3 fragments, technical difficulties have hampered a similar analysis of surface-bound fragments. The use of MoAbs, however, has overcome these difficulties, and they have been shown to be more sensitive in probing conformational changes in proteins than the low-resolution spectroscopic methods (COLLAWN et al. 1988). Especially in studying C3, the MoAbs have provided a wealth of data on the specific epitopes expressed during the degradation of C3; the identification of these epitopes using the other methods mentioned above would not be possible. The reactivity of several MoAbs with $C3(H_2O)$ as well as with C3b but not with native C3 (Table 1) has suggested that similar conformational changes are induced when $C3(H_2O)$ or C3b are generated (HACK et al. 1988). This is in agreement with the similar conformations observed spectroscopically (ISENMAN et al. 1981; PANGBURN 1987) as well as with the similar functional activities expressed by these molecules; the $C3(H_2O)$ acquires "C3b-like activities" by expressing sites which are recognized by C3b binding proteins

(PANGBURN et al. 1981 and Volanakis, this volume). The epitopes recognized by these antibodies are located within the different fragments of the C3 molecule, namely the C3c, C3dg, or C3a fragments (Table 1). Despite the observed similarities between C3(H$_2$O) and C3b, epitopes expressed only in C3 degradation fragments C3a and C3b but not in C3(H$_2$O) have also been described (Table 1), thus demonstrating that some conformational differences exist between these two molecules.

Cleavage of fluid-phase C3b to iC3b by factor I in the presence of one of its several cofactor molecules (BECHERER et al. 1989a) is accompanied by further conformational changes in the molecule. This has been shown using MoAbs that are specific for fluid-phase or surface-bound iC3b fragments of C3 and agrees with the appearance and disappearance of ligand binding sites upon this cleavage (for review see ROSS and MEDOF 1985 and BECHERER et al. 1989a; as used in the present review, the term ligand is defined as those complement proteins which bind to C3, whether they are found in serum or on the surface of cells). The newly exposed sites in iC3b include the conglutinin, the CR2, and the CR3 binding sites (LAMBRIS 1988; BECHERER et al. 1989b). The iC3b-specific MoAbs react with the iC3b and/or C3dg/C3d fragments but not with C3, C3(H$_2$O), or C3b (Table 1) (TAMARIUS et al. 1985; LACHMANN et al. 1982; IIDA et al. 1987a). Interestingly, most of the iC3b-specific antibodies that react with neoantigenic sites recognize the C3dg/C3d fragment of C3. This fragment contains the binding site for CR2 and one of the H sites (LAMBRIS et al. 1985, 1988).

In addition to neoantigenic sites expressed in both fluid-phase and surface-bound C3 fragments, epitopes expressed solely in surface-bound C3 fragments have also been described (Table 1). It appears that most of these epitopes are also expressed by denatured C3 (NILSSON et al. 1982, 1986, 1987, 1989a, 1989b), a fact which has facilitated their localization within the C3 molecule (see below).

3.2 Analysis of the C3 Functions

Due to their remarkable specificity, MoAbs have been used as tools to identify functionally important sites in C3. The inhibition of C3 binding to a given ligand is a good starting point for identifying the ligand interaction site(s) in C3. Although several points of concern are associated with such inhibitory effects (steric inhibition, change in C3 coformation upon antibody binding, inability to localize the recognized epitope due to its dependence on the conformation, etc.) several interesting observations have been made using anti-C3 MoAbs. One example is the MoAb 130. This antibody was found to inhibit the binding of C3d to CR2, an observation which has assisted the localization of the CR2 binding site in C3d (LAMBRIS et al. 1985). Another finding is that MoAbs to either C3c (C3-9, Ab 12, and Ab 72) or C3d (311, Ab 14) inhibit H binding to C3b (Table 1; TAMERIUS et al. 1982; BECHERER et al. 1989c) which, together with the results derived using MoAbs to H (ALSENZ and LAMBRIS 1988), suggests that these two molecules interact via multiple binding sites. Furthermore, the observation that several MoAbs to C3 (C3-9, 311, H11, H2) inhibit H, CR1, CR2, and/or B binding to C3 suggests that these ligands may bind to common or proximal sites in C3 (Table 1; BECHERER et al. 1989c; WÖRNER et al. 1989; Koistinen et al. 1989). This assumption is in agreement with the recent localization of the CR1, B, and H binding sites in C3. By using synthetic peptides, two H interaction sites

Table 1. Summary of the monoclonal anti-C3 antibodies recognizing neoantigens and/or influencing C3-ligand interactions

MoAb	Fragment recognized	Neoantigen expressed in	Inhibition of the C3 interaction with						References
			H	B	P	CR1	CR2	gC	
C3-5	C3a	C3*/C3a							HACK et al. 1988
4SD 17.1	C3a	C3a							NILSSON et al. 1988
H453, H454	C3a	C3a							BURGER et al. 1988
C3-1	C3c	C3*/C3b/C3c							HACK et al. 1988
C3-9	C3c	C3*/C3b/C3c	+	+		+	+	–	HACK et al. 1988; BECHERER et al. 1989c; HUEMER et al. 1989
C-5G	C3c	C3b/C3c	+	–		+	–		IDA et al. 1987a
bH6	C3c	C3b/iC3b/C3c						+	GARRED et al. 1988b, HUEMER et al. 1989, BECHERER et al. 1989c
7D326.1, 7D331.1	C3c	sb⁺ C3b/iC3b	–	–			+		NILSSON et al. 1989a
MoAb 130	C3d	iC3b/C3dg/C3d	–	–		–	+		TAMERIUS et al. 1982, LAMBRIS et al. 1985
G-3E	C3d	iC3b/C3dg/C3d	–	–		–	+		IDA et al. 1987a
clone 9	C3dg	iC3b/C3dg/C3g			+				LACHMANN et al. 1980
C3-11	C3dg	C3*/C3b/C3dg							HACK et al. 1988
7D323.1, 7D84.1, 7D264.6	C3dK	sb iC3b							NILSSON et al. 1989b
105	C3c	–	–	+	–	–	–	+	BURGER et al. 1982, HUEMER et al. 1989
111	C3c	–	–	+	+	–	–	+	BURGER et al. 1982, HUEMER et al. 1989
Ab 12	C3c	–	+	–	–	–	–		TAMERIUS et al. 1982
498	C3c	–	–	+		–	–		BECHERER et al. 1989c
Ab 84	C3c	–	–	–	+	–	–		TAMERIUS et al. 1982
311	C3d	–	+	–		–	+	+	BECHERER et al. 1989c. HUEMER et al. 1989
31	C3d	–	+	–	+	–	–		BECHERER et al. 1989c
Ab 14, Ab 72	C3d	–	+	–		–			TAMERIUS et al. 1982
4C2	C3d	–	+	+	–	+	–		KOISTINEN et al. 1989
H11	β-chain	–	–	+	–	+	–	+	WÖRNER et al. 1989, HUEMER et al. 1989
H2	β-chain	–	–	+	–	+	–	+	WÖRNER et al. 1989, HUEMER et al. 1989

C3* = C3(H₂O); sb⁺ = surface bound.
Other anti-C3 MoAbs reacting with:
C3: anti-FG11 (MA et al. 1985); HAV4-1 (KOCH and BEHRENDT 1986)
C3a: H13 (BURGER et al. 1987); MoAb 868 (KLOS et al. 1988)
C3b: MoAb 755 (KLOS et al. 1988)
C3c: clone 4 (LACHMANN et al. 1980); N-7A (IDA et al. 1987a); WM-1 (WHITEHEAD et al. 1981); MoAb-BRL (AGUADO et al. 1985); anti-C3c (22 different MoAb) (DOBBIE et al. 1987)
C3d: Fc 112, Fc 280, Fc 283, Ortho (CHAPLIN and MONROE 1986); anti-C3d (nine different MoAbs) (DOBBIE et al. 1987); 3D4H3 (RUDDY et al. 1983); BRIC 8

have been identified in C3b, one in C3d and another in C3c (GANU and MÜLLER-EBERHARD 1985; LAMBRIS et al. 1988; BECHERER et al. 1989a; LAMBRIS et al. 1989; BECHERER et al. 1989c). The H binding site in C3d is associated with the CR2 site while the site in C3c is within the domain of C3 containing the B and CR1 binding sites. It is interesting that the latter site contains a sequence (between residues 744 and 755) that is similar to the sequence of the CR2 binding site in C3d (for more details see BECHERER et al. 1989a). Furthermore, a third sequence similar to the CR2 binding site has been identified in the β-chain of C3 (between residues 295 and 306), and a peptide spanning this sequence has been found to bind to CR2 (ESPARZA et al. 1989). Although a direct binding of CR1, H, or B to this peptide has not yet been determined, this may explain the finding that MoAbs to the β-chain of C3 have also been found to inhibit H, CR1, and B binding to C3b (WÖRNER et al. 1989).

The selective inhibition by MoAbs of either B (498, 105, 111) or H binding (311, Ab 12, 14, 72, and 84) to C3b (TAMERIUS et al. 1982; BURGER et al. 1982) suggests the presence of distinct binding sites for B and H on C3b. This is not necessarily in contrast to the above findings and might be attributed to the interaction of these molecules with C3b via multiple sites. In this context, it has been observed that MoAb 130 inhibits CR2 but not H binding to C3d; both sites are located within residues 1187–1249 of C3. The latter fact further supports the findings with synthetic peptides which have shown that these two sites are related but are not the same (LAMBRIS et al. 1985, 1988). Of interest also is the recent finding that MoAbs to C3 which binding H were also found to inhibit the binding of glycoprotein C (gC) of herpes simplex virus type 1 (HSV-1) to C3b (HUEMER et al. 1989). However, the inhibition of H but not gC binding to C3b by other antibodies (Table 1) may explain the functional differences between these two molecules; H has both cofactor and decay-accelarating activity while gC has only the latter.

The interaction of properdin with C3b, as investigated by MoAbs to either C3c or C3d (Table 1; TAMERIUS et al. 1982), suggests a two-side interaction between these two molecules. Since the site in C3c for P has recently been localized to residues 1402–1435 of C3 (DAOUDAKI et al. 1988), it will be of interest to see whether these anti-C3c antibodies react with this region of C3, or whether the observed inhibition is indirect (steric or allosteric effects, etc.). The presence of a second P interaction site in C3b remains to be determined.

The epitopes recognized by several other MoAbs (Table 1) have not yet been localized, and this makes it difficult to correlate the structural elements involved in antibody binding to those involved in ligand binding. In addition, MoAbs which bind close to a functional site in C3 (see also Fig. 1) were fount not to inhibit its reactivity with the ligand. However, this is not necessarily unexpected since it has been shown that a peptide, 24 amino acids in length, is able to bind two different antibodies simultaneously (JACKSON et al. 1988).

3.3 Mapping of the Epitopes Recognized by the Monoclonal Antibodies

A major breakthrough in understanding antibody-antigen interactions was provided by the crystallographic studies of Fab fragments complexed to the protein antigens lysozyme (AMIT et al. 1986) and influenza virus neuraminidase (COLMAN et al. 1987).

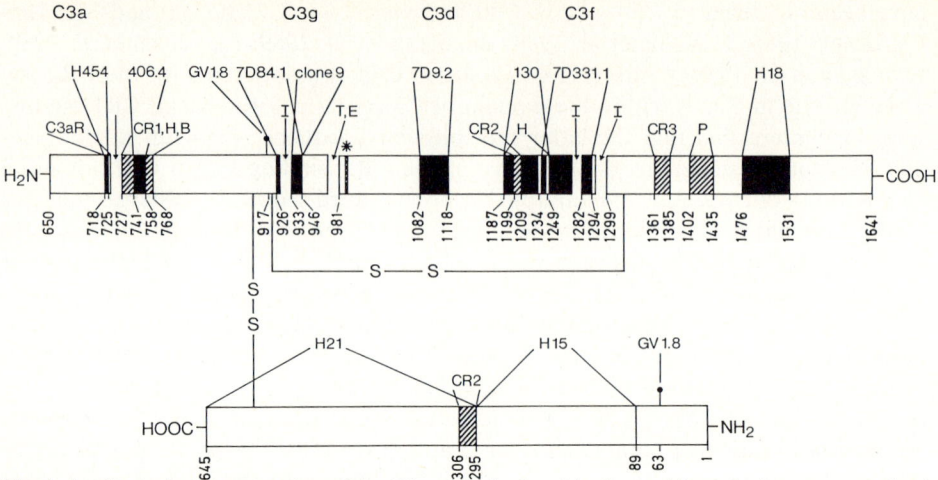

Fig. 1. A schematic representation of the C3 molecule showing the position of epitopes recognized by the anti-C3 MoAbs and ligand binding sites in C3. The thioester site in C3d (*), the carbohydrate structure (—●), and the sites of C3 cleaved by the C3 convertases (between residues 726 and 727), factor I (I), elastase (E), and trypsin (T) are indicated by *arrows*. Other MoAbs with specificities similar to the ones shown are listed in Table 2. The numbering of the residues of C3 is based on the reported amino acid sequence of C3 (DE BRUIJN and FEY 1985) after subtracting the signal peptide

However, since the three-dimensional structure for every antibody-antigen complex — in our case an antibody-C3 complex — is not practical, other methods have been employed to study the structural requirements of antigen-antibody interactions (COLLAWN et al. 1988). Although the information obtained using these methods is not as detailed as that from crystallographic data, it is nonetheless beneficial in the understanding of structural and functional aspects. The strategy that is used in our laboratory to study several aspects of C3 structure and functions approaches the problem at hand from three different directions and is outlined in Fig. 2. The first approach is to study the problem on the protein level. C3 is degraded enzymatically or chemically into small fragments which are then analyzed for binding to the various MoAbs or C3 binding proteins. Following purification of the fragment of interest, its sequence is determined in order to localize its position within the C3 sequence. Overlapping peptides within the identified area are then synthesized and analyzed for binding to the MoAb or ligand of interest. The second approach, which orignates at the DNA level, involves construction of an expression minilibrary from the cDNA spanning the entire coding sequence of C3 (kindly provided by G. FEY, Scripps Clinic, LA JOLLA, California, USA; DE BRUIJN and FEY 1985). In this case, the cDNA is first digested with DNAase, fragments 200–300 bases in length are cloned into an expression vector (λgt11 (GROSSBERGER et al. 1988), and the library is then screened with the monoclonal anti-C3 antibodies (GROSSBERGER et al. 1988; NILSSON et al. 1989a; WÖRNER et al. 1989) or with various proteins that bind C3. The reactive clones are then sequenced to map the specific site. This approach can be a fast and efficient method for screening numerous monoclonal antibodies.

A. *PROTEIN*

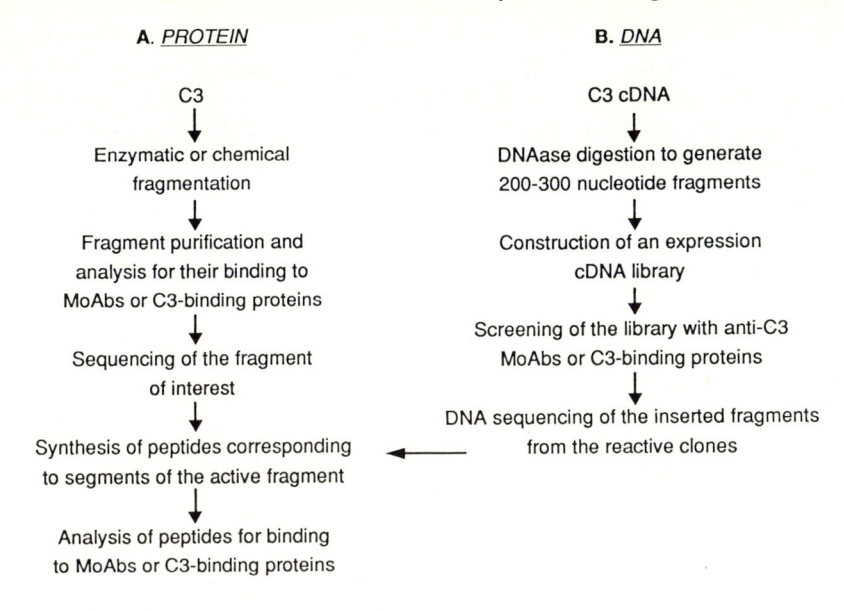

C3
↓
Enzymatic or chemical fragmentation
↓
Fragment purification and analysis for their binding to MoAbs or C3-binding proteins
↓
Sequencing of the fragment of interest
↓
Synthesis of peptides corresponding to segments of the active fragment ⟵
↓
Analysis of peptides for binding to MoAbs or C3-binding proteins

B. *DNA*

C3 cDNA
↓
DNAase digestion to generate 200-300 nucleotide fragments
↓
Construction of an expression cDNA library
↓
Screening of the library with anti-C3 MoAbs or C3-binding proteins
↓
DNA sequencing of the inserted fragments from the reactive clones

C. *CONSERVATION OF THE LIGAND BINDING SITES AND MoAb – RECOGNIZED EPITOPES IN C3*

Analysis of the MoAb binding to C3 from different species or of the C3-ligand interactions between different species.
↓
Comparison of the amino acid sequence of C3 and other homologous proteins from different species.
↓
Analysis of the amino acid similarities, in conjuction with the results from approach A and B, for determining the structural elements in C3 important for binding of MoAb and C3-binding proteins.

Fig. 2. Schematic representation of a general strategy to localize MoAb-recognized epitopes and ligand binding sites in C3

The third approach deals with the conservation of antigenic and functional sites within C3 from different species and other homologous proteins such as C4, C5, and α_2-macroglobulin. First, C3 from different species is purified and tested for its ability to bind to MoAbs or to the various human C3 ligands (ALSENZ et al. 1989; BECHERER et al. 1987). These proteins can then be sequenced at the DNA level, thus allowing analysis of the conservation of epitopes or binding sites between different species. This, together with the conserved sequences of other homologous proteins that do or do not bind the MoAb or ligand of interest, offers a wealth of information on the structural features of the C3 molecule. Concerning this third approach, the complete amino acid sequences of human and mouse C3 (DE BRUIJN and FEY 1985; LUNDWALL et al. 1984; WETSEL et al. 1984) and the partial sequences of rabbit (KUSANO et al. 1986) and *Xenopus* C3 (GROSSBERGER et al. 1989) have been resolved recently.

Using the above approaches, various epitopes recognized by anti-C3 MoAbs have been localized (Table 2). Several of the antibodies recognize epitopes located within or close to functional regions of C3 and therefore have been used as tools to analyze its functions. For example, the MoAb 130 was localized to a discontinuous epitope within residues 1192–1249 of C3 (LAMBRIS et al. 1989). This MoAb inhibits the binding of C3d to CR2 and assists in the localization of the CR2 binding site within this fragment to residues 1199–1210 (LAMBRIS et al. 1985).

Table 2. Summary of the monoclonal anti-C3 antibodies whose epitopes in C3 have been mapped

Antibodies	C3-fragment(s) recognized	Residues	References
H11[a,h]	C3b/C3c/β-chain	1–645	WÖRNER et al. 1989
GV1.8[b,h]	β-chain/CHO	63*	GRIER et al. 1987
GV1.10[b,h]	α/β-chain, CHO	63*, 917*	GRIER et al. 1987
H7[a,h]	C3b/C3c/β-chain	89–645	WÖRNER et al. 1989
H15[a,h]	C3b/C3c/β-chain	98–293	WÖRNER et al. 1989
H2[a,h]	C3b/C3c/β-chain	98–293	WÖRNER et al. 1989
H21[a,h]	C3b/C3c/β-chain	294–645	WÖRNER et al. 1989
H453, H454[c,i]	C3a	718–725	BURGER et al. 1988
398.1[d,i]	sb C3b	741–758	BECHERER, NILSSON, LAMBRIS unpublished
406.4[d,i]	sb C3b	741–758	BECHERER, NILSSON, LAMBRIS unpublished
595.2[d,i]	sb C3b	741–758	BECHERER, NILSSON, LAMBRIS unpublished
615.1[d,i]	sb C3b	741–758	BECHERER, NILSSON, LAMBRIS unpublished
7D 84.1[d,i]	sb C3b/iC3b	926–946	NILSSON et al. 1989a
7D 264.6[d,i]	sb C3b/iC3b	926–936	NILSSON et al. 1989a
7D 323.1[d,i]	sb C3b/iC3b	926–946	NILSSON et al. 1989a
Clone 9[e,i]	iC3b/C3dg	933–946	MYONES et al. 1989
7D 9.2[d,i]	sb C3b	1082–1118	NILSSON et al. 1989a
MoAb 130[f,i]	iC3b/C3dg/C3d	1192–1249	LAMBRIS et al. 1985, 1989
7D 326.1[d,j]	sb C3b/iC3b	1234–1294	NILSSON et al. 1989a
7D 331.1[d,j]	sb C3b/iC3b	1234–1294	NILSSON et al. 1989a
4SD 11.1[d,j]	sb C3b	1476–1510	NILSSON et al. 1989a
4SD 18.1[d,j]	sb C3b	1476–1510	NILSSON et al. 1989a
H18[a,j]	C3c	1476–1531	WÖRNER et al. 1989
H6b[a,j]	C3c	1476–1531	WÖRNER et al. 1989
H3[a,j]	C3c	1476–1531	WÖRNER et al. 1989
H206[d,j]	C3c	1476–1531	WÖRNER et al. 1989
H215[g,j]	C3c	1476–1531	WÖRNER et al. 1989

sb = surface bound;
* The MoAb binds to the carbohydrate moiety linked to this residue.

Immunization with:
[a] native C3;
[b] cobra venom factor;
[c] synthetic peptide C3$^{69-76}$ coupled to KLH;
[d] SDS-denatured C3;
[e] inulin-fixed C3b/iC3b;
[f] trypsin-generated C3b;
[g] expressed C3 fragment.

Mapped by:
[h] enzymatic fragmentation;
[i] synthetic peptides;
[j] expressed C3 fragments

The segment of C3 spanning residues 929–946 contains at least four different overlapping epitopes (Table 2). These epitopes are expressed by surface-bound and/or fluid-phase iC3b and are within the segment of C3 which also contains: (a) the carbohydrate moiety mediating conglutinin binding (HIRANI et al. 1985), (b) the factor I cleavage site(s) (DAVIS and HARRISON 1982 and BECHERER et al. 1989a), and (c) the leukocytosis-inducing activity of the C3 molecule (MEUTH et al. 1983). Using these antibodies to localize the fragments generated upon cleavage of iC3b by factor I, it was found that factor I cleaves C3 within this region at three different positions (NILSSON-EKDAHL et al. 1989; BECHERER et al. 1989a). The existence of several epitopes within such a limited amino acid sequence is not surprising since a segment as small as 19 amino acids has been shown to accommodate three different epitopes (FIESER et al. 1987).

Other epitopes in C3 have been mapped using either synthetic peptides or expressed C3 fragments (Table 2). These epitopes — several being expressed solely by surface-bound C3 fragments — have been located to residues 741–758 (four MoAbs), 1234–1294 (two MoAbs), and 1476–1531 (seven MoAbs) of the C3 sequence (for references see Table 2). The localization of these epitopes has been greatly facilitated by their expression in denatured C3 also. The possibility that some of these antibodies recognize the same epitope has not been excluded.

Related to the third approach (conservation of epitopes and binding sites), anti-C3 antibodies have been shown to cross-react with C3 from other species. Cross-reactivity studies using MoAbs have facilitated the identification of which residues are involved in the binding, and which MoAbs recognize the same or different epitopes. For example, from three MoAbs which recognize epitopes within residues 929–946 of human C3, only MoAb 7D323.1 and not 7D84.1 or 7D264.6 react with rabbit C3. Based on their reactivity with rabbit C3 and overlapping synthetic peptides, it was predicted that these MoAbs recognize different epitopes, and that residues Arg^{929}, Arg^{932}, and Glu^{933} are essential for the binding of MoAbs 7D84.1 and 7D264.6 but not that of 7D323.1 (NILSSON et al. 1989b).

4 Analysis of the Complement Activation C3 Products

The selective binding of MoAbs to either native C3 or its degradation products, whether fluid-phase or surface-bound, has rendered possible the development of sensitive assays detecting these products in biological fluids. Most of these are simple (e.g., enzyme-linked immunosorbent assay, or ELISA) and available to laboratories that are not specialized in dealing with some of the tedious and complicated assays used in analyzing the complement system. To date three MoAbs have been found to react specifically with the C3a fragment of C3 (Table 1). The detection of this fragment is relevant for the diagnosis and/or prognosis of adult respiratory distress syndrome and various other diseases (e.g., rheumatoid arthritis, systemic lupus erythematosus; SLE) (for review see BITTER-SUERMANN 1988 and HUGLI 1989). Two versions of ELISA have been described for detecting C3a. In a competition ELISA using MoAb 4SD 17.1, which recognizes C3a but not native C3 (NILSSON et al. 1988), the amount of C3a in the test sample is quantitated based on its ability to inhibit the binding of this MoAb to microtiter plate-fixed C3a. In the ELISA reported by BURGER et al. (1988), the C3a-containing sample is first incubated with an immobilized polyclonal anti-C3a antibody, and the bound C3a is detected by the MoAb H453. The sensitivity of both assays briefly described above is approximately 1 ng/ml.

In addition to the above antibodies which detect complement activation based on the generation of the C3a fragment, the MoAbs 130 (KANAYAMA et al. 1986), clone 9 (MOLLNES and LACHMANN 1987), and bH6 (GARRED et al. 1988a), all of which recognize neoantigens (Table 1), have also been used to develop assays that detect complement activation. These assays use either MoAbs fixed to microtiter plates (MOLLNES and LACHMANN 1987; GARRED et al. 1988a) or competition assays similar to that described for C3a.

The reactivity of several MoAbs with surface-bound, but not fluid-phase, C3 fragments led to the development of assays detecting and quantitating these fragments fixed to immune complexes (AGUADO et al. 1985; IIDA et al. 1987b) or particles and micro-organisms (NEWMAN and MIKUS 1985). A highly sensitive ELISA measuring C3 fragments in immune complexes has been developed using MoAbs 130 and 105. This assay detects as little as 6.26 µg aggregated human Ig/ml serum (AGUADO et al. 1985) and is comparable or superior to the most commonly used Raji (THEOFILOPOULOS et al. 1976) and C1q assays (ZUBLER et al. 1976). Using this assay, increased levels of complement-fixing immune complexes were detected in plasma of patients with autoimmune diseases (rheumatoid arthritis, Sjögrens syndrome, systemic lupus erythematosus) and paracoccidioidomycosis (AGUADO et al. 1985). Interestingly, patients with paracoccidioidomycosis were found to have a markedly decreased degradation of immune complex bound C3b/iC3b to C3dg compared to patients with autoimmune diseases. This observation resulted from the differential specificities of MoAb 105, an antibody which binds preferentially to surface-bound C3b, and MoAb 130, an antibody specific for iC3b. In addition to the different fragments of C3 found on immune complexes, differences were also observed, using MoAb clone 9, in the fragments bound to the surface of complement activators (NEWMAN and MIKUS 1985). These findings emphasize the importance of monoclonal antibodies in distinguishing the C3 activation products in biological fluids.

5 Conclusions

MoAbs have been useful tools for studying the C3-ligand interactions as well as the conformational changes associated with its degradation. A great deal of information on the functional aspects of C3 has been obtained based on the ability of MoAbs differentially to inhibit certain C3 interactions. Since the results obtained with MoAbs are often due to steric or allosteric effects, and since most antibodies recognize conformation-dependent epitopes which are difficult to localize by present techniques, other approaches are necessary to confirm functional sites identified only by MoAbs. For diagnostic purposes, MoAbs directed against particular fragments of C3 have facilitated the development of assays detecting these fragments in biological fluids, and their use will augment our understanding of complement involvement, particularly the role of C3, in various diseases.

Acknowledgements. We thank Drs. D. Goundis and H. Huemer for helpful discussions and critical reading of the manuscript. Also, we thank Drs. J. Tamerius and B. Kolb (Cytotech) for generously providing serveral of the MoAbs used in our studies.

References

Aguado MT, Lambris JD, Tsokos GC, Burger R, Bitter-Suermann D, Tamerius JD, Dixon FJ, Theofilopoulos AN (1985) Monoclonal antibodies against complement 3 neoantigens for detection of immune complexes and complement activation. Relationship between immune complex levels, state of C3, and numbers of receptors for C3b. J Clin Invest 76: 1418–1426

Alsenz J, Lambris JD (1988) Modulation of factor H (H) functional activities by monoclonal antibodies (MoAbs) to H. Complement 5: 202–203 (Abstract)

Alsenz J, Becherer JD, Esparza I, Daoudaki ME, Avila D, Oppermann S, Lambris JD (1989) Structure and function analysis of C3 from different species. Complement Inflamm 6: 307 (Abstract)

Amit AG, Mariuzza RA, Phillips SEV, Poljak RJ (1986) Three-dimensional structure of an antigen-antibody complex at 2.8A resolution. Science 233: 747–753

Becherer JD, Daoudaki ME, Lambris JD (1987) Conservation of the C3 ligand binding sites within different species. Fed Proc 46: 771 (Abstract)

Becherer JD, Alsenz J, Lambris JD (1989a) Molecular aspects of C3 interactions and structural/functional analysis of C3 from different species. Curr Top Microbiol Immunol 153: 45–72

Becherer JD, Alsenz J, Servis C, Myones BL, Lambris JD (1989b) Cell surface proteins reacting with activated complement components. Complement Inflamm 6: 142–165

Becherer JD, Alsenz J, Hack CE, Drakopulou E, Lambris JD (1989c) Identification of common binding domains in C3b for members of the complement family of C3b-binding proteins. Complement Inflamm 6: 313 (Abstract)

Bitter-Suermann D (1988) The anaphylatoxins. In: Rother K, Till GO (ed) The complement system. Springer Berlin Heidelberg New York, pp 367–395

Burger R, Deubel U, Hadding U, Bitter-Suermann D (1982) Identification of functionally relevant determinants on the complement component C3 with monoclonal antibodies. J Immunol 129: 2042–2050

Burger R, Bader A, Kirschfink M, Rother U, Schrod L, Wörner I, Zilow G (1987) Functional analysis and quantification of the complement C3 derived anaphylatoxin C3a with a monoclonal antibody. Clin Exp Immunol 68: 703–711

Burger R, Zilow G, Bader A, Friedlein A, Naser W (1988) The C terminus of the anaphylatoxin C3a generated upon complement activation represents a neoantigenic determinant with diagnostic potential. J Immunol 141: 553–558

Chaplin H Jr, Monroe MC (1986) Comparisons of pooled polyclonal rabbit anti-human C3d with four monoclonal mouse anti-human C3ds. I. Preparation, purification and binding properties. Vox Sang 50: 42–51

Collawn JF, Wallace CJA, Proudfoot AEI, Paterson Y (1988) Monoclonal antibodies as probes of conformational changes in protein-engineered cytochrome c. J Biol Chem 263: 8625–8634

Colman M, Laver WG, Varghes JN, Baker AT, Tulloch PA, Air GM, Webster RG (1987) Three-dimensional structure of a complex of antibody with influenza virus neuraminidase. Nature 326: 358–363

Daoudaki ME, Becherer JD, Lambris JD (1988) A 34-amino acid peptide of the third component of complement mediates properdin binding. J Immunol 140: 1577–1580

Davis AEIII, Harrison RA (1982) Structural characterization of factor I mediated cleavage of the third component of complement. Biochemistry 21: 5745–5749

De Bruijn MHL, Fey GH (1985) Human complement component C3: cDNA coding sequence and derived primary structure. Proc Natl Acad Sci USA 82: 708–712

Dobbie D, Brazier DM, Gardner B, Holbum AM (1987) Epitope specificities and quantitative and serologic aspects of monoclonal complement (C3c and C3d) antibodies. Transfusion 27: 453–459

Esparza I, Vilbois F, Becherer JD, Lambris JD (1989) Multiple sites of interaction in C3 for CR2/EBV-receptor: functional implications. Complement Inflamm 6: 334 (Abstract)

Fieser TM, Tainer JA, Geysen HM, Houghten RA, Lerner RA (1987) Influence of protein flexibility and peptide conformation on reactivity of monoclonal anti-peptide antibodies with a protein α-helix. Proc Natl Acad Sci USA 84: 8568–8572

Ganu VS, Müller-Eberhard HJ (1985) Inhibition of factor B and factor H binding to C3b by synthetic peptide corresponding to residues 749–789 of human C3. Complement 2: 27 (Abstract)

Garred P, Mollnes TE, Lea T (1988a) Quantification in enzyme-liked immunosorbent assay of a C3 neoepitope expressed on activated human complement factor C3. Scand J Immunol 27: 329–335

Garred P, Mollnes TE, Lea T, Fischer E (1988b) Characterization of a monoclonal antibody MoAb bH6 reacting with a neoepitope of human C3 expressed on C3b, iC3b, and C3c. Scand J Immunol 27: 319–327

Grier AH, Schultz M, Vogel CW (1987) Cobra venom factor and human C3 share carbohydrate antigenic determinants. J Immunol 139: 1245–1252

Grossberger D, Riegert P, Becherer JD, Nilsson B, Nilsson-Ekdahl K, Nilsson UR, Lambris JD (1988) Monoclonal anti-C3 antibodies specific for either bound or fluid phase C3: mapping of their antigenic sites in C3 by the use of an expression minilibrary. FASEB J A1643 (Abstract)

Grossberger D, Marcuz A, Du Pasquier L, Lambris JD, (1989) Conservation of structural and functional domains in complement component C3 of *Xenopus* and mammals. Proc Natl Acad Sci USA 86: 1323–1327

Hack CE, Paardekooper J, Smeenk RJT, Abbink J, Eerenberg AJM, Nuijens JH (1988) Disruption of the internal thioester bond in the third component of complement (C3) results in the exposure of neodeterminants also present on activation products of C3. An analysis with monoclonal antibodies. J Immunol 141: 1602–1609

Hirani S, Lambris JD, Müller-Eberhard HJ (1985) Localization of the conglutinin binding site on the third component of human complement. J Immunol 134: 1105–1109

Holt PDJ, Donaldson C, Judson PA, Johnson P, Parsons SF, Anstee DJ (1985) NBTS/BRIC 8. A monoclonal anti-C3d antibody. Transfusion 25: 267–269

Huemer HP, Burger R, Garred P, Cohen G, Eisenberg R, Friedman H, Esparza I, Dierich MP, Lambris JD (1989) The interaction sites in C3b for HSV-1 gC and factor H are related. Complement Inflamm 6: 348 (Abstarct)

Hugli TE (1989) Structure and function of C3a anaphylatoxin. Curr Top Microbiol Immunol 153: 181–208

Iida K, Mitomo K, Fujita T, Tamura N (1987a) Characterization of three monoclonal antibodies against C3 with selective specificities. Immunology 62: 413–417

Iida K, Mitomo K, Fujita T, Tamura N (1987b) A solid-phase anti-C3 assay for detection of immune complexes in six distinguished forms. J Immunol Methods 98: 23–28

Isenman DE (1983) Conformational changes accompanying proteolytic cleavage of human complement protein C3b by the regulatory enzyme factor I and its cofactor H. Spectroscopic and enzymological studies. J Biol Chem 258: 4238–4244

Isenman DE, Cooper NR (1981) The structure and function of the third component of human complement 1. The nature and extent of conformational changes accompanying C3 activation. Mol Immunol 18: 331–339

Isenman DE, Kells DIC, Cooper NR, Müller-Eberhard HJ, Pangburn MK (1981) Nucleophilic modification of human complement protein C3: correlation of conformational changes with acquisition of C3b-like functional properties. Biochemistry 20: 4458–4467

Jackson DC, Poumbouries P, White DO (1988) Simultaneous binding of two monoclonal antibodies to epitopes separated in sequence by only three amino acid residues. Mol Immunol 25: 465–471

Kanayama Y, Kurata Y, McMillan R, Tamerius JD, Negoro N, Curd JG (1986) Direct quantitation of activated C3 in human plasma with monoclonal anti-iC3b-C3d-neoantigen. J Immunol Methods 88: 33–36

Klos A, Ihrig V, Messner M, Grabbe J, Bitter-Suermann D (1988) Detection of native human complement components C3 and C5 and their primary activation peptides C3a and C5a (anaphylatoxic peptides) by ELISAs with monoclonal antibodies. J Immunol Methods 111: 241–252

Koch C, Behrendt N (1986) A novel polymorphism of human complement component C3 detected by means of a monoclonal antibody. Immunogenetics 23: 322–325

Koistinen V, Wessberg S, Leikola J (1989) Common binding region of complement factors B, H and CR1 on C3b revealed by monoclonal anti-C3d. complement Inflamm 6: 270

Kusano M, Choi NH, Tomita M, Yamamoto K, Migita S, Sekiya T, Nishimura S (1986) Nucleotide sequence of cDNA and derived amino acid sequence of rabbit complement component C3 alpha-chain. Immunol Invest 15: 365–378

Lachmann PJ, Oldroyd RG, Milstein C, Wright BW (1980) Three rat monoclonal antibodies to human C3. Immunology 41: 503–515

Lachmann PJ, Pangburn MK, Oldroyd RG (1982) Breakdown of C3 after complement activation. Identification of a new fragment, C3g, using monoclonal antibodies. J Exp Med 156: 205–216

Lambris JD (1988) The multifunctional role of C3, the third component of complement. Immunol Today 9: 387–393

Lambris JD, Ganu VS, Hirani S, Müller-Eberhard HJ (1985) Mapping of the C3d receptor (CR2)-binding site and a neoantigenic site in the C3d domain of the third component of complement. Proc Natl Acad Sci USA 82: 4235–4239

Lambris JD, Avilla D, Becherer JD, Müller-Eberhard HJ (1988) A discontinuous factor H binding site in the third component of complement as delineated by synthetic peptides. J Biol Chem 263: 12147–12150

Lambris JD, Becherer JD, Daoudaki ME, Servis C, Alsenz J (1989) Use of synthetic peptides in exploring and modifying complement reactivities. In: Sim RB (ed) Activators and inhibitors of complement activation. Kluer Academic, Lancaster (in press)

Lundwall A, Wetsel RA, Domdey H, Tack BF, Fey GH (1984) Structure of murine complement component C3. I. Nucleotide sequence of cloned complementary and genomic DNA coding for the β chain. J Biol Chem 259: 13851–13862

Ma D, Sessler MJ, Meyer TF, Schrod L, Hänsch GM, Burger R (1985) Expression of polypeptide segments of the human complement component C3 in *E. coli*: genetic and immunological characterization of cDNA clones specific for the β-chain of C3. J Immunol 135: 3398–3402

Meuth JL, Morgan EL, DiScipio RG, Hugli TE (1983) Suppression of T lymphocyte functions by human C3 fragments. I. Inhibition of human T cell proliferative responses by a kallikrein cleavage fragment of human iC3b. J Immunol 130: 2605–2611

Molenaar JL, Helder AW, Müller MAC, Goris-Mulder M, Jonker LS, Brouwer M, Pondman KW (1975) Physico-chemical and antigenic properties of human C3. Immunochemistry 12: 359–364

Mollnes TE, Lachmann PJ (1987) Activation of the third component of complement (C3) detected by a monoclonal anti-C3′g′ neoantigen antibody in a one-step enzyme immunoassay. J Immunol Methods 101: 201–207

Myones BL, Avila D, Lachmann PJ, Lambris JD (1989) Localization of the epitopes recognized by the anti-C3G (clone 9) monoclonal antibody using synthetic peptides. Complement Inflamm 6: 373 (Abstract)

Newman SL, Mikus LK (1985) Deposition of C3b and iC3b onto particulate activators of the human complement system. Quantitation with monoclonal antibodies to human C3. J Exp Med 161: 1414–1431

Nilsson UR, Nilsson B (1982) Analogous antigenic alterations elicited in C3 by physiologic binding and by denaturation in the presence of sodium dodecylsulfate. J Immunol 129: 2594–2597

Nilsson B, Nilsson UR (1986) SDS denaturation of complement factor C3 as a model for allosteric modifications occurring during C3b binding: demonstration of a profound conformational change by means of circular dichroism and quantitative immunoprecipitation. Immunol Lett 13: 11–14

Nilsson B, Svensson KE, Borwell P, Nilsson UR (1987) Production of mouse monoclonal antibodies that detect distinct neoantigenic epitopes on bound C3b and iC3b but not on the corresponding soluble fragments. Mol Immunol 24: 487–494

Nilsson B, Svensson KE, Inganäs M, Nilsson UR (1988) A simplified assay for the detection of C3a in human plasma employing a monoclonal antibody raised against denatured C3. J Immunol Methods 107: 281–287

Nilsson B, Grossberger D, Nilsson-Ekdahl K, Riegert P, Becherer JD, Nilsson U, Lambris D (1989a) Localization of neoantigenic epitopes in C3b and iC3b. Complement Inflamm 6: 376 (Abstract)

Nilsson B, Nilsson-Ekdahl K, Avila D, Nilsson UR, Lambris JD (1989b) Conformational changes in C3 as detected by monoclonal antibodies: mapping of the recognized epitopes by synthetic peptides. J Biol Chem (submitted for publication)

Nilsson-Ekdahl K, Nilsson B, Becherer JD, Nilsson UR, Lambris JD (1989) Further studies on the degradation of complement factor C3 by factor I. Inhibition of factor I with DFP. Seventh International Congress of Immunology Gustav Fischer Verlag Stuttgart p. 131 (Abstract 22–24)

Pangburn MK (1987) A fluorimetric assay for native C3. The hemolytically active form of the third component of human complement. J Immunol Methods 102: 7–14

Pangburn MK, Schreiber RD, Müller-Eberhard HJ (1981) Formation of the initial C3 convertase of the alternative pathway. Acquisition of C3b-like activities by spontaneous hydrolysis of the putative thioester in native C3. J Exp Med 154: 856–867

Perkins SJ, Sim RB (1986) Molecular modelling of human complement C3 and its fragments by solution scattering. Eur J Biochem 157: 155–168

Ross GD, Medof (1985) Membrane complement receptors specific for bound fragments of C3. Adv Immunol 37: 217–267

Ruddy S, Moxley GF, Purkall DB (1983) Inhibition of classic and alternative pathway convertases by rat monoclonal antibody to human C3. Immunobiology 164: 291 (Abstract)

Tamerius JD, Pangburn MK, Müller-Eberhardt HJ (1982) Selective inhibition of functional sites of cell-bound C3b by hybridoma-derived antibodies. J Immunol 128: 512–514

Tamerius JD, Pangburn MK, Müller-Eberhard HJ (1985) Detection of a neoantigen on human C3bi and C3d by monoclonal antibody. J Immunol 135: 2015–2019

Theofilopoulos AN, Wilson CB, Dixon FJ (1976) The Raji cell radioimmune assay for detecting immune complexes in human sera. J Clin Invest 57: 169–182

Wetsel RA, Lundwall A, Davidson F, Gibson T, Tack BF, Fey GH (1984) Structure of murine complement component C3. II. Nucleotide sequence of clones complementary DNA coding for the alpha chain. J Biol Chem 259: 13857–13862

Whitehead AS, Sim RB, Bodmer WF (1981) A monoclonal antibody against human complement component C3: the production of C3 by human cells in vitro. Eur J Immunol 11: 140–146

Wörner I, Burger R, Lambris JD (1989) Localization and functional characterization of epitopes on α and β-chains of C3. Complement Inflamm 6: 416 (Abstract)

Zubler RH, Lange G, Lambert PH, Miescher PA (1976) Detection of immune complexes in untreated sera by a modified ^{125}I-C1q binding test. J Immunol 116: 232–235

Subject Index

α_2-Macroglobulin 46, 49, 58–59, 78
Anaphylatoxins 181–208
Antipeptide antibodies 51, 56, 200, 236

β-glucan 106–108
Bilharziasis 172

C3
- binding sites 50–58, 237–239
- biochemical characteristics 23–24
- biosynthesis 25–29
- – leader peptide 25
- – precursor processing and intracellular transport 25–26
- – glycosylation 26
- – tetra-arginine linker 25
- C3 from
- – axolotl 49, 60, 63
- – chicken 49, 61
- – cobra 46, 60–62
- – feline 60, 61
- – guinea pig 49, 60, 61
- – human 45–72
- – lamprey 64
- – mouse 46, 49, 55, 59
- – porcine 60, 61
- – quail 49, 61
- – rabbit 46, 49, 55, 59–60
- – rat 60, 61
- – trout 49, 60, 63
- – *Xenopous* 46, 49, 55, 60, 62
- C3 (H$_2$O) 5, 236–237
- C3a 181–208
- fragments 46–48, 181–208, 235, 243–244
- gene structure 31–35
- – correlation of protein domains with gene structure 33–35
- – intron/exon organization 31–35
- – structural map 33
- gene transcription 29
- – truncated message 29
- glycosylation 24, 46, 52, 59–63
- monoclonal antibodies, MoAb 200, 235–245

- – analysis of C3 functions 237–239
- – analysis of conformational changes 236–237
- – analysis of complement activation 243–244
- – epitope mapping 239–243
- – detection of C3 fragments 200, 236–238, 243
- – generation 235–236
- – neoepitopes 200, 238
- neoepitopes 51, 200, 238
- polymorphism 36
- sites of synthesis 26–29
- – astrocytes 26
- – endothelial 28
- – epithelial 28–29
- – hepatocytes 26–27
- – monocyte/macrophage 27–28
- – Raji cells 29
- transcriptional and translational regulation 29–31
- – acute phase response 29, 31
- – stimulation by
    glucocorticoids 30
    hormones 31
    interferons 30–31
    interleukins 30–31
    lipopolysaccharide (LPS) 30
    phorbol esters 30
    tumor necrosis factor (TNF) 30–31
C3 convertase
- alternative pathway 2–3, 12–13, 40
- – amplification 6
- – initiation 5
C3a receptors
- C3a 197–200
- C3a functional aspects
- – C3a 190–196
- – immunomodulation 194–195
- – isolated tissues 192, 194
- – mast cells 195
- – microvascular 191–192
- – platelet activation 195–196

Subject Index

– – skin responses   191
– – vasoconstriction   192, 194
– cross-linking receptor-ligand   198–199
– granulocytes   197
C4   46, 49, 58, 59, 64
C4 binding protein   47, 48, 56
C5   47–49, 58–59, 61, 63
C5 convertase
– alternative pathway   2–3, 13–14
– classical pathway   2–3, 14–15
Candida albicans   165, 173–175
Carboxypeptidase N   181, 183, 190–192
Chagas' disease   170
Chemical cross-linking   196–199
Cobra Venom factor, CVF   49, 57, 61–63, 229, 231
Complement receptor type 1, CR1
– allotypes   85–86
– binding site(s) in C3   50–52, 56, 59–60, 237–239
– biosynthesis   87
– functions   47, 48, 90–92
– glycosylation   87
– sites of expression   87
– structure   83–85, 212–214, 217
Complement receptor type 2, CR2
– binding site(s) in C3   53–56, 59–60, 237–239, 242
– biosynthesis   89
– functions   47, 48, 53–54, 90–92
– glycosylation   90
– sites of expression   89–90
– structure   88, 212–214, 217
Complement receptor type 3, CR3
– binding site(s) in C3   55–56, 103–109
– functions   109–112
– ligands   55–56, 103–109, 174
– surface expression   110–112
– structure   100–103, 214, 216, 218
Complement receptor type 4, CR4 see p150,95
Conformation
– C3a   187–190
– – Circular dichroism   187–188
– – NMR   189
– – X-ray crystallography   188
– C3 conformational changes   46–48, 236–237
Conglutinin   48, 52, 58, 217
CR1 see Complement receptor type 1
CR2 see Complement receptor type 2
CR3 see Complement receptor type 3
CVF see Cobra Venum factor

DAF see Decay-Accelerating factor
Decay-Accelerating factor, DAF
– activities   131–132
– alternative forms   127–128
– biosynthesis   124–125
– blood groups   133
– cDNA   128–131
– gene   131
– glycophospholipid anchor   126
– glycosylation   124–125
– guinea pig   133
– paroxysmal nocturnal hemoglobinuria, PNH   132
– physiological role   131–132
– purification   124
– rabbit   133
– sites of expression   127–128
– *Trypanosoma cruzi*   170, 175
Deficiencies
– C2   225, 228
– C3   36–37, 223–232
– – acquired   231
– – inherited   223, 227, 229
– – human   224–226
– – guinea pig   226
– – dog   229
– – rabbit   230
– – biological consequences   224–226
– – infections   224, 230
– – membranoproliferative glomerulonephritis   224, 230
– – immune response   225, 228, 230
– C3a receptor   227
– C4   225, 228
– C8   230
– factor H   151
– Leucocyte adhesion dificiency (LAD)   112–114
– – animal model   114

EBV, see Epstein-Barr virus
ELAM-1   104, 214
Epstein-Barr virus, EBV   50, 52–55, 89–90, 92–93, 168–169

Factor B   6–9, 47–51, 56–57, 61, 215, 217, 237
Factor D   10, 61
Factor H
– allelic variants   152
– alternative splicing   155, 157
– binding characteristics   56–57, 60–63, 149, 150, 237
– carbohydrates   151
– cofactor activity   47–51, 148

– decay accelerating activity   148
– gene linkage   154
– genomic organization   153
– regulatory sequences   154
– structure   212–214, 217
Factor I   47–50, 56, 57, 59–60, 216–217, 243
Factor X   56, 105–107
Fibrinogen   55–56, 105–106
Fibronectin   105

Herpes Simplex virus, HSV   50, 164–169
– glycoprotein C   165, 166–168, 239
Histoplasma Capsulatum   108
HSV, see Herpes Simplex virus

ICAM-1   218
Integrins   100–101
– leucocyte integrins   100–103
– α-subunits   100–103
– – Ca^{++} binding motifs   102, 216
– – L-domain   102, 215
– β-subunit
– – cysteine-rich octet   101–102
– – gene location   112
– – high homology region   102
– – RFLP   112

Leishmania   55–56, 105–108
– gp63   49, 55–56, 105–107
Leukocytosis   47, 49
LFA-1   100–101
Lipopolysaccharide (LPS)   106, 108–109

Malaria   171
MCP see Membrane cofactor protein
Membrane cofactor protein, MCP
– biosynthesis   136
– cDNA   136–138
– gene   138
– glycosylation   136
– physiological role   47–48, 139
– purification   133
– sites of expression   139
– structural variations   134–136
– structure   212–214

NeF   231

p150,95   99–122, 218
Phagocytosis   100–110

Plasmodium falciparum   171
Plasmodium knowlesi   171
Plasmodium vivax   171
Primary structure
– C3a   183
– C3   49
– DAF   129
– MCP   137
Properdin
– binding site(s) in C3   47, 48, 57–63, 239
– functions   11–12
– structure   214, 218

Respiratory distress syndrome   243
RGD-tripeptide   55, 62, 105
Rheumatoid arthritis   243, 244

Schistosoma mansoni   165, 172–173
Serine proteinase   9, 10, 215
Staphylococcus epidermidus   108
Structural motifs
– metal binding domain   102, 216
– factor I/C7 repeat   102, 216
– serine protease domain   7, 215
– LDLR domain   215
– thrombospondin repeat   11, 214
– Von Willebrand factor repeat   9, 214
– complement control protein repeat, CCP   7, 84, 88, 140, 153, 213–215
Synthetic peptides
– epitope mapping   236, 240, 241, 243
Synthetic peptides binding to
– CR1   51
– CR2   53–54, 239
– CR3   55, 105
– H   56
– Properdin   59
– C3a receptor
– – LGLAR   183–186
– – C3a analogs   184–187
– – hydrophobic groups   185–187
Systemic lupus erythrematosus   243, 244

Thiolester bond   5, 25–26, 46–47, 60–64, 73–80, 219–220
Trypanosoma cruzi   165, 170–171

Vaccinia virus   165, 169–170
Von Willebrand factor   9, 214

Zymosan   108